577.472 (26)

AF749115

Developments in Hydrobiology 202

Series editor

K. Martens

Challenges to Marine Ecosystems

Proceedings of the 41st European Marine Biology Symposium

Edited by

John Davenport, Gavin M. Burnell, Tom Cross,
Mark Emmerson, Rob McAllen, Ruth Ramsay
& Emer Rogan

University College Cork, Ireland

Reprinted from Hydrobiologia, Volume 606 (2008)

52902

Springer

Library of Congress Cataloging-in-Publication Data

A C.I.P. Catalogue record for this book is available from the Library of Congress.

ISBN-13: 978-1-4020-8807-0

Published by Springer,
P.O. Box 17, 3300 AA Dordrecht, The Netherlands

Cite this publication as Hydrobiologia vol. 606 (2008).

Cover illustration: Lough Hyne Marine Reserve, Southwest Ireland. Photo credit: Dr R. McAllen.

Printed on acid-free paper

All Rights reserved
© 2008 Springer

No part of this material protected by this copyright notice may be reproduced or utilized in any form or by any means, electronic or mechanical, including photocopying, recording or by any information storage and retrieval system, without written permission from the copyright owner.

Printed in the Netherlands

TABLE OF CONTENTS

CLIMATE CHANGE AND MARINE ECOSYSTEMS

SUSTAINABLE FISHERIES/AQUACULTURE

Hydrobiologia (2008) 606:1–5
DOI 10.1007/s10750-008-9352-x

Foreword

© Springer Science+Business Media B.V. 2008

This volume presents a representative sample of contributions to the 41st European Marine Biology Symposium held on 4–8 September 2005 in Cork, Ireland. The theme of the symposium was 'Challenges to Marine Ecosystems', and this was divided into four sub-themes: Genetics, Marine Protected Areas, Global Climate Change and Marine Ecosystems, Sustainable Fisheries and Agriculture. The symposium attracted 308 participants from 33 countries and, over the 5 days, 92 oral presentations were given as well as 175 posters presented.

The world's marine ecosystems face multiple challenges, some natural, but many resulting from humankind's activities. Global climate change, driven by influences of energy usage and industrial practices, is a reality now accepted by most of the world's scientists, media and political establishments. Warming seas and rising sea levels are regarded as threats, while visionaries consider deep ocean carbon disposal as a technological opportunity. Exploitation of the seas continues apace, with repeated concerns over the impact of over-fishing, plus reservations about the environmental effects of marine aquaculture. We need to understand how resilient organisms and ecosystems are to these challenges, while responding by protecting biologically meaningful areas of the oceans. The sub-themes of the 41st European Marine Biology Symposium address all these matters.

The symposium was hosted by the Department of Zoology, Ecology and Plant Science, University College Cork, and we would like to thank all members of the Department who assisted in the running of the event. We would also like to thank all our sponsors for their vital financial support.

Thanks are also due to all the referees for their efforts in evaluating and improving the manuscripts that were submitted for publication in this volume.

Local organising committee

Anne Marie Power	Mark Emmerson
Dave McGrath	Padriag Whelan
Dave Morritt	Rob McAllen
Emer Rogan	Ruth Ramsey
Gavin Burnell	Sandy O'Driscoll
Jim Wilson	Sarah Culloty
John Davenport	Tasman Crowe
John Joyce	Tom Cross
Julia Davenport	Tom Kelly

Guest editors: J. Davenport, G. Burnell, T. Cross, M. Emmerson, R. McAllen, R. Ramsay & E. Rogan
Challenges to Marine Ecosystems

List of participants

Australia
Maria Schreider
Thomas Wernberg
Jacqui Pocklington

Austria
Joerg Ott
Bettina Riedel
Marlene Thaler

Belgium
Gregory Maes
Thomas Remerie

Canada
Tony J Pitcher

Chile
Günter Försterra
Vreni Häussermann

China
Jiang Hong

Croatia
Zrinka Buric
Marija Despalatovic
Daniel Lyons
Sanja Matic-Skoko
Slavica Matijevic
Jakica Njire
Lorena Peric
Tomislav Radic
Gorenka Sinovcic
Sanda Skejić
Ana Travizi
Damir Vilicic

Denmark
Finn Andersen
Thomas Kirk Sørensen

Falklands
Daniel A Smale

Finland
Erk Bonsdorff
Jenni Jäänheimo
Alf Norkko
Joanna Norkko
Ari Ruuskanen

France
Claire Gollety
Christian Hily
Jean-Baptiste Ledoux
Jérémy Lobry
Núria Teixidó
Marie Voisin

Germany
Jeff Ardron
Friedrich Buchholz
Jennifer Dannheim
Ute Jacob
Mark Lenz
Rafael Panusch

Greece
Stelios Katsanevakis
Persefoni Megalofonou
Chryssi Mytilineou
Artemis Nicolaidou
Maria Salomidi
Miltiadis Seferlis
Maria Thessalou-Legaki
Vassiliki Vassilopoulou

Iceland
Sigridur Kristinsdottir

Iran
Fatemeh Abbassi
Majid Sampour

Ireland
Alice Antoniacomi
Javier Atalah
Alan Berry
Deirdre Brophy
Gavin Burnell
Edward A. Codling
Solene Connan
Paul Connolly
Ilaria Coscia
James Coughlan
Tom Cross
Tasman Crowe
Sarah Culloty
Amy Dale
John Davenport
Julia Davenport
Carys Ann Davies
Mark Emmerson
Louise Firth
Elizabeth Gosling
Mark Holmes
Sandra Irwin
Mark Jessopp
John Joyce
Eamonn Kelly
Tom Kelly
Sandrine Laurand
David Lyons
Alexia Massa-Gallucci
Rob McAllen
Fionnuala McBreen
Philip McGinnity
Dave McGrath
Luca Mirimin
Liam Morrison
Olwyen Mulholland
Vanessa Murphy
Roisin Nash
Sophie Nicol
Brendan O'Connor
Nessa O'Connor
Brian O'Loan
Aengus Parsons
Katja Peijnenburg
Anne Marie Power
Ruth Ramsay
Emer Rogan
Maria Sala
Samuel Shephard
Elizabeth Sides
Susan Steele
Rob Thomas
Anna Was
Padriag Whelan
James D Wilson

Israel
Bella Galil

Italy
Alberto Addis
Rita Cannas

Leonardo Cannizzoaro
Giulia Ceccherelli
Aldo Corriero
Roberta Cupido
Giovanni D'Anna
Doris De Vito
Rossana Di Donato
Floriana Di Stefano
Adele Fabbrocini
Maria Cristina Follesa
Maria Francesca Fornasier
Gianuluca Franceschini
Christina Gambi
Simona Genovese
Adriana Giangrande
Folco Giomi
Basilone Gualtiero
Monica Landi
Giuseppe Magnifico
Sara Marzialetti
Carlotta Mazzoldi
Martina Milanese
Carlo Pipitone
Giulio Relini
Lidia Orsi Relini
Giovanni Sansone
Giovanni Santangelo
Antonio Sará
Gianluca Sará
Renata Zaccone

Latvia
Atis Minde
Didzis Ustups

Mexico
Diego Lercari
Armando Ortega-Salas
Manuel Zetina-Rejón
Ana Maria Amaya
Maria Teresa Barreiro
Gisele Signoret
Martha Signoret
Pedro Ardisson

Morocco
Abdellatif Bayed

Netherlands
Carlo Heip
Herman Hummel
António Pagarete
Karin Troost

New Zealand
Bill Ballantine
Judi Hewitt
Drew Lohrer
Islay Diane Marsden
Simon Thrush

Norway
Asbjøm Aanonsen
Torjan Bodvin
Geir Dahle
Tim Dempster
Knut E Jørstad
Daniela Pampanin

Poland
Lidia Dzierzbicka-Glowacka
Monika Kedra
Rafal Lasota
Maciej Wolowicz

Portugal
Adelaide Almeida
Nuno Vaz Álvaro
Andreia Braga-Henriques
Isabel Caçador
Joana Campos
Susana Carvalho
Rui Coelho
Marina Cunha
Joana Fernandez-Carvalho
Wallenstein Francisco
Vânia Freitas
Catarina Grilo
Ruth Higgins
David Jacinto
Sara Leston
Fernando P Lima
Frederico Lyra
Filipe Martinho
Irene Martins
Virgínia Alves Martins
Telmo Morato
Carlos Moura
Henrique Queiroga
Pedro Ribeiro
Paulo Santos
Tiago Verdelhos
Ivan Viegas
Raquel Xavier

Russia
Dmitri Aristov
Eugene Genelt-Yanovsky
Ksenia Shunkina
Elena Ponomareva

Slovenia
Janja France
South Africa
Christopher McQuaid
Vincent Nakin

Spain
Elena Acero-Sánchez
Adolfo Carballo Penela
Silvia de Juan Mohan
David Díaz
Marta Díaz-Valdés
Raquel Maestre Conejero
Concepción Marcos Diego
Ángel Pérez-Ruzafa
Clara Pintado Gorordo
Adoración Sánchez-Mata
Arabella G. Sánchez-Mata
Vanessa Stelzenmüller
Damian Fernandez Jover

Sweden
Mikael Dahl
Johanna Gardeström
Tuuli Mäkinen
Karl Norling
Pia Pettersson
Anssi Vainikka
Olivia Langhamer

Taiwan
Benny Chan

United Kingdom
Jonathan Adey
Louise Allcock
Roger Bamber
Andy Beaumont
Andrew Blight
Helen Bloomfield
Paul Brazier
Robert Brownlow
Gavin Burns

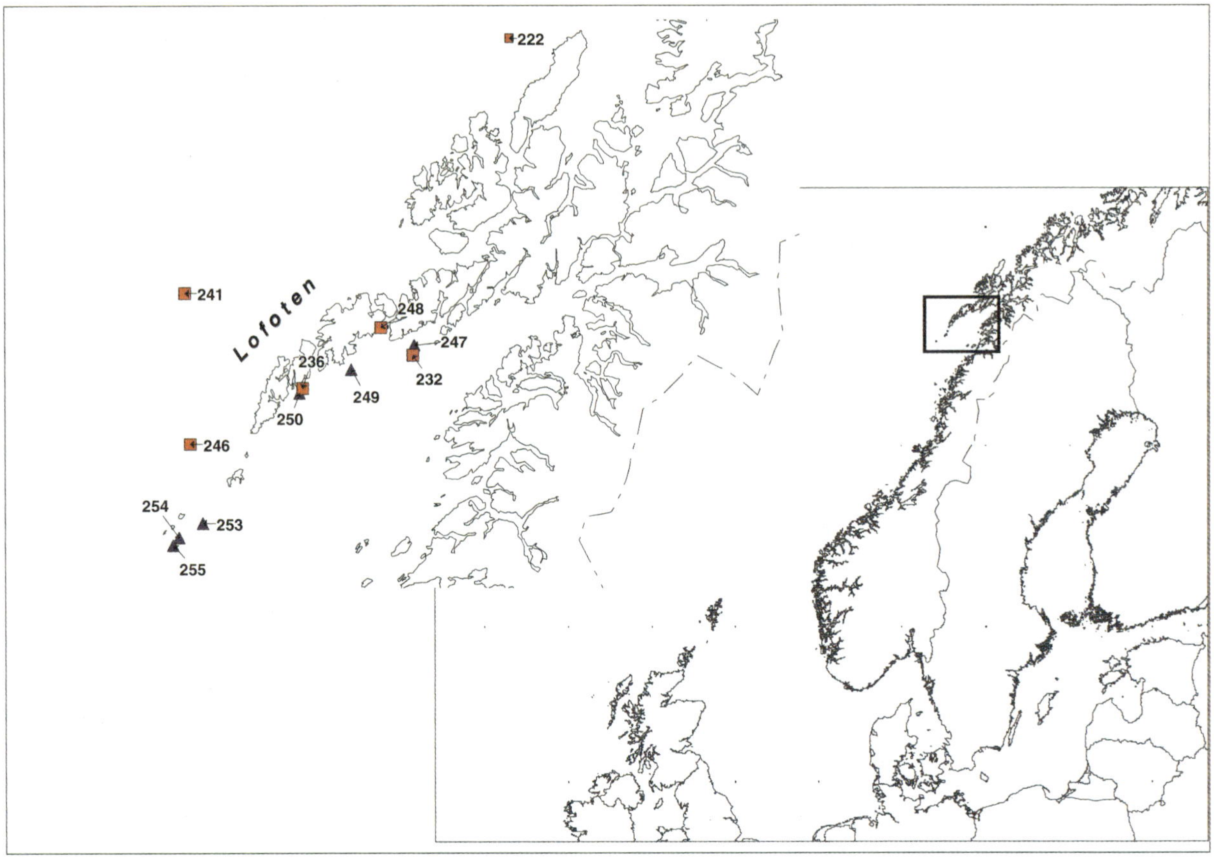

Fig. 1 Sampling stations in 2002 (black triangles) and 2003 (red squares)

Table 1 Location, date collected, number of individuals in samples, otolith classification of individuals and average age at sampling locations. Latitudes are given in degrees N, and longitudes are degrees E

Station no	Date	Latitude	Longitude	No fish	Otolith classification					# classified	Av. age (years)
					1	2	3	4	5		
247	27.03.2002	68.158	14.457	80	33	4	0	6	16	59	5.7
249	29.03.2002	68.067	13.815	51	26	4	0	3	18	51	5.6
250	30.03.2002	67.978	13.289	51	24	3	0	5	19	51	6.2
253	30.03.2002	67.495	12.31	60	6	2	0	10	29	47	6.9
254	31.03.2002	67.443	12.078	58	6	3	1	6	42	57	6.8
255	31.03.2002	67.413	12.017	36	0	3	0	5	28	36	6.8
222	24.03.2003	69.3	15.4	95	2	1	0	9	24	36	6.3
232	29.03.2003	68.12	14.44	96	45	14	0	11	11	81	6.4
236	30.03.2003	67.998	13.328	96	15	10	0	18	53	96	6.4
241	02.04.2003	68.352	12.137	96	3	1	0	10	76	90	6.8
246	03.04.2003	67.792	12.19	95	1	3	0	14	74	92	6.8
248	05.04.2003	68.223	14.115	96	72	14	0	6	3	95	5.6

uncertain NCC, respectively), while otolith type 4 and 5 correspond to the NEAC group (uncertain NEAC and certain NEAC, respectively). Otoliths not attributable to NCC of NEAC are typed as 3. This classification approach has been used for decades in cod stock management in the Northeast Atlantic.

Genetic analyses

Due to lack of stability, the blood samples had to be analysed within 3 to 4 days. After sedimentation of the blood cells, distilled water was added and the cells were disrupted. The samples were then analysed using the agar gel electrophoresis approach, first described by Sick (1961) and later modified by Jørstad (1984). The different banding patterns corresponding to the most common genotypes have earlier been described by Sick (1961). More detailed descriptions are also given in Dahle & Jørstad (1993) and Husebø et al. (2004). All blood sample analyses were conducted on fresh samples on the research vessel during the ongoing survey.

The samples of white muscle tissue were stored at low temperature (−80 °C) at the molecular biology laboratory of IMR in Bergen until analyses. The samples were analysed for 3 to 5 months after collection at sea, using standard starch gel electrophoresis with modifications as described by Jørstad (1984). Five different tissue enzymes were analysed, including lactate dehydrogenase (LDH), isocitrate dehydrogenase (IDH), glycerophosphate dehydrogenase (GPD), phosphoglucomutase (PGM) and phosphoglucose isomerase (GPI). These allozymes have earlier been described as polymorphic in cod (Mork et al., 1985; Jørstad & Nævdal, 1989).

Fin clips were stored (4°C) in the laboratory in Bergen until DNA extraction, which was carried out by using Qiagen DNeasy 96 Tissue kit (Qiagen). DNA quality was controlled by running the extractions on ready-to-run 1.2% agarose gels (Amersham Biosciences). The analyses of the *Pan* I locus followed the procedures described by Fevolden & Pogson (1997). The samples obtained from the 2002 survey were analysed in the laboratory at the Norwegian College of Fishery Science, University of Tromsø, while the 2003 collections were analysed in the laboratory in Bergen. The samples were analysed for six microsatellite loci (Gmo2; Gmo3; Gmo34; Gmo35; Gmo132; Tch11), all well established in cod studies and used in a number of other population studies in cod (Brooker et al., 1994; Miller et al., 2000; O'Reilly et al., 2000). Primer labels were according to ABI 3100 requirements, and the PCR reactions were mainly run separately for each primer-set to ensure sufficient amplification. The PCR products were mixed in appropriate groups according to primer labelling and fragment size and co-loaded on an ABI 3100 genetic analyser. Size standard GeneScan 500 LIZ (Applied Biosystems) was added to each sample. Fragment sizing was conducted by using the ABI Genotyper computer program package, with manual control of the automatically scored peaks.

Statistical analysis

Allele frequencies, F_{ST} values and exact tests of population difference between population pairs, tests for linkage disequilibrium and deviations from Hardy-Weinberg equilibrium were calculated in the program GENEPOP 3.4 (Raymond & Rousset, 1995). Nei's (1978) genetic distance and heterozygosities were computed in TFPGA (Miller, 1997). This program was also used to compute the UPGM dendrogram, and the % of bootstraps supporting the nodes based upon Nei's (1978) genetic distance. Allelic richness was calculated in FSTAT 2.9.3.2 (Goudet, 2001).

For the assignment and mixed-stock analysis of the samples collected in 2002 & 2003, two different approaches were used. First, base populations for presumed NCC and NEAC were constructed by pooling all individuals from stations 232 to 248 as a baseline for NCC and all individuals from stations 222, 241 and 246 as the NEAC baseline. Based on otolith classification these two groups of stations appeared to be dominated by NCC and NEAC, respectively, although individuals with otoliths of NEAC type were present in the NCC baseline and individuals with CC type otoliths were present in the NEAC baseline. The mixed samples consisted of stations 236, 247, 249, 250, 253, 254 and 255. Nineteen individuals in the mixed samples had alleles not present in the two baseline samples, and blanks were substituted for alleles at affected loci for these individuals before conducting the analysis (Gmo 2; 1 ind., Gmo35; 2 ind., Gmo132; 5 ind., Tch11; 5 ind. GPI-1; 3 ind., and GPD; 3 ind.). No individual in the mixed sample had more than one allele not represented in the baseline.

The program GENECLASS2 (Piry et al., 2004) was used for individual assignment analyses, and mixed stock analysis (MSA) was conducted in the program BAYES (Pella & Masuda, 2001). MSA and individual assignment analysis was conducted on all samples from 2002, and on the sample from station 236 from

2003, using various combinations of loci. The assignment power of the loci was evaluated using the program WhichLoci (Banks et al., 2003). This program uses successive assignment trials from one locus at a time, on a test data set generated by random sampling of the alleles according to their frequency in the populations in the given data set, to generate a ranking of the loci's efficiency for correct population assignment. In the present study, the population size parameter for the randomly generated populations was set to 500. The program Genetix (Belkhir et al., 2004) was used to conduct a factorial analysis of the baseline populations constructed for the mixed stock analysis.

In order to verify the validity of this approach, with defining baseline populations from geography, a second analysis of the data was conducted in the program Structure (Pritchard et al., 2000). In this analysis, no prior information of morphology (otoliths) or geography (sampling locality) was used, and individuals were partitioned into putative groups based on genetic information alone. In the runs, the burn in length was set to 30,000, and the run length to 500,000. Possible values of K from 1 to 4 were evaluated, with three parallel runs at each value of K. The samples from 2002 and 2003 were run in separate tests. For all analyses the admixture model was used.

Results

Otolith types and age structure at the sampled stations

According to otolith classification, all stations in both years had individuals of NCC and NEAC type, but in different proportions (Table 1). The trawl stations in 2002 were distributed in increasing distance from the mainland in a southwesterly transect along the Lofoten Islands, with rising station numbers (see Fig. 1). The distribution of otolith types show a decreasing trend in the frequencies of NCC (types 1 and 2) and a corresponding increase in NEAC (type 4 and 5) along this transect. At the innermost trawl station (247), the fraction of NCC according to otolith classification was 62%, while it was reduced to only 8% at station 255. Ages from 1 to 15 years were represented in the samples, with age-classes from 4 to 8 being most abundant. There was also a slight increase in average age from the inner to the outer stations (Table 1).

The trawl stations sampled in 2003 were distributed somewhat differently. Three stations, 222, 241 and 246, were located offshore from the mainland, on the shelf banks west of the Lofoten islands, while the other three stations were situated closer to shore (Fig. 1) within the Vestfjorden area. The distribution of otolith types shows that the three outer stations were almost totally dominated by NEAC (92–96%), while the two innermost stations appear to be dominated by NCC (73% and 91%). One station, 236, appears to consist of a more even mix of individuals of both types. Age-classes from three to ten were represented, with ages from five to eight being most abundant. The average age of individuals was quite uniform among the stations, except for station 248 where the average age was lower (Table 1).

Allelic variation among loci and samples

The number of alleles across samples ranged from two at the *Pan* I locus to 33 at the Gmo132 locus, and mean heterozygosity among loci varied from 0.017 to 0.932 (Table 2). Allelic richness among loci across samples varied from 1.0 at the IDH locus at several stations to 16.8 at the Gmo132 locus at station 232 in 2003 (Table 2).

In general, heterozygosity and allelic richness were much higher at the microsatellite loci than at the *Pan* I, haemoglobin and allozyme loci. Allele numbers across samples at microsatellite loci ranged from 8 to 33. Lower numbers of alleles at the microsatellite loci Gmo34 and Gmo132 were observed in the samples from the outer stations. This difference between inner and outer stations in allelic richness at these two loci was apparent in both years, with a declining trend in number of alleles at the Gmo34 locus with increasing distance from the mainland at the stations sampled in 2002. Average heterozygosity at individual loci across samples varied from 0.176 at the Gmo3 locus to 0.932 at the Tch11 locus. The allelic distribution at four of the loci was similar across samples, but the loci Gmo34 and Gmo132 displayed a clear difference between samples dominated by individuals with otoliths of NEAC and NCC type.

Table 2 Summary statistics for the 13 loci in the 12 samples.Abbreviations are as follows: number of individuals (*N*), scoring percentage achieved at locus (*S*), number of alleles at locus (N_A), allelic richness (AR) F_{ST} in individual loci across samples (F_{ST}), and Nei's unbiased heterozygosity (*H*)

	Location		*Pan* I	Gmo2	Gmo3	Gmo34	Gmo35	Gmo132	Tch11	Hem	GPD	IDH	PGM	GPI-1	LDH-3	Average
2002	St. 247	*N*	77	77	78	77	77	78	71	59	59	59	59	59	59	68.38
		S	96	96	98	96	96	98	89	74	74	74	74	74	74	85.61
		N_A	2	11	4	8	8	22	18	2	2	2	3	3	2	6.69
		AR	2.000	9.288	3.138	6.205	7.257	14.289	15.199	2.000	1.931	1.407	2.287	2.931	2.000	5.37
		H	0.446	0.872	0.201	0.506	0.829	0.724	0.927	0.374	0.082	0.017	0.082	0.483	0.504	0.47
	St. 249	*N*	51	49	51	48	47	49	51	50	51	51	51	51	48	49.84
		S	100	96	100	94	92	96	100	98	100	100	100	100	94	97.69
		N_A	2	12	4	7	7	15	19	3	2	1	2	4	3	6.23
		AR	2.000	9.601	3.118	5.843	6.510	11.074	16.302	2.732	1.722	1.000	1.962	3.396	2.500	5.21
		H	0.502	0.797	0.131	0.352	0.829	0.659	0.934	0.408	0.039	0.000	0.094	0.405	0.503	0.43
	St. 250	*N*	51	51	51	49	47	51	51	50	51	51	51	51	50	50.38
		S	100	100	100	96	92	100	100	98	100	100	100	100	98	98.77
		N_A	2	13	5	7	8	18	20	3	2	2	2	3	2	6.69
		AR	2.000	10.867	3.644	6.200	7.708	13.749	17.019	2.732	1.722	1.471	1.856	2.722	2.000	5.67
		H	0.502	0.872	0.184	0.477	0.852	0.760	0.938	0.377	0.039	0.020	0.058	0.522	0.498	0.47
	St. 253	*N*	56	59	56	59	55	58	57	58	60	60	60	60	60	58.31
		S	93	98	93	98	92	97	95	97	100	100	100	100	100	97.15
		N_A	2	14	3	6	7	13	21	2	2	2	2	3	3	6.15
		AR	2.000	10.981	2.104	4.449	6.812	7.843	16.334	2.000	1.642	1.400	1.875	2.400	2.642	4.81
		H	0.446	0.874	0.053	0.329	0.812	0.497	0.927	0.265	0.033	0.017	0.065	0.459	0.497	0.41
	St. 254	*N*	54	49	56	51	51	58	50	58	58	58	58	58	58	55.15
		S	93	84	97	88	88	100	86	100	100	100	100	100	100	95.08
		N_A	2	10	4	5	7	10	17	3	3	2	3	4	3	5.62
		AR	2.000	8.561	2.674	4.039	6.924	7.062	14.980	2.658	1.828	1.414	2.216	3.216	2.802	4.64
		H	0.316	0.839	0.087	0.235	0.839	0.453	0.930	0.395	0.034	0.017	0.068	0.489	0.484	0.40
	St. 255	*N*	35	35	36	35	35	36	33	24	36	36	36	36	36	34.54
		S	97	97	100	97	97	100	92	67	100	100	100	100	100	95.92
		N_A	2	12	5	3	7	8	20	2	1	2	2	3	2	5.31
		AR	1.998	10.412	4.299	2.896	6.904	6.871	18.095	2.000	1.000	1.667	1.990	2.667	2.000	4.83
		H	0.135	0.858	0.277	0.162	0.846	0.468	0.943	0.284	0.000	0.028	0.106	0.504	0.468	0.39

Table 2 continued

	Location		*Pan* I	Gmo2	Gmo3	Gmo34	Gmo35	Gmo132	Tch11	Hem	GPD	IDH	PGM	GPI-1	LDH-3	Average
2003	St. 222	*N*	87	95	96	95	89	95	94	94	96	96	96	96	96	94.23
		S	91	99	100	99	93	99	98	98	100	100	100	100	100	98.23
		N_A	2	15	5	4	8	14	23	3	3	2	3	2	3	6.69
		AR	2.000	11.278	2.952	3.233	7.318	7.424	17.506	2.251	1.500	1.438	2.198	2.000	2.438	4.89
		H	0.239	0.860	0.120	0.141	0.825	0.489	0.939	0.207	0.021	0.021	0.109	0.469	0.487	0.38
	St. 232	*N*	71	81	81	80	81	81	81	81	81	81	81	81	81	80.15
		S	88	100	100	99	100	100	100	100	100	100	100	100	100	99.00
		N_A	2	15	5	8	8	23	19	3	2	2	3	3	2	7.31
		AR	2.000	11.606	3.810	6.107	7.228	15.009	16.804	2.506	1.296	1.654	2.450	2.832	2.000	5.79
		H	0.336	0.879	0.266	0.557	0.833	0.798	0.937	0.459	0.012	0.037	0.118	0.479	0.481	0.48
	St. 236	*N*	96	94	96	96	96	96	96	96	96	96	96	96	95	95.76
		S	100	98	100	100	100	100	100	100	100	100	100	100	99	99.76
		N_A	2	11	7	7	10	19	18	3	3	1	2	5	2	6.92
		AR	2.000	8.872	4.429	5.082	7.614	10.480	14.890	2.250	2.161	1.000	1.871	3.405	2.000	5.08
		H	0.476	0.827	0.265	0.348	0.834	0.589	0.928	0.308	0.061	0.000	0.071	0.446	0.483	0.43
	St. 241	*N*	84	89	90	88	89	90	90	90	90	90	90	90	90	89.23
		S	93	99	100	98	99	100	100	100	100	100	100	100	100	99.15
		N_A	2	16	7	5	8	11	18	3	2	2	4	4	3	6.53
		AR	1.993	11.515	3.860	3.403	7.187	6.354	14.946	2.264	1.792	1.463	2.817	2.981	2.468	4.85
		H	0.154	0.860	0.149	0.152	0.837	0.403	0.930	0.191	0.054	0.022	0.117	0.460	0.512	0.37
	St. 246	*N*	90	92	93	93	93	93	91	93	93	93	93	93	93	92.53
		S	97	99	100	100	100	100	98	100	100	100	100	100	100	99.53
		N_A	2	12	8	5	9	11	20	2	3	2	3	3	3	6.38
		AR	1.998	9.056	3.849	3.188	7.230	6.094	15.508	2.000	1.709	1.451	2.171	2.779	2.451	4.57
		H	0.190	0.826	0.154	0.124	0.814	0.479	0.923	0.250	0.032	0.021	0.093	0.471	0.512	0.38
	St. 248	*N*	94	96	95	94	95	94	96	96	96	96	96	96	96	95.38
		S	98	100	99	98	99	99	100	100	100	100	100	100	100	99.46
		N_A	2	17	7	7	9	25	20	2	2	1	4	4	3	7.92
		AR	2.000	11.356	4.230	5.927	7.465	15.588	15.987	2.000	1.438	1.000	2.375	3.198	2.250	5.75
		H	0.277	0.877	0.224	0.601	0.833	0.842	0.930	0.415	0.021	0.000	0.071	0.505	0.489	0.47
		$TotN_A$	2	21	10	8	12	33	28	3	4	2	5	6	3	
		AvH	0.335	0.853	0.176	0.332	0.832	0.597	0.932	0.328	0.036	0.017	0.088	0.474	0.493	
		AvAR	1.999	10.185	3.443	4.604	7.154	9.569	16.144	2.271	1.682	1.326	2.191	2.926	2.308	
		F_{ST}	0.330	0.008	0.007	0.050	0.000	0.035	−0.001	0.029	0.001	−0.003	−0.003	0.006	0.001	

The greatest differences in allele frequencies between samples were found at the two-allele *Pan* I locus, with frequencies of the *Pan* I^A allele ranging from 0.83 (station 248) to 0.07 (station 255) In general, the frequencies decreased with increasing distance from the mainland. The distribution of *Pan* I genotypes was also closely related to the distribution of otolith types, with the *Pan* I^A being most common in individuals with otolith type 1 and 2, and the *Pan* I^B allele in individuals with otolith types 4 and 5. The fraction of homozygotes for the *Pan* I^A allele decreases from over 70% in individuals with otolith type 1 (certain NCC) to around 5% in individuals with otolith type 5 (certain NEAC). The fraction of *Pan* I^{AB} heterozygotes is somewhat higher in individuals of otolith type 2 and 4 (uncertain NCC and uncertain NEAC, respectively), 29% and 45%, than in individuals of otolith type 1 and 5, which had heterozygote fractions of 24% and 17%, respectively.

The frequencies of haemoglobin variants varied between stations, and followed a similar pattern as the *Pan* I locus, with different frequency distributions in stations close to the mainland and further out. The frequency of the most common allele HbI^2, varied from around 0.7 at the stations dominated by individuals with otolith type 1 and 2 to around 0.85 to 0.90 at stations dominated by individuals with NEAC otolith types. Three alleles in all were found at this locus.

The number of alleles in allozymes varied from 2 (IDH) to 6 (GPI-1). In general, there was little variation between stations in both years, and no apparent trend in allele distributions at any locus. Some stations were monomorphic for IDH.

Hardy–Weinberg and linkage disequilibrium tests

The overall sample collection was tested according to Hardy–Weinberg equilibrium for each of the thirteen loci investigated. Five of the loci revealed significant departure from HW and all these were due to heterozygote deficiency. These loci included *Pan* I (P <0.0001); Gmo132 ($P = 0.034$); Tech11 (P <0.0001), Gmo3 ($P < 0.0001$) and HbI (P = 0.0054). The overall test including all loci and all samples demonstrated highly significant ($P < 0.0001$) departure from HW due to deficiency of heterozygotes.

The two-tailed test for Hardy–Weinberg equilibrium, considered for each population and with sequential Bonferroni correction for multiple tests (Rice, 1989), showed 11 significant deviations from equilibrium. Deficiency of heterozygotes was observed at locus *Pan* I at stations 222, 249 and 250, at haemoglobin at station 249, at locus Tch11 at stations 232, 236 and 246, at locus Gmo2 at stations 247, 254, and 246, and at locus Gmo3 at station 246. No instances of heterozygote excess were observed.

In the tests for linkage disequilibrium of the samples from 2002 to 2003 three significant deviations from equilibrium were found after Bonferroni correction for 78 tests; between *Pan* I and Gmo34, between *Pan* I and Gmo132 and between *Pan* I and Haemoglobin.

Genetic differentiation

The sample collection was tested for homogeneity at the various loci analysed and significant differentiation among samples following sequential Bonferroni correction was found at all loci except Gmo35, Tch11, GPD, IDH, PGM, GPI and LDH.

The exact tests for pairwise genetic differentiation showed significant differences between most stations in both years (see Table 3), but there were also stations that were not significantly different. The largest differences in 2002 were found between the innermost and outermost stations. The inner (stations 232 and 248) and outer stations (stations 222, 242, 246) in 2003, consisting of mostly of individuals with coastal and oceanic type otoliths, respectively, formed two groups with small and not significant differences within the groups, and large differences between them (Table 3). F_{ST} values between stations ranged from zero to 0.12. A test of genetic differentiation between year classes 6, 7 and 8 of individuals pooled from outer stations 222, 241 and 246, with 85, 82 and 33 individuals, respectively, showed no significant differences across loci ($P = 0.55$), nor at any individual locus (P-values ranging from 0.11 to 0.86), demonstrating temporal stability and homogeneity within the samples from these stations.

The greatest genetic differentiation among the microsatellite loci was observed at loci Gmo34 and Gmo132, while the other four loci showed little or no differentiation measured as F_{ST} (see Table 4). Pairwise F_{ST} between stations calculated for microsatellites alone ranged from zero to 0.0387 across loci (data not shown, available from the authors upon request), with stations 241 and 248 being most

Table 3 Matrix of exact tests of genic differentiation at the 13 loci (above diagonal), and genetic differentiation estimated as F_{ST} (below diagonal). Bold indicates values that remain significant after sequential Bonferroni correction

	(1)	(2)	(3)	(4)	(5)	(6)	(7)	(8)	(9)	(10)	(11)	(12)
St 222 (1)		**<0.0001**	**<0.0001**	0.4588	0.5349	**<0.0001**	**<0.0001**	**<0.0001**	**<0.0001**	0.0295	0.1524	0.5162
St 232 (2)	0.0861		**<0.0001**	**<0.0001**	**<0.0001**	0.5188	**0.0006**	**<0.0001**	0.0508	**<0.0001**	**<0.0001**	**<0.0001**
St 236 (3)	0.0144	0.0337		**<0.0001**	**<0.0001**	**<0.0001**	**<0.0001**	**<0.0001**	**0.0010**	0.2836	0.0126	**<0.0001**
St 241 (4)	0.0000	0.1014	0.0202		0.6405	**<0.0001**	**<0.0001**	**<0.0001**	**<0.0001**	**<0.0001**	0.0126	0.8710
St 246 (5)	0.0001	0.0942	0.0165	-0.0008		**<0.0001**	**<0.0001**	**<0.0001**	**<0.0001**	**<0.0001**	0.0200	0.4327
St 248 (6)	0.1015	-0.0005	0.0442	0.1174	0.1106		0.0169	**<0.0001**	**<0.0001**	**<0.0001**	**<0.0001**	**<0.0001**
St 247 (7)	0.0642	0.0064	0.0193	0.0748	0.0699	0.0065		0.0363	0.2406	**<0.0001**	**<0.0001**	**<0.0001**
St 249 (8)	0.0351	0.0300	0.0093	0.0415	0.0419	0.0354	0.0099		0.1399	**0.0011**	0.0080	**<0.0001**
St 250 (9)	0.0397	0.0109	0.0117	0.0532	0.0479	0.0165	0.0031	0.0111		0.00125	**<0.0001**	**<0.0001**
St 253 (10)	0.0073	0.0475	0.0009	0.0123	0.0130	0.0581	0.0260	0.0117	0.0149		0.4098	0.0098
St 254 (11)	0.0017	0.0689	0.0082	0.0072	0.0062	0.0828	0.0475	0.0166	0.0287	0.0026		0.3221
St 255 (12)	-0.0009	0.0945	0.0193	-0.0001	0.0024	0.1108	0.0715	0.0422	0.0438	0.0132	0.0018	

Table 4 Relative score of individual loci in individual assignment evaluated in WHICHLOCI

Rank	Locus	% (Relative score)
1	PAN	20.8
2	Gmo132	19.1
3	Gmo34	17.4
4	HEM	16.0
5	Tch11	9.8
6	Gmo2	8.5
7	Gmo3	2.9
8	Gmo35	2.3
9	PGM	1.5
10	PGI-1	1.0
11	GPD	0.4
12	LDH-3	0.2
13	IDH-2	0.2

differentiated. Pairwise F_{ST} values for the *Pan* I locus were high between inshore and offshore stations, with the highest observed value of 0.68 found between stations 232 and 241. The highest value for the haemoglobin locus of 0.14 was also found between these two stations. The pairwise F_{ST} values for the allozyme loci were in general low. A regression analysis of the correlation between F_{ST} values at the *Pan* I locus and the microsatellite loci was highly significant ($r^2 = 0.82$, $P < 0.0001$).

The three UPGMA dendrograms of the samples from 2003 presented in Fig. 2 show a very similar topology. In all three the stations 232 and 248 form a very tight cluster, as do the stations with dominance of NEAC otolith types; 222, 241 and 246. Station 236 takes an intermediate position. Excluding the *Pan* I and haemoglobin loci from the marker set did not have any influence on the tree topology.

Mixed stock analysis and individual assignment analysis

(a) Using baselines defined from geographical distribution of samples

The two baseline data sets constructed from outer (222, 241 and 246) and inner (232, 247) stations in 2003 were highly differentiated with an F_{ST} value of 0.11 across 13 loci. The largest differentiation was observed at the *Pan* I locus, which had an F_{ST} value of 0.68. Figure 3 shows the graphical distribution of individuals from the factorial analysis conducted in GENETIX. There is little overlap between the two groups, but the spread of individuals of NCC type is larger than for those of NEAC type. The self-assignment test conducted in GENECLASS2 resulted in correct assignment of 402 of the 456 individuals in the two baseline groups, corresponding to 88.2%. The analysis of the performance of the different loci in a self-assignment test on the two baselines conducted in WHICHLOCI, showed that the *Pan* I locus contributed most to correct assignment, followed by Gmo132, Gmo34 and Haemoglobin. The other loci contributed little or nothing (allozymes) to the assignment. The ranking and relative score of the loci are given in Table 4.

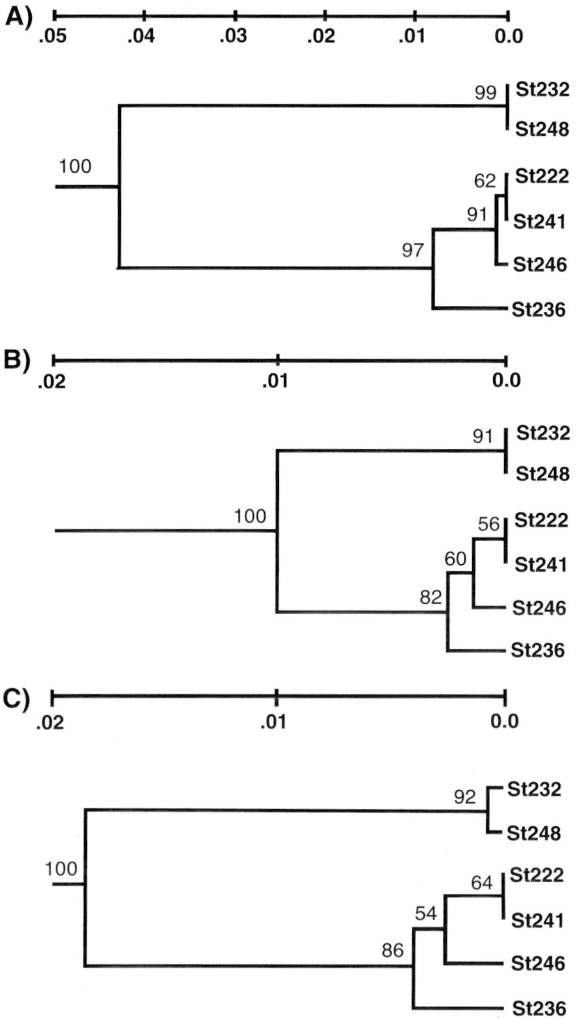

Fig. 2 UPGM dendrograms of the samples collected in 2003, based on Nei's unbiased genetic distance. (**A**) Including all 13 loci, (**B**) Based on five allozymes and six microsatellites (excluding *Pan* I and haemoglobin) (**C**) Based on six microsatellites. Numbers by nodes indicate percentage of similar replicates in 1000 bootstraps

The results from the classification of the individuals of the mixed samples (236, 247, 249, 250, 253, 254 and 255) in the individual assignment tests and MSA analysis are presented in Fig. 4. The correlation between classification based on otolith types, and the classification based both on the individual assignment test and the MSA was very high when all 13 loci were used in the analysis. The difference in the estimate of stock proportions between MSA and otolith classification ranged from 1.9 to 9.5%, with an average difference of 5.9%. A regression analysis showed that the correlation between otolith classification and MSA analysis was highly significant ($r^2 = 0.97$, $P = 0.0003$). The results achieved when the MSA was based on microsatellite loci alone, were similar, with deviations from the otolith classification ranging from 2.7 to 13.2%, average 6.4%. When the stock proportions were estimated from the distribution of *Pan* I genotypes differences increased somewhat; ranging from 1.6 to 19.5%, average 8.7%. Apparently, the differences between estimates based on otoliths and *Pan* I increased with increasing fraction of NEAC in the samples.

(b) Mixed stock analysis based on genetic information alone, using Structure

The analyses in Structure, using various combinations of markers, gave somewhat differing results as to possible number of genetic groupings within the sampled individuals. Using all 13 markers, there were strong indications in both years of the presence of two groups within the sampled individuals (Table 5). When the analysis was run using only allozymes and Haemoglobin, no stabilization of the value of ln Pr(X|K) was observed when K = 1...4 was simulated. Similar results were obtained when the analysis was run using only microsatellites, with no apparent stabilisation of the ln Pr(X|K) value, and thus no definite indication of the value of K. The simulations of $K > 2$, for both allozymes/haemoglobin and microsatellites also resulted in fairly even distribution of individuals within samples to the clusters, indicating that no structure was detected. The analysis using microsatellites with K = 2 however, resulted in a fairly uneven partitioning of individuals largely corresponding to what was observed using all 13 markers.

The stock proportions estimated at different stations by Structure (assuming K = 2), were in the same range as those estimated above by the other methods (Fig. 4). The greatest difference between the estimates of the NEAC fraction by otoliths and Structure were observed at the outermost stations. The differences in estimates ranged from 1.8 to 17.6%, on average, the difference in estimates was 9.7%. Further, a comparison of the classification of 432 individual fish in Geneclass and Structure, showed that 94.9% of the individuals were assigned to the same stock by both methods.

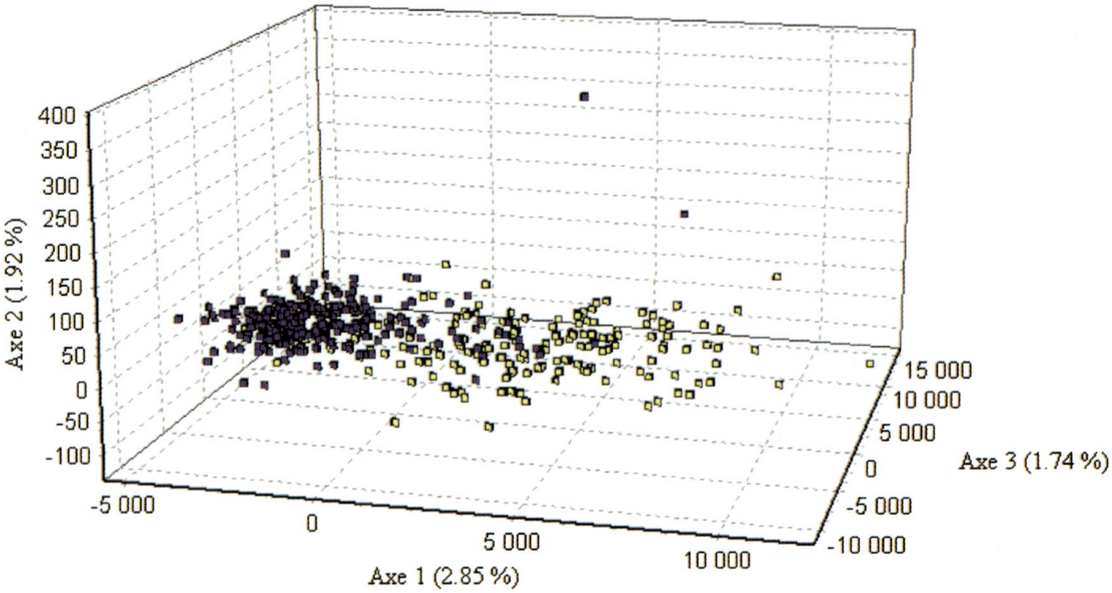

Fig. 3 Diagram of the factorial analysis conducted in Genetix showing the separation individuals in the two baseline populations. Dark blue points represent NEAC, and yellow points NCC

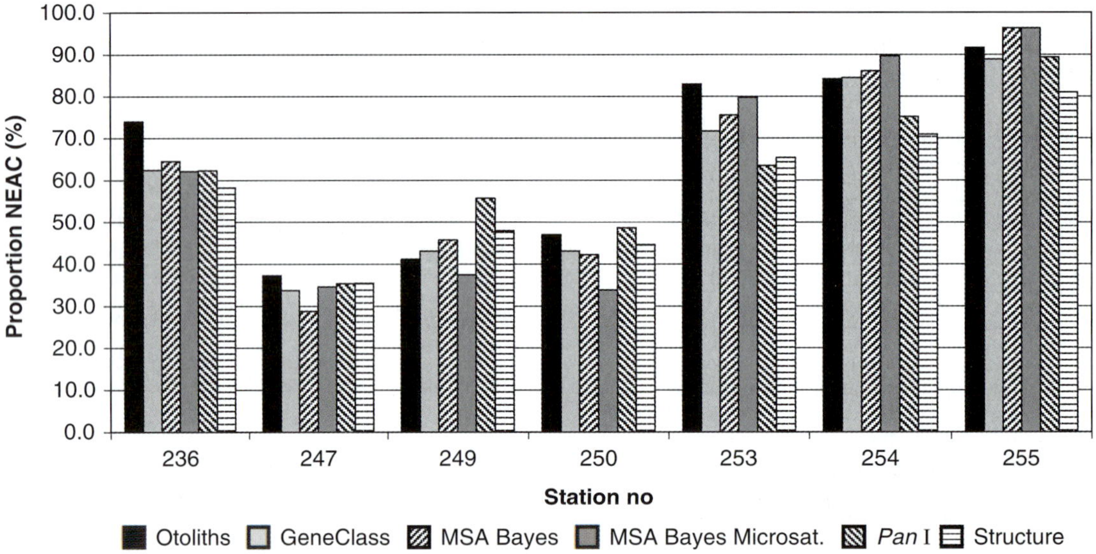

Fig. 4 Estimation of fraction of NEAC in samples by different methods and different combinations of genetic markers. Otoliths: Proportion of otolith types 4 & 5, GeneClass: Proportion individuals assigned to NEAC baseline, MSA Bayes: Stock proportions estimated from 13 loci, MSA Bayes microsat: stock proportions estimated from six microsatellites alone, Structure: stock proportions estimated using all markers in Structure

Table 5 Estimated posterior probabilities of K for different combinations of markers

K	2002			2003		
	$Pr(K/X_{all})$	$Pr(K/X_{enz})$	$Pr(K/X_{micro})$	$Pr(K/X_{all})$	$Pr(K/X_{enz})$	$Pr(K/X_{micro})$
1	~0.000	~0.000	~0.000	~0.000	~0.000	~0.000
2	~1.000	~0.000	~0.000	~1.000	~0.000	~0.000
3	~0.000	~0.000	~0.000	~0.000	~0.000	~0.000
4	~0.000	~0.000	~0.000	~0.000	~0.000	~0.000

Discussion

The results both from otolith classification and the genetic analyses confirmed earlier reports of concurrent presence of more than one group of cod present at the Lofoten spawning grounds (Rollefsen, 1954; Hylen, 1964; Møller, 1966; Nordeide, 1998). The model based clustering conducted in STRUCTURE, using all markers, indicated that the most likely number of groups was two, though further structuring within these groups cannot be excluded. Also, the STRUCTURE analysis using only allozymes and haemoglobin, and only microsatellites, did not give conclusive results as to the number of groups present within the samples, though there were indications that the most likely value of K was 2 when using microsatellites. In the spawning area under investigation, the Lofoten area, it is reasonable to expect the presence of spawners from several of the fjords in close proximity of the main spawning ground. This might explain the apparent absence of structuring in the samples when analysing the allozyme and microsatellite data separately. Both these markers types have preciously shown differentiation between different fjord locations (Jørstad & Nævdal, 1989, Knutsen et al., 2003).

The two main groups identified were, however, somewhat unevenly distributed. In both sampling years, cod of otolith classes 1 and 2 (NCC) predominated in the most inshore stations, whereas cod of otolith classes 4 and 5 predominated in the offshore sampling stations. Geographically intermediate sampling stations comprised a mixture of individuals of all otolith classes.

Three different classes of genetic markers (*Pan* I, microsatellites and HbI) revealed significant differences between the two groups of cod. The different classes of genetic markers applied in this study discriminated between samples of putative coastal and oceanic cod to varying extents. The differences in allele frequencies observed at the *Pan* I locus among groups of different otolith classes were substantial, and in accordance with earlier reports (Berg et al., 2005). Significant differences between the two groups were also observed at two of the microsatellite loci (Gmo34 and Gmo132). These loci showed little or no differentiation among samples within the two groups of cod. Allele frequencies between cohorts within the NEAC group at the microsatellite loci displayed temporal stability, in contrast to previous reports of temporal fluctuations in allele frequencies of Gmo2 and Gmo132 between cohorts in a local NCC population in Trondheimsfjorden, middle Norway (Karlson & Mork, 2005).

The level of the observed differentiation at *Pan* I is remarkably high compared to other classes of markers. The pairwise F_{ST} values observed for the microsatellites between the two groups are also high compared to those reported between offshore and inshore cod off Newfoundland (Ruzzante et al., 1997), but comparable to those observed between North Sea cod and Barents Sea cod (Hutchinson et al., 2001). The pairwise F_{ST} values observed between NCC and NEAC in the present study are of a higher magnitude than those observed in other studies of small to medium scale genetic differentiation in cod (Hutchinson et al., 2001; Knutsen et al., 2003).

The distribution of haemoglobin allele frequencies was also significantly different in the two groups, in accordance with earlier studies (Møller, 1966; Dahle & Jørstad, 1993; Nordeide & Pettersen, 1998). Thus the present data have given strong support for the suggestion that two genetically different groups of Atlantic cod spawn in the Lofoten area. The relatively strong bootstrap support of the clusters of the UPGMA dendrograms, based on different combinations of genetic markers, underline this (Fig. 2), as does the separation of individuals in the baseline populations in the factorial analysis (Fig. 3). Among the markers used herein only allozymes were unable to differentiate between the alleged NCC and NEAC samples.

The results of the mixed stock analyses, and individual assignment tests, employing different combinations of markers, showed that the stock proportions estimated from otoliths and genetic markers were highly correlated. Moreover, the stock proportion estimates based on *Pan* I analyses and microsatellites were also in good concordance, and differentiation among sampling stations measured as F_{ST} based on the two marker classes were highly correlated. The estimates based on individual assignment and MSA did not differ much (Fig. 4). Koljonen et al. (2005) found that the estimates between individual assignment and MSA differed most when the genetic differentiation between baseline populations was small. In the present study, the differentiation between the baseline populations was substantial, and resulted in similar performance of the two methods. The high correct assignment of 88% achieved in the self-assignment test conducted in GENECLASS2, also demonstrated that the individual assignment tests performed well with the present level of differentiation. The good concordance between results from the assignment of individuals in STRUCTURE, without using information of morphology or sampling locality, and the assignment conducted in GENECLASS2 confirmed the strong association between morphology, distribution of spawning groups and genetics. Among the greatest deviations in the group proportion estimates were between the estimates based on *Pan* I and otoliths at stations with a high fraction of NEAC, where the *Pan* I estimates of NEAC fractions were lower than the other estimates. Such a single locus estimate is, however, much more sensitive to sampling error in the baseline populations than methods based on multiple loci. For reliable estimates of allele frequencies of *Pan* I alleles in NCC and NEAC, larger sample sizes of baseline populations than those used herein should be analysed. The NEAC fraction estimated in STRUCTURE was also lower than the estimate obtained by other methods at the same stations. However, the results of this study suggest that otolith classification can in fact be used as a reliable method for estimating stock proportions of NCC and NEAC in the area studied, in support of Berg et al. (2005).

Though the relative effects of environment and genetic background on otolith shape and structure is not clear, as there are to our knowledge no published studies on this, otoliths may serve as an indicator of life history pattern, and thus indirectly of genetic background of individuals. Løken & Pedersen (1996) investigated the effects of parent type and temperature on vertebrae number, another morphological trait differing between NCC and NEAC, and found that the vertebrae numbers were at least partly genetically determined. Using genetic methods in combination with otolith classification provide useful tools for managing this mixed stock fishery, rendering possible differential regulation of fishing mortality on NCC and NEAC in the Lofoten area.

The distinction of cod groups in the northeast Atlantic has been a matter of debate and conflicting views have existed as to whether NEAC and NCC really represent two distinct sub-units with limited gene flow between them (Arnason & Palsson, 1996; Fevolden and Pogson, 1997; Borisov et al., 1999; Pogson & Fevolden, 2003; Sarvas & Fevolden, 2005). The selective neutrality of some of the markers that show the greatest divergence between the two groups of cod has been questioned. The early studies by Sick (1965) and Møller (1966, 1968) indicated that haemoglobin could be used as a reliable marker for the separation of cod populations. Later studies have questioned the consistency of haemoglobin variation due to possible temperature-induced selection on its genotype frequencies (e.g. Karpov & Novikov, 1980; Brix et al., 2004; Petersen & Steffensen, 2003). Notwithstanding, Dahle & Jørstad (1993) examined haemoglobin variation in more than 5000 individuals and did not find the data consistent with the model of a temperature induced clinical variation as proposed by Karpov & Novikov (1980). Used with caution, and as in the present study in combination with other markers, we believe haemoglobin provides valuable information in attempts to discriminate the two groups in question.

In numerous papers it has been suggested that also the *Pan* I locus is influenced by selectional constraints (in cod; Pogson, 2001; Beacham et al., 2002; Fevolden & Pogson, 1997; Karlson & Mork, 2003; Pogson & Mesa, 2004; Case et al., 2005, 2006, and in walleye pollock; Canino & Bentzen, 2004; Canino et al., 2005). Although the dynamic of selection on *Pan* I is far from being completely understood, the marker's role as a valid contributor to elucidate stock structuring has been recognized (Fevolden & Pogson, 1997; Pogson & Fevolden, 2003; Case et al. 2005). Pogson & Fevolden (2003) suggested that the differences observed today between NCC and NEAC in distribution of the *Pan* I^A and *Pan* I^B allele is the result of recent diversifying selection more than historical isolation. In recently segregated populations genetic differences will first appear in loci subjected to selection, if the selective forces are different for the populations. This could explain why more neutral mtDNA markers have shown less divergence among populations of Atlantic cod (Arnason & Palsson, 1996).

Large variation in the magnitude of F_{ST} estimates among the microsatellites employed in this study was observed. The highest differentiation between NCC and NEAC was observed at the loci Gmo34 and Gmo132, while pairwise F_{ST} values at the other four microsatellites were insignificant. Higher F_{ST} values at certain loci, relative to other comparable loci, have been interpreted as indicative of non-neutrality (Beaumont, 2005 and references therein). Low *P*-values across samples were observed in the test of linkage disequilibrium between Gmo132, Gmo34 and *Pan* I. Linkage disequilibrium can be caused by natural selection, physical linkage between loci, genetic drift in a finite population, or population subdivision (Waples & Smouse, 1990). Looking at individual samples, there was no indication that instances of significant linkage disequilibrium were more frequent at stations with a more even mixture of individuals with NCC and NEAC otolith types than at stations dominated by either type. These results may indicate that selection is operating somewhere in the genomic region(s) in which these markers reside, and that the selection differs between inshore and offshore areas. That selection may be influencing the allelic distribution at the Gmo132 locus was confirmed by a recent study (Nielsen et al., 2006). In their study, which included samples of cod from the Baltic, North and Barents Seas, and Newfoundland, they suggested that this locus is subjected to hitch-hiking selection. However, Ruzzante et al. (1996) did not find significant differences in the distribution of alleles at the Gmo132 locus in their comparison of inshore and offshore cod off Newfoundland, and in a comparative study of microsatellite and *Pan* I variation in cod linkage disequilibrium between Gmo132 and *Pan* I was not observed (Skarstein et al., 2006). The lower number of alleles observed at Gmo132 and Gmo34 in NEAC compared to NCC could be indicative of directional selection in operation, but could also be explained by alternative models. Varying mutations rates at different microsatellite loci can result in different timescales for build-up of genetic differences. However, the analyses applied by Nielsen et al. (2006) takes this possibility into account, indicating that varying mutation rates would not be sufficient to explain the observed differences among loci. The observed lower diversity at some loci in NEAC was also noted by Hutchinson et al. (2001). They used five microsatellite loci, including one common to the present study (Gmo2), and found lower genetic diversity in a Barents Sea sample compared to samples from the North Sea. They suggested that the lower genetic diversity in the Barents Sea could be a result of a recent bottleneck and barriers to gene flow.

The lower genetic diversity observed in NEAC may indicate that NCC represents the ancestral life history pattern of cod and at some stage a group of cod adopted a migratory life history. Pogson (2001) also suggested that the *Pan* I^A allele class, predominant in coastal cod, is the older of the two allele classes in the *Pan* I gene, and showed more intra-allele variation than the more recently developed *Pan* I^B allele.

The lack of differentiation at four microsatellite loci is in itself not evidence of absence of reproductive isolation. The mutational processes generating allelic variation in microsatellites are still not fully understood, and may not affect all microsatellites in the same way. O'Reilly et al. (2004) used 14 microsatellite loci to investigate population structuring in walleye pollock in the Pacific ocean. In their study, they found that there was a significant negative correlation between measures of locus polymorphism (allelic richness and expected heterozygosity) and estimates of F_{ST}. They suggested that this might be a result of the mutational properties of highly variable microsatellites, resulting in homoplasy masking population differences. In the present study however, the highest values for F_{ST} among microsatellites were found at the Gmo132 locus, which also had the highest number of alleles; 33. The second largest value was observed at the Gmo34 locus, which had a moderate number of alleles; 8. Other loci, with both high and moderate numbers of alleles, did not differentiate the two groups of cod, and there was no significant correlation between number of alleles and differentiation measured as F_{ST} in the microsatellite markers employed in this study.

As to the reproductive isolation between the two groups of cod, the markers used herein may not provide fully conclusive results. However, lack of detectable differences at selectively neutral loci is in itself inconclusive regarding the amount of divergence among population units. Even if populations are indistinguishable at such loci, undetected differences may exist at loci of adaptive significance. In contrast, differentiation displayed in neutral loci is always conclusive with respect to divergence of at least part of the genome and implies the potential for differences throughout the entire genome (Ryman et al., 1995). We believe, as have been suggested by others (Fevolden & Pogson, 1997; Pogson & Fevolden, 2003; Sarvas & Fevolden, 2005), that the large differences observed at the *Pan* I locus, where NCC and NEAC are almost fixed for different alleles, cannot result from selection alone. A very likely contributor to the divergence must be some degree of reproductive isolation. Moreover, such a genetic division of inshore and offshore cod is not exceptional to the NE Atlantic but has been shown to exist in the NW Atlantic at Newfoundland (Ruzzante et al., 1997). Tagging studies at Greenland have demonstrated the existence of offshore migratory cod and local non-migratory cod in fjords there (Storr-Paulsen et al., 2004), although their genetic divergence has not yet been confirmed. The results of this study suggest that the two groups of cod should be managed independently and further that the effects of the mixed stock fishery in the Lofoten area on the local NEAC populations contributing to the NCC group should be evaluated.

Acknowledgements Biological sampling and analyses were conducted by the staff of Institute of Marine Research (IMR) during the two research vessel surveys, providing the biological data used in this study. These surveys were partially linked to the ongoing Norwegian/Russian genetic studies on Atlantic cod, and G.G. Novikov and A. Semenova participated in the research surveys, including blood sampling and HbI analyses. We are also indebted to technical assistance in *Pan* I analyses at the Norwegian College of Fishery Science, Tromsø (Judith Volkers, Tanja Hanebrekke) and IMR (Eva Farestveit) and microsatellite DNA analyses (Eva Farestveit and Ole Ingar Paulsen). We are also grateful for the assistance of Michele Masuda in preparing input files for the BAYES program.

References

Arnason, E. & S. Palsson, 1996. Mitochondrial cytochrome B DNA sequence variation of Atlantic cod *Gadus morhua*, from Norway. Molecular Ecology 5: 715–724.

Arnason, E., S. Palsson & A. Arason, 1992. Gene flow and lack of differentiation in Atlantic cod, *Gadus morhua* L., from Iceland, and comparison of cod from Norway and Newfoundland. Journal of Fish Biology 40: 751–770.

Arnason, E., P. H. Petersen, K. Kristinsson, H. Sigurgislason & S. Palsson, 2000. Mitochondrial cytochrome B DNA sequence variation of Atlantic cod from Iceland and Greenland. Journal of Fish Biology 56: 409–430.

Banks, M. A. & W. Eichert, 2000. Whichrun (Version 3.2): A computer program for population assignment of individuals based on multilocus genotype data. Journal of Heredity 91: 87–89.

Banks, M. A., W. Eichert & J. B. Olsen, 2003. Which genetic loci have greater population assignment power?. Bioinformatics 19: 1436–1438.

Beacham, T. D., J. Brattey, K. M. Miller, K. D. Le & R. E. Withler, 2002. Multiple stock structure of Atlantic cod (Gadus morhua) off Newfoundland and Labrador determined from genetic variation. ICES Journal of Marine Science 59: 650–665.

Beaumont, M. A., 2005. Adaptation and speciation: what can F_{ST} tell us?. Trends in Ecology & Evolution 20: 435–440.

Belkhir K., P. Borsa, L. Chikhi, N. Raufaste & F. Bonhomme, 2004. GENETIX 4.05, logiciel sous Windows TM pour la génétique des populations. Laboratoire Génome, Populations, Interactions, CNRS UMR 5171, Université de Montpellier II, Montpellier (France).

Berg, E., T. Sarvas, A. Harbitz, S. E. Fevolden & A. B. Salberg, 2005. Accuracy and precision in stock separation of north-east Arctic and Norwegian coastal cod by otoliths – comparing readings, image analysis and a genetic method. Marine and Freshwater Research 56: 753–762.

Borge, A., Å. Fotland, H. Gjøsæter & H. Mjanger 2002. Manual for Sampling of Fish and Crustaceans, version 1.0. Institute of Marine Research.

Borisov, V. M., V. P. Ponomarenko & N. A. Yaragina, 1999. A critical review of the population status in coastal cod (*Gadus morhua*) from Barents Sea region and northern Norway. Journal of Ichthyology 39: 18–28.

Brix, O., S. Thorkildsen & A. Colosimo, 2004. Temperature acclimation modulates the oxygen binding properties of the Atlantic cod (*Gadus morhua* L.) genotypes–HbI*1/1, HbI*1/2, and HbI*2/2–by changing the concentrations of their major haemoglobin components (results from growth studies at different temperatures). Comparative Biochemistry and Physiology—Part A: Molecular & Integrative Physiology 138: 241–251.

Brooker, A. L., D. Cook, P. Bentzen, J. M. Wright & R. W. Doyle, 1994. Organization of microsatellites differs between mammals and cold-water teleost fishes. Canadian Journal of Fisheries and Aquatic Sciences 51: 1959–1966.

Cadrin, S. X., K. D. Friedland. & J. R. Waldman, 2005. Stock Identification Methods—Applications in Fishery Science. Elsevier Academic Press, Burlington, MA. pp 719.

Canino, M. F. & P. Bentzen, 2004. Evidence for positive selection at the Pantophysin (*Pan* I) locus in walleye pollock, *Theragra chalcogramma*. Molecular Biology and Evolution 21: 1391–1400.

Canino, M. F., P. T. O'Reilly, L. Hauser & P. Bentzen, 2005. Genetic differentiation in walleye pollock (*Theragra chalcogramma*) in response to selection at the pantophysin (*Pan* I) locus. Canadian Journal of Fisheries and Aquatic Sciences 62: 2519–2529.

Carvalho, G.R. & L. Hauser, 1994. Molecular genetics and the stock concept in fisheries. Reviews in Fish Biology and Fisheries 4: 326–350.

Case, R. A. J., W. F. Hutchinson, L. Hauser, C. Van Oosterhout & G. R. Carvalho, 2005. Macro- and micro-geographic variation in pantophysin (*Pan* I) allele frequencies in NE Atlantic cod *Gadus morhua*. Marine Ecology-Progress Series 301: 267–278.

Case, R. A. J., W. F. Hutchinson, L. Hauser, V. Buehler, C. Clemmesen, G. Dahle, O.S. Kjesbu, E. Moksness,

H. Otterå, H. Paulsen, T. Svåsand, A. Thorsen & G. Carvalho, 2006. Association between growth and *Pan* I* genotype within Atlantic cod full-sibling families. Transactions of the American Fisheries Society 135: 241–250.

Cornuet, J.- M., S. Piry, G. Luikart, A. Estoup & M. Solignac, 1999. New methods employing multilocus genotypes to select or exclude populations as origins of individuals. Genetics 153: 1989–2000.

Dahle, G., 1991. Cod, *Gadus morhua* L, populations identified by mitochondrial DNA. Journal of Fish Biology 38: 295–303.

Dahle, G. & K. E. Jørstad, 1993. Haemoglobin variation in cod – a reliable marker for Arctic cod (*Gadus morhua* L.). Fisheries Research 16: 301–311.

Dawson, K. J. & K. Belkhir, 2001. A Bayesian approach to the identification of panmictic populations and the assignment of individuals. Genetical Research 78: 59–77.

Fevolden, S. E. & G. H. Pogson, 1997. Genetic divergence at the synaptophysin (*Syp*I) locus among Norwegian coastal and north-east arctic populations of Atlantic cod. Journal of Fish Biology 51: 895–908.

Fournier, D. A., T. D. Beacham, B. E. Riddell & C. A. Busack, 1984. Estimating stock composition in mixed stock fisheries using morphometric, meristic, and electrophoretic characteristics. Canadian Journal of Fisheries and Aquatic Sciences 41: 400–408.

Frydenberg, O., D. Møller, G. Nævdal & K. Sick, 1965. Haemoglobin polymorphism in Norwegian cod populations. Hereditas-Genetiskt Arkiv 53:257.

Frydenberg, O., J. T. Nielsen & V. Simonsen, 1969. The maintenance of haemoglobin polymorphism of the cod. Japanese Journal of Genetics 44: 160–165.

Goudet, J., 2001. FSTAT: A Program to Estimate and Test Gene Diversities and Fixation Indices. Version 2.9.3.2 Available from http://www.unil.ch/izea/softwares/fstat.html.

Hansen, M. M., E. Kenchington & E. Nielsen, 2001. Assigning individual fish to populations using microsatellite markers. Fish and Fisheries 2: 93–112.

Husebø, Å., A. K. Imsland & G. Nævdal, 2004. Haemoglobin variation in cod: a description of new variants and their geographical distribution. Sarsia 89: 369–378.

Hutchinson, W. F., G. R. Carvalho & S. I. Rogers, 2001. Marked genetic structuring in localised spawning populations of cod *Gadus morhua* in the North Sea and adjoining waters, as revealed by microsatellites. Marine Ecology-Progress Series 223: 251–260.

Hylen, A., 1964. Coastal cod and skrei in the Lofoten area. Fiskeridirektoratets Skrifter Serie Havundersøkelser 13: 27–42.

Jonsdottir, O. D. B., A. K. Danielsdottir & G. Nævdal, 2001. Genetic differentiation among Atlantic cod (*Gadus morhua* L.) in Icelandic waters: temporal stability. ICES Journal of Marine Science 58: 114–122.

Jørstad, K. E., 1984. Genetic analyses of cod in Northern Norway. Flødevigen Rapportserie 1: 745–760.

Jørstad, K. E. & G. Nævdal, 1989. Genetic variation and population structure of cod, *Gadus morhua* L, in some fjords in northern Norway. Journal of Fish Biology 35: 245–252.

Karlsson, S. & J. Mork, 2003. Selection-induced variation at the Pantophysin locus (*Pan* I) in a Norwegian fjord population of cod (*Gadus morhua* L.). Molecular Ecology 12: 3265–3274.

Karlsson, S. & J. Mork, 2005. Deviation from Hardy-Weinberg equilibrium, and temporal instability in allele frequencies at microsatellite loci in a local population of Atlantic cod. ICES Journal of Marine Science 62: 1588–1596.

Karpov, A.K. & G. G. Novikov, 1980. Haemoglobin alloforms in cod, *Gadus morhua* (Gadiformes, Gadidae), their functional characteristics and occurrence in populations. Journal of Ichthyology 20: 45–49.

Knutsen, H., P. E. Jorde, C. Andre & N. C. Stenseth, 2003. Fine-scaled geographical population structuring in a highly mobile marine species: the Atlantic cod. Molecular Ecology 12: 385–394.

Koljonen, M. L., J. J. Pella & M. Masuda, 2005. Classical individual assignments versus mixture modeling to estimate stock proportions in Atlantic salmon (*Salmo salar*) catches from DNA microsatellite data. Canadian Journal of Fisheries and Aquatic Sciences 62: 2143–2158.

Løken, S. & T. Pedersen, 1996. Effect of parent type and temperature on vertebrae number in juvenile cod, *Gadus morhua* (L), in Northern Norway. Sarsia 80: 293–298.

Millar, R. B., 1987. Maximum-likelihood-estimation of mixed stock fishery composition. Canadian Journal of Fisheries and Aquatic Sciences 44: 583–590.

Miller, M. P., 1997. Tools for population genetic analyses (TFPGA)1.3. Available from: http://www.marksgenetic software.net/tfpga.htm.

Miller, K. M., K. D. Le & T .D. Beacham, 2000. Development of tri- and tetranucleotide repeat microsatellite loci in Atlantic cod (*Gadus morhua*). Molecular Ecology 9: 238–239.

Mork, J. & M. Giæver, 1999. Genetic structure of cod along the coast of Norway: results from isozyme studies. Sarsia 84: 157–168.

Mork, J., N. Ryman, G. Ståhl, F. Utte & G. Sundnes, 1985. Genetic variation in Atlantic cod (*Gadus morhua*) throughout its range. Canadian Journal of Fisheries and Aquatic Sciences 42: 1580–1587.

Møller, D., 1966. Genetic differences between cod groups in Lofoten area. Nature 212: 824.

Møller, D., 1968. Genetic diversity in spawning cod along the Norwegian coast. Hereditas-Genetiskt Arkiv 60: 1.

Nei, M., 1978. Estimation of average heterozygosity and genetic distance from a small number of individuals. Genetics 89: 583–590.

Nielsen, E. E. & E. Kenchington, 2001. A new approach to prioritizing marine fish and shellfish populations for conservation. Fish and Fisheries 2: 328–343.

Nielsen, E.E., M.M. Hansen, D.E. Ruzzante, D. Meldrup & P. Gronkjaer, 2003. Evidence of a hybrid-zone in Atlantic cod (*Gadus* morhua) in the Baltic and the Danish Belt Sea revealed by individual admixture analysis. Molecular Ecology 12: 1497–1508.

Nielsen, E. E., M. M. Hansen & D. Meldrup, 2006. Evidence of microsatellite hitch-hiking selection in Atlantic cod (*Gadus morhua* L.): implications for inferring population structure in nonmodel organisms. Molecular Ecology 15: 3219–3229.

Nordeide, J. T., 1998. Coastal cod and north-east arctic cod—do they mingle at the spawning grounds in Lofoten? Sarsia 83: 373–379.

Nordeide, J.T. & E. Kjellsby, 1999. Sound from spawning cod at their spawning grounds. ICES Journal of Marine Science 56:326–332.

Nordeide, J. T. & I. H. Pettersen, 1998. Haemoglobin frequencies and vertebral numbers of cod (*Gadus morhua* L.) off northern Norway - Test of a population structure hypothesis. Ices Journal of Marine Science 55: 134–140.

O'Reilly, P. T., M. F. Canino, K. M. Bailey & P. Bentzen, 2000. Isolation of twenty low stutter di- and tetranucleotide microsatellites for population analyses of walleye pollock and other gadoids. Journal of Fish Biology 56: 1074–1086.

O'Reilly, P. T., M. F. Canino, K. M. Bailey & P. Bentzen, 2004. Inverse relationship between F_{ST} and microsatellite polymorphism in the marine fish, walleye pollock (*Theragra chalcogramma*): implications for resolving weak population structure. Molecular Ecology 13: 1799–1814.

Paetkau, D., W. Calvert, I. Stirling & C. Strobeck, 1995. Microsatellite analysis of population-structure in Canadian polar bears. Molecular Ecology 4: 347–354.

Pampoulie, C., D. E. Ruzzante, V. Chosson, T. D. Jorundsdottir, L. Taylor, V. Thorsteinsson, A. K. Danielsdottir & G. Marteinsdottir, 2006. The genetic structure of Atlantic cod (*Gadus morhua*) around Iceland: Insight from microsatellites, the *Pan* I locus, and tagging experiments. Canadian Journal of Fisheries and Aquatic Sciences 63: 2660–2674.

Pella, J. J. & M. Masuda, 2001. Bayesian methods for analysis of stock mixtures from genetic characters. Fishery Bulletin 99: 151–167.

Pella, J. J. & G. B. Milner, 1987. Use of genetic marks in stock composition analysis. In Ryman, N. & F. Utter (eds), Population Genetics and Fishery Management. University of Washington Press, Seattle, WA, USA, 247–276.

Petersen, M. F. & J. F. Steffensen, 2003. Preferred temperature of juvenile Atlantic cod *Gadus morhua* with different haemoglobin genotypes at normoxia and moderate hypoxia. Journal of Experimental Biology 206: 359–364.

Piry, S., A. Alapetite, J.M. Cornuet, D. Paetkau, L. Baudouin & A. Estoup, 2004. Geneclass2: a software for genetic assignment and first-generation migrant detection. Journal of Heredity 95: 536–539.

Pogson, G. H., K. A. Mesa & R. G. Boutilier, 1995. Genetic population structure and gene flow in the Atlantic cod *Gadus morhua* - a comparison of allozyme and nuclear RFLP loci. Genetics 139: 375–385.

Pogson, G. H., 2001. Nucleotide polymorphism and natural selection at the Pantophysin (*Pan* I) locus in the Atlantic cod, *Gadus morhua* (L.). Genetics 157: 317–330.

Pogson, G. H. & K. A. Mesa, 2004. Positive Darwinian selection at the pantophysin (*Pan* I) locus in marine Gadid fishes. Molecular Biology and Evolution 21: 65–75.

Pogson, G. H. & S. E. Fevolden, 2003. Natural selection and the genetic differentiation of coastal and Arctic populations of the Atlantic cod in northern Norway: a test involving nucleotide sequence variation at the pantophysin (*Pan* I) locus. Molecular Ecology 12: 63–74.

Pritchard, J. K., M. Stephens & P. Donnelly, 2000. Inference of population structure using multilocus genotype data. Genetics 155: 945–959.

Rannala, B. & J. L. Mountain, 1997. Detecting immigration by using multilocus genotypes. Proceedings of the National Academy of Sciences of the United States of America 94: 9197–9201.

Raymond, M. & F. Rousset, 1995. Genepop (Version-1.2) - Population-genetics software for exact tests and ecumenicism. Journal of Heredity 86: 248–249.

Rice, W. R., 1989. Analyzing tables of statistic tests. Evolution 43: 223–225.

Rollefsen, G., 1933. The otoliths of cod. Fiskeridirektoratets skrifter, serie Havundersokelser 4: 1–14.

Rollefsen, G., 1954. Observations on the cod and cod fisheries of Lofoten. Rapports et Procés-Verbaux des Réunions du Conseil International Pour l'Exploration de la Mer 133: 40–47.

Ruzzante, D. E., C. T. Taggart, C. Cook & S. Goddard, 1996. Genetic differentiation between inshore and offshore Atlantic cod (*Gadus morhua*) off Newfoundland: Microsatellite DNA variation and antifreeze level. Canadian Journal of Fisheries and Aquatic Sciences 53: 634–645.

Ruzzante, D. E., C. T. Taggart, D. Cook & S. V. Goddard, 1997. Genetic differentiation between inshore and offshore Atlantic cod (*Gadus morhua*) off Newfoundland: a test and evidence of temporal stability. Canadian Journal of Fisheries and Aquatic Sciences 54: 2700–2708.

Ruzzante, D. E., C. T. Taggart & D. Cook, 1999. A review of the evidence for genetic structure of cod (*Gadus morhua*) populations in the NW Atlantic and population affinities of larval cod off Newfoundland and the Gulf of St. Lawrence. Fisheries Research 43: 79–97.

Ruzzante, D. E., C. T. Taggart, S. Lang & D. Cook, 2000. Mixed-stock analysis of Atlantic cod near the Gulf of St. Lawrence based on microsatellite DNA. Ecological Applications 10: 1090–1109.

Ryman, N., F. Utter & L. Laikre, 1995. Protection of intraspecific biodiversity of exploited fishes. Reviews in Fish Biology and Fisheries 5: 417–446.

Sarvas, T. H. & S. E. Fevolden, 2005. The scnDNA Locus *Pan* I reveals concurrent presence of different populations of Atlantic cod (*Gadus morhua* L.) within a single fjord. Fisheries Research 76: 307–316.

Sick, K., 1961. Haemoglobin polymorphism in fishes. Nature 35: 894–896.

Sick, K., 1965. Haemoglobin polymorphism of cod in the North Sea and the North Atlantic Ocean. Hereditas 54: 49–69.

Skarstein, T. H., J. I Westgaard. & S. E. Fevolden, 2006. Comparing microsatellite variation in north-east Atlantic cod (*Gadus morhua* L.) to genetic structuring as revealed by the pantophysin (*Pan* I) locus. Journal of Fish Biology 69(Suppl. C): 261–262.

Smith, P. J., A .J. Birley, A. Jamieson & C. A. Bishop, 1989. Mitochondrial DNA in the Atlantic cod, *Gadus morhua*: lack of genetic divergence between eastern and western populations. Journal of Fish Biology 34: 369–373.

Stenevik, E. K. & S. Sundby, 2005. Buoyancy of eggs of Norwegian coastal cod from different areas along the coast. GLOBEC International Newsletter 11(1): 23.

Storr-Paulsen, M., K.Wieland, H. Hovgard & H. J. Ratz, 2004. Stock structure of Atlantic cod (*Gadus morhua*) in West Greenland waters: Implications of transport and migration. Ices Journal of Marine Science 61: 972–982.

Vikebø, F., S. Sundby, B. Ådlandsvik & Ø. Fiksen, 2005. The combined effect of transport and temperature on distribution and growth of larvae and pelagic juveniles of Arcto-Norwegian cod. ICES Marine Science Symposia 62: 1375–1386.

Waples, R. S. & P. E. Smouse, 1990. Gametic disequilibrium analysis as a means of identifying mixtures of salmon populations. American Fisheries Society Symposium 7: 439–458.

Ward, R. D., 2000. Genetics in fisheries management. Hydrobiologia 420: 191–201.

Waser, P. M. & C. Strobeck, 1998. Genetic signatures of interpopulation dispersal. Trends in Ecology & Evolution 13: 43–44.

Xu, S. Z., C. J. Kobak & P. E. Smouse, 1994. Constrained least-squares estimation of mixed population stock composition from MtDNA haplotype frequency data. Canadian Journal of Fisheries and Aquatic Sciences 51: 417–425.

Hydrobiologia (2008) 606:27–33
DOI 10.1007/s10750-008-9350-z

Contrasting levels of genetic differentiation among putative neutral microsatellite loci in Atlantic herring *Clupea harengus* populations and the implications for assessing stock structure

Phillip C. Watts · David O'Leary · Mary C. Cross · Jamie Coughlan · Eileen Dillane · Suzanne M. Kay · Suzanne Wylde · René Stet · Richard D. M. Nash · Emma M. C. Hatfield · Thomas F. Cross

© Springer Science+Business Media B.V. 2008

Abstract Microsatellite DNA loci, when used in population genetic studies, are usually assumed to be neutral (unaffected by natural selection, either directly or as a result of tight linkage), but this assumption is rarely tested. Here, the assumption of neutrality is examined using established methods, principally that based on the expected relationship between F_{ST} and heterozygosity, at 12 putative neutral microsatellite loci utilised in a study of Atlantic herring *Clupea harengus* in the north east Atlantic (west of Great Britain and around Ireland) and in the Baltic Sea. All but two of these loci demonstrate relationships that suggest that they may be regarded as neutral genetic markers. Of the other two loci, however, one shows a relationship suggestive of the action of directional selection and the other of balancing natural selection, though other locus-specific effects may operate. Thus, the latter two loci may provide inaccurate inference if used in phylogeographic studies and also demonstrate the danger of assuming neutrality at all microsatellite loci without explicit testing. However, such loci, particularly those affected by directional as opposed to balancing selection, may be of great use in stock discrimination studies, and selected loci in general, have considerably potential in studies of adaptation.

Keywords F_{ST} · Microsatellite · Neutrality · Selection

Guest editors: J. Davenport, G. Burnell, T. Cross, M. Emmerson, R. McAllen, R. Ramsay & E. Rogan
Challenges to Marine Ecosystems

P. C. Watts (✉) · S. M. Kay
Marine and Freshwater Biology Research Group, The Biosciences Building, School of Biological Sciences, University of Liverpool, Crown Street, Liverpool L69 7ZB, UK
e-mail: p.c.watts@liv.ac.uk

D. O'Leary · M. C. Cross · J. Coughlan · E. Dillane · S. Wylde · T. F. Cross
Department of Zoology, Ecology & Plant Science, Aquaculture & Fisheries Development Centre, National University of Ireland, Cork, Ireland

R. Stet
University of Aberdeen, Aberdeen, UK

R. D. M. Nash
Institute of Marine Research, PO Box 1870, Nordnes, 5817 Bergen, Norway

E. M. C. Hatfield
FRS Marine Laboratory, Aberdeen, UK

Introduction

Molecular genetic markers offer a convenient solution to the problem of inferring connectivity among populations of species that are difficult or impossible to track directly. Accordingly, genetic techniques have been used widely to study dispersal in aquatic organisms, with particular emphasis on defining stock boundaries for fisheries management. Although not able to provide a migration rate per se (see Whitlock & McCauley, 1999), Wright's (1951) estimator of genetic differentiation among populations (F_{ST}) is broadly correlated with dispersal capability (Bohonak, 1999; Kinlan & Gaines, 2003; Watts & Thorpe, 2006) and is the most widely employed measure of population connectivity. The basic premise behind the use of this (and also many other genetic statistics) is that the spatial distribution of allele frequencies is a consequence of the effects of migration, genetic drift, mutation and selection. For an estimate of migration, it is necessary to make two important assumptions: (1) mutation has a negligible effect during the contemporary timescale of the study and (2) the loci used for genetic investigations are (relatively) unaffected by selection, so-called neutral markers.

Typically, mutation is sufficiently rare to be ignored when quantifying dispersal at scales relevant to interacting populations, whereas the general presumption of marker neutrality is more contentious. For example, many studies of marine species present evidence for a possible influence of selective processes operating at certain genetic loci (e.g. Hilbish & Koehn, 1985; Pogson et al., 1995; Lemaire et al., 2000; Karlsson & Mork, 2003; Case et al., 2005). Notably, the assumption of neutrality is seldom questioned at putative non-coding markers such as microsatellites because they are expected to respond to selective processes only rarely, for instance when they are linked to a functional gene, i.e. "hitch-hiking" selection. A crucial point is that different evolutionary processes leave characteristic signatures on the genome: migration and genetic drift are expected to affect all loci more or less equally while selective processes should operate on a specific subset of gene regions. This suggests a framework to test for selective neutrality among a panel of genetic markers, with neutral loci demonstrating broadly similar levels of F_{ST} and balancing or directional (spatial) selection at a locus indicated by atypically low or high values of genetic differentiation respectively. Failure to recognise and identify these effects can lead to misinterpretation of the evolutionary processes that generate spatial genetic structure.

Lewontin & Krakauer (1973) proposed that directional selection could be detected by significantly large observed variance in values of F_{ST} (s_F^2) compared with a theoretically expected variance of F_{ST} among loci (σ_F^2) that is equal to $k\overline{F}_{ST}{}^2/(n-1)$, where k is a constant specific to the underlying distribution of allele frequencies among populations, $\overline{F}_{ST}$ is an average (over all loci) and n is the number of populations sampled. This test has been criticised, particularly because it is sensitive to the model of population structure and also correlated allele frequencies among populations (Robertson, 1975; Nei & Maruyama, 1975), and, as a consequence, is considered unreliable and rarely used. Beaumont & Nichols (1996) developed an alternative method to detect outlying loci, which is based on an expected distribution of F_{ST} conditional on heterozygosity (H_e), and several alternative statistical tests for selection have been proposed subsequently (reviewed by Guinand et al., 2004). It is apparent, however, that there is a general lack of an explicit inspection for locus-specific effects, with heterogeneity among loci explained as a stochastic (sampling) effect (under the *untested* assumption of locus neutrality) or overlooked when the data set is presented as multi-locus averages. With this in mind, Baer (1999) used the Lewontin-Krakauer test to examine 102 published allozyme data sets on fish and concluded that 'locus-specific' processes were not sufficiently strong enough to mask real population structures.

Given the widespread use of genetic markers for fisheries research (e.g. see EU-funded projects: CODTRACE, www.ucd.ie/codtrace/index.htm; HERGEN, www.hull.ac.uk/hergen/; WESTHER, www.clupea.net/westher/), it is of substantial importance that outlying loci are identified during a genetic analysis of population genetic structure or history. In this paper we examine the extent of variability among a panel of 12 microsatellite loci that were used to characterise the level of genetic differentiation among putative stocks of Atlantic herring *Clupea harengus*. Atlantic herring stocks in the North and Baltic Seas are characterised by substantial differences between the North Sea and the Baltic (Ruzzante et al., 2005), but weak if any population structure within the North Sea (Mariani et al., 2005) or Irish Sea, Celtic Sea and west of Ireland (WESTHER EU consortium, unpublished data).

Here, we quantify the level of heterogeneity in allele frequencies at 12 putative neutral microsatellite loci using several methods that were developed to identify outlying loci that may be responding to selective processes. Under a shared, neutral demographic history, all genetic markers should engender equivalent values of F_{ST}. However, we uncover significant heterogeneity in the level of spatial genetic differentiation among loci, suggesting that selective neutrality cannot be assumed within a panel of microsatellite loci.

Methods

Sample collection and genotyping

Spawning herring were sampled between 2003 and 2005 from 10 locations in the North Atlantic: six sites to the west of Great Britain and Ireland, two sites off the Norwegian Coast, one area in the Baltic Sea and one from Hudson Canyon in the western Atlantic (Fig. 1). Liver tissue or fin clips were taken from the fish and stored in absolute ethanol until sample processing. Total genomic DNA was extracted using either Chelex-100 (Walsh et al., 1991) or high-salt (Aljanabi & Martinez, 1997) standard protocols. PCR was performed using standard reagents and thermal cycling conditions for 12 polymorphic, unlinked microsatellite loci (*Cpa*101, *Cpa*111, *Cpa*112, *Cpa*114, *Cha*1020, *Cha*1202, *Cpa*4, *Cpa*6, *Cha*1017, *Cha*1059, *Cpa*107 and *Cpa*113) described by McPherson et al. (2001a), Miller et al. (2001) and Olsen et al. (2002): these loci were selected from a larger group of microsatellites as a panel that were unlinked and did not have null alleles (P. C. Watts & D. O'Leary, unpublished). Microsatellite alleles were separated on either a LiCor4200 or an ABI3100, with fragment sizes quantified against custom-made size ladders or by comparing mobilities with a GENESCAN LIZ-500 size-standard (Applied Biosystems) and scored using GENEMAPPER (Applied Biosystems) software.

Statistical analyses

Basic measures of diversity (numbers of alleles, N_a; gene diversity, H_e) were calculated by FSTAT v.2.9.3 (Goudet, 1995). Every locus was tested for departure from expected Hardy–Weinberg equilibrium (HWE) conditions for each sample using the randomisation tests (5,000 permutations) implemented by FSTAT. Individual locus values of F_{ST} over all samples were calculated using Weir & Cockerham's (1984) unbiased estimator of F_{ST} using FSTAT, with standard errors for the single-locus estimates generated by jackknifing over populations and 95% confidence intervals about the multi-locus estimates made by bootstrapping over loci.

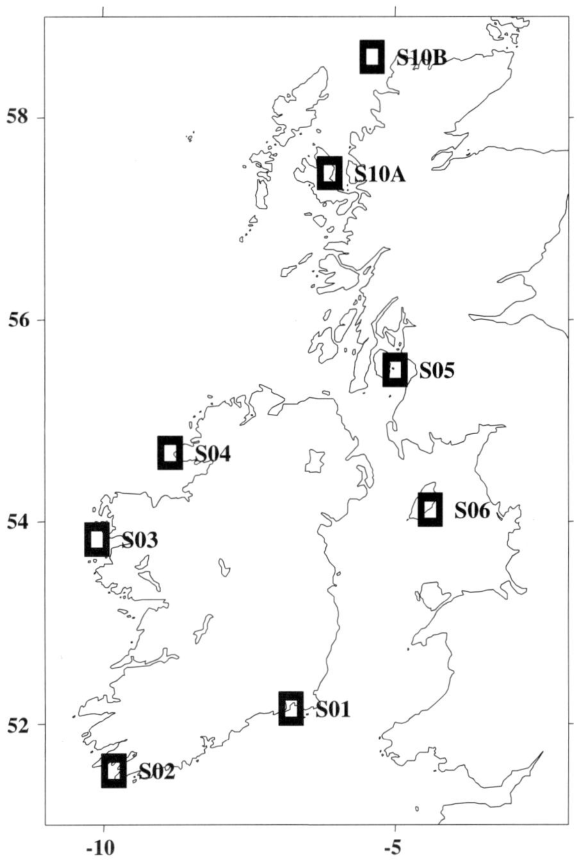

Fig. 1 Sample locations of spawning herring west of Great Britain and around Ireland

We performed Lewontin-Krakauer tests for two values of k (= 2 and 7.6) by calculating the theoretically expected variance of F_{ST} (σ_F^2) using the equation described in the *Introduction*. After simulating several allele frequency distributions, Lewontin & Krakaeur (1973) suggested that $k \leq 2$ for loci that are governed by neutral processes. To account for some of the criticisms of this test, Baer (1999) employed a more stringent threshold value of $k = 7.6$, which is based on a skewed allele frequency distribution (see Baer, 1999 for further details). The significances of any differences between σ_F^2 and s_F^2 were tested using a standard variance-ratio test.

Outlying values of genetic differentiation at specific loci were detected using the procedure described by Beaumont & Nichols (1996). Briefly, FDIST2 (http://www.rubic.rdg.ac.uk/~mab/software.html) runs coalescent simulations to generate an expected distribution of F_{ST} with heterozygosity (H_e), that is based on the average genetic differentiation among samples (over all samples and loci) for the test data set. The distribution of F_{ST} as a function of H_e is then described, with 0.025, 0.5 and 0.975 quantiles (i.e. 95% confidence limits and median) against which the locus-specific values were plotted to detect anomalous loci.

Results

As typical for this species, most *C. harengus* microsatellite loci investigated here were highly polymorphic. Average gene diversity (H_e) over all samples varied from 0.36 up to 0.92 and the mean number of alleles (N_a) ranged between seven and 30, with only one locus (*Cpa*107) having an average of less than 10 alleles (Table 1). Just four out of the 156 tests for departure from expected HWE conditions were significant ($P < 0.05$, Bonferroni correction, Rice, 1989, applied for $k = 13$ tests per sample) (data not shown). While this is indicative of non-random mating within samples, it should be noted that only four loci (*Cpa*112, *Cha*1017, *Cha*1020, *Cha*1059) demonstrated significant ($P < 0.05$, $k = 13$) deviations from expected HWE conditions over all samples.

Average genetic differentiation ($F_{ST} = 0.007$) in the study area was weak but significantly different from zero (95% confidence intervals: 0.003, 0.012). Individual locus estimates of F_{ST} varied considerably, from zero (*Cha*1202) up to ~0.028 (*Cpa*107, *Cpa*112). Rather than displaying a random distribution of values of F_{ST} among loci there is a marked dichotomy to the locus-specific pattern of differentiation: (1) most ($n = 9$) markers are characterised by very weak differentiation over all samples ($F_{ST} < 0.01$), while (2) three loci (*Cpa*6, *Cpa*107, *Cpa*112) possess mean values of F_{ST} greater than or equal to 0.02 (Table 1). Given this heterogeneity among loci it is not surprising, therefore, that the Lewontin-Krakaeur tests are significant for $k = 2$ ($\sigma_F^2/s_F^2 = 79.62$, d.f. = 12, $P << 0.001$), although the test is not significant at the more stringent value of $k = 7.6$ ($\sigma_F^2/s_F^2 = 20.95$, d.f. = 12, $P \sim 0.06$).

Table 1 Locus-specific and average estimates of genetic differentiation and genetic variability at 12 microsatellite loci from samples of Atlantic herring, *C. harengus*. F_{ST}, Wright's (1951) estimator of genetic differentiation

Locus	F_{ST}	SE	H_e	N_a
*Cpa*101	0.006	0.003	0.913	29
*Cpa*111	0.007	0.005	0.414	10
*Cpa*112	0.027	0.016	0.880	30
*Cpa*114	0.000	0.001	0.910	23
*Cha*1020	0.004	0.003	0.916	23
*Cha*1202	0.001	0.004	0.752	13
Cpa4	0.003	0.002	0.873	24
Cpa6	0.020	0.014	0.400	10
*Cha*1017	0.003	0.003	0.826	14
*Cha*1059	0.004	0.004	0.715	15
*Cpa*107	0.028	0.009	0.363	7
*Cpa*113	0.006	0.003	0.925	20
	F_{ST}	+95% CI	−95% CI	
All loci	0.008	0.013	0.004	
9 loci	0.004	0.005	0.002	
3 loci	0.025	-	-	

SE, standard error of F_{ST}; H_e, expected heterozygosity; N_a, number of alleles; 95% CI, 95% confidence intervals

When values of F_{ST} were plotted against heterozygosity (Fig. 2) we note that F_{ST} is depressed with higher values of this statistic, which is an expected consequence of high polymorphism (see Hedrick, 1999). With respect to the loci themselves, two features are evident: (1) most loci fall within the 95% envelope of neutral expectations of F_{ST} conditional on heterozygosity and (2) only two loci are identified as outliers, *Cpa*112 that lies above the upper 95% confidence interval and *Cpa*114 that is less differentiated than expected if it were a neutrally evolving locus.

Discussion

In this paper we uncovered significant heterogeneity to the overall level of population differentiation within a panel of supposedly neutral microsatellite loci. While we do not characterise the population structure of Atlantic herring in detail—these data will be published in more detail elsewhere—it is evident that this species is weakly, but significantly,

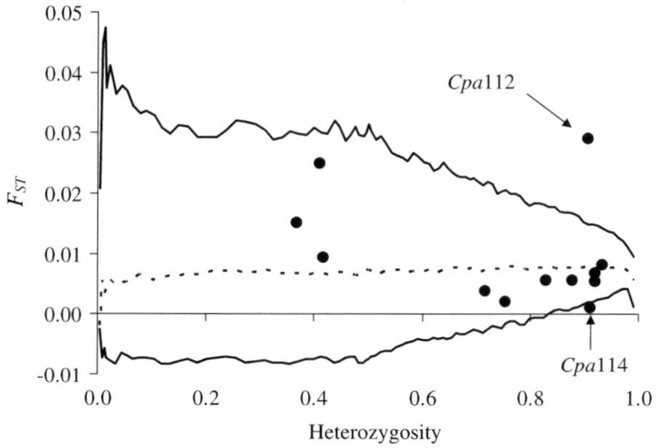

Fig. 2 Values of average F_{ST} for 12 microsatellite loci in Atlantic herring *C. harengus* against heterozygosity. Lines indicate estimated quantiles (solid = 95% confidence intervals, dashed = median) for the expected distribution of F_{ST} (= 0.008) with heterozygosity

differentiated at microsatellite loci throughout a large portion of its distribution. This is concordant with previous research (Shaw et al., 1999; Mariani et al., 2005; Ruzzante et al., 2005). However, the results presented here highlight an important, but often neglected, finding that all loci are not necessarily independent replicates of the same evolutionary processes. Thus, the level of spatial genetic structure uncovered can depend on the choice of genetic markers used.

Much of the statistical framework to assess the level of genetic differentiation among loci is presented as an appraisal of selective processes operating at one or few loci. However, interpreting significant inter-locus variation as evidence for selection itself using the Lewontin-Krakauer test is problematic because this method is sensitive to other demographic processes, such as isolation by distance, genetic structure and variation in temporal separation. These processes produce correlated values of F_{ST} among populations (Nei & Maruyama, 1975; Robertson, 1975). The potential confounding effect of such processes on our data is not known at present, although, for example, evidence for significant isolation by distance in European *C. harengus* populations is equivocal (Mariani et al., 2005). Perhaps more relevant, Baer (1999) raised the point that even without the action of selection, certain other locus-specific processes, including differences in mutation rate and the presence of null alleles, can generate significant inter-locus heterogeneity. Accordingly loci used in the present study were selected from a larger panel of microsatellites because, after preliminary data analysis, they did not appear to have null alleles, deviate from expected HWE conditions, etc. Nonetheless, substantial differences in mutation rates do seem plausible given the heterogeneity in variability among loci (Table 1), although the level of polymorphism per se is not directly associated with the value of F_{ST} (Table 1, Fig. 2). From a pragmatic standpoint, it may not be as important to identify the specific processes (i.e. selection versus mutation) that are operating, as much as actually uncovering the presence of significant inter-locus heterogeneity in the first place. The latter observation affords a more informed perspective on whether the panel of loci can be interpreted in terms of neutral evolutionary processes. As such, the Lewontin-Krakauer test presents a convenient, preliminary indicator of the presence of locus-specific effects among a panel of loci, though this should be evident from the pattern of variation in single-locus estimates of F_{ST} (Table 1), and significant results can be further explored using other computationally more intensive methods, such as that proposed by Beaumont & Nichols (1996).

In contrast with previous studies that failed to find heterogeneity among microsatellite loci (e.g. de Leon et al., 1997), two 'neutral' microsatellite loci fell outside (one above and one below) the 95% limits of an expected neutral distribution of F_{ST} against heterozygosity (Fig. 2). One interpretation is that these loci are, or are at least linked to regions that are, subject to directional and balancing selection respectively. Addressing the major criticisms laid at the Lewontin-Krakaeur test, Beaumont & Nichols (1996) tested the effects of a variety of factors that may confound the distribution of F_{ST} for neutral markers. They noted that while some of these effects can broaden the distribution of F_{ST} among loci, the model is surprisingly robust and under most circumstances the model should be valid. However, asymmetrical colonisation can have a particularly substantial effect, so analyses using this test alone should be interpreted with caution. Moreover, since Beaumont & Nichols (1996) recommend that more than 20 loci be used for their method, additional loci should be incorporated into this analysis to either confirm or refute the robustness of this result. Despite these concerns, there is substantial variation in values of F_{ST} among loci

(Table 1) and the Beaumont and Nichols test highlighted two unusual loci, irrespective of the underlying processes. While various other analyses may be used to detect specific outlier loci (e.g. Vitalis et al., 2001; Schlotterer, 2002; Beaumont & Balding, 2004; reviewed by Guinand et al., 2004) a simulation study demonstrated that the method of Beaumont and Nichols performs as well as an alternative Bayesian method of detecting selection at a locus (Beaumont & Balding, 2004). Interestingly, during their evaluation, Beaumont and Balding (*ibid.*) established that the two methods to detect outlier loci could be expected to identify adaptive (spatial) variation but lack power to uncover balancing selection. Given the particularly large number of alleles at *Cpa*114, it would not be surprising if the significantly low level of genetic differentiation was the outcome of a very high mutation rate and concomitant allelic homplasy, rather than the action of a selective process per se. Despite this, if the action of balancing selection can be confirmed, then it would be extremely interesting to identify the environmental correlates and functional genomic region(s) involved.

Elucidating the stock structure of Atlantic herring is an applied problem that has received considerable scientific attention (e.g. Shaw et al., 1999; McPherson et al., 2001b, 2003; Mariani et al., 2005; Ruzzante et al., 2005). With the promise of affording high-resolution stock discrimination, almost 40 microsatellite loci have been characterised to study the population dynamics of this species (McPherson et al., 2001a; Miller et al., 2001; Olsen et al., 2002). Similarly large numbers of microsatellite loci are available for many other commercially exploited marine teleosts. Because of financial and logistical constraints, however, only a subset of markers can be used for population genetic analyses. Our identification of significant heterogeneity in the level of differentiation among microsatellite loci impacts on the common assumption that this class of marker is always neutral and crucially raises the question: which loci should be employed to characterise population-genetic structure? As an illustration, a re-analysis of the data but excluding the three most differentiated loci provides a low but still significant average F_{ST} of 0.004 (0.002, 0.005 ±95% confidence intervals). Using just the three loci with the largest F_{ST}s (*Cpa*6, *Cpa*107, *Cpa*112) provides an estimate of average differentiation that is almost an order of magnitude greater ($F_{ST} = 0.025$; three loci are too few loci to generate confidence intervals by re-sampling) (Table 1). It is clear, therefore, that estimation of parameters of population structure and history under a neutral of model of evolution (e.g. migration and genetic drift) can be significantly biased without prior identification and removal of outlying marker loci. Baer's (1999) meta-analysis argued against this effect being widespread, but our data demonstrate that exceptions can occur. Moreover, Baer (1999) limited his evaluation to allozyme loci. Given the widespread use of microsatellite loci to quantify population structures of marine teleosts it would be informative to repeat this analysis and contrast the different categories of loci. On the other hand, genetic markers may be used as discriminatory characters, for example to assign the origins(s) of samples or for mixed stock analyses, even if selection is implicated (Beacham et al., 2004). Under such circumstances, a preliminary analysis of locus-specific heterogeneity in F_{ST}s would explicitly identify a subset of highly discriminatory markers that would be better suited to achieve this aim, and in a more cost-effective manner, than by using many loci. As demonstrated by studies that exploit different information from different markers (e.g. Pogson et al., 1995; Lemaire et al., 2000; Ward, 2000), the most powerful approach to uncovering important evolutionary and demographic processes will be to first identify and then use a combination of neutral and selective genetic markers.

Acknowledgements The WESTHER project would not have been possible without a huge sampling and coordination effort that involved many people: we are grateful to the crews of FRS Scotia, RV Roagan, RV Lough Foyle, RV Celtic Voyager and the various commercial fishing vessels that provided access to samples. Also, we thank J. Boyd, N. Campbell, J. Chubb, C. Collins, M. A. Cross, C. O. Cunningham, A. J. Geffen, F. Gell, S. Jansen, C. Kelly, K. MacKenzie, J. Schlickeisen, P.-J. Schön, D. Wolfenden, C. Zimmermann and A. Zuur for collecting samples and informative discussions. M. Burt kindly provided the samples from Hudson Canyon. Funding was provided by the EU 5th Framework RTD contract QLRT-2001-01056.

References

Aljanabi, S. M. & I. Martinez, 1997. Universal and rapid salt-extraction of high quality genomic DNA for PCR-based techniques. Nucleic Acids Research 22: 4692–4693.

Baer, C. F., 1999. Among-locus variation in F_{ST}: fish, allozymes and the Lewontin-Krakauer test revisited. Genetics 152: 653–659.

Beacham, T. D., M. Lapointe, J. R. Candy, B. McIntosh, C. MacConnachie, A. Tabata, K. Kaukinen, L. Deng, K. M. Miller & R. E. Withler, 2004. Stock identification of Fraser river sockeye salmon using microsatellites and major histocompatibility complex variation. Transactions of the American Fisheries Society 133: 1117–1137.

Beaumont, M. A. & D. J. Balding, 2004. Identifying adaptive genetic divergence among populations from genome scans. Molecular Ecology 13: 969–980.

Beaumont, M. A. & R. A. Nichols, 1996. Evaluating loci for use in the genetic analysis of population structure. Proceedings of the Royal Society of London B 263: 1619–1626.

Bohonak, A. J., 1999. Dispersal, gene flow and population structure. Quarterly Review of Biology 74: 21–45.

Case, R. A. J., W. F. Hutchinson, L. Hauser, C. Van Oosterhout & G. R. Carvalho, 2005. Macro- and micro-geographic variation in pantophysin (*Pan*I) allele frequencies in NE Atlantic cod *Gadus morhua*. Marine Ecology Progress Series 301: 267–278.

Goudet, J., 1995. FSTAT, Version 1.2: a computer program to calculate *F*-statistics. Journal of Heredity 86: 485–486.

Guinand, B., C. Lemaire & F. Bonhomme, 2004. How to detect polymorphisms undergoing selection in marine fishes? A review of methods and case studies, including flatfishes. Journal of Sea Research 51: 167–182.

Hedrick, P. W., 1999. Perspective: highly variable loci and their interpretation in evolution and conservation. Evolution 53: 313–318.

Hilbish, T. J. & R. Koehn, 1985. The physiological basis of natural selection at the *LAP* locus. Evolution 39: 1302–1317.

Karlsson, S. & J. Mork, 2003. Selection-induced variation at the pantophysin locus (*Pan*I) in a Norwegian fjord population of cod (*Gadus morhus* L.). Molecular Ecology 12: 3265–3274.

Kinlan, B. P. & S. D. Gaines, 2003. Propagule dispersal in marine and terrestrial environments: a community perspective. Ecology 84: 2007–2020.

Lemaire, C., G. Allegrucci, M. Naciri, L. Bahri-Sfar, H. Kara & F. Bonhomme, 2000. Do discrepancies between microsatellite and allozyme variation reveal differential selection between sea and lagoon in the sea bass (Dicentrarchus labrax)? Molecular Ecology 9: 457–467.

de Leon, F. G., L. Chikhi & F. Bonhomme, 1997. Microsatellite polymorphism and population subdivision in natural populations of sea bass *Dicentrarchus labrax* L. 1758. Molecular Ecology 6: 51–62.

Lewontin, R. C. & J. Krakauer, 1973. Distribution of gene frequency as a test of the theory of selective neutrality. Genetics 74: 175–195.

Mariani, S., W. F. Hutchinson, E. M. C. Hatfield, D. E. Ruzzante, E. J. Simmonds, T. G. Dahlgren, C. Andre, J. Brigham, E. Torstensen & G. R. Carvalho, 2005. North Sea herring population structure revealed by microsatellite analysis. Marine Ecology Progress Series 303: 245–247.

McPherson, A. A., P. T. O'Reilly, T. L. McParland, M. W. Jones & P. Bentzen, 2001a. Isolation of nine novel tetranucleotide microsatellites in Atlantic herring (*Clupea harengus*). Molecular Ecology Notes 1: 31–32.

McPherson, A. A., R. L. Stephenson, P. T. O'Reilly, M. W. Jones & C. T. Taggart, 2001b. Genetic diversity of coastal Northwest Atlantic herring populations: implications for management. Journal of Fish Biology 59: 356–370.

McPherson, A. A., R. L. Stephenson & C. T. Taggart, 2003. Genetically different Atlantic herring *Clupea harengus* spawning waves. Marine Ecology Progress Series 247: 303–309.

Miller, K. M., K. Laberee, A. D. Schulze & K. H. Kaukinen, 2001. Development of microsatellite loci in Pacific herring (*Clupea pallasi*). Molecular Ecology Notes 1: 131–132.

Nei, M. & T. Maruyama, 1975. Lewontin-Krakauer test for neutral genes. Genetics 80: 395.

Olsen, J. B., C. J. Lewis, E. J. Kretschmer, S. L. Wilson & J. E. Seeb, 2002. Characterization of 14 tetranucleotide microsatellite loci derived from Pacific herring. Molecular Ecology Notes 2: 101–103.

Pogson, G. H., K. A. Mesa & R. G. Boutilier, 1995. Genetic population structure and gene flow in the Atlantic cod *Gadus morhua*: a comparison of allozyme and nuclear RFLP loci. Genetics 139: 375–385.

Rice, W. R., 1989. Analyzing tables of statistical tests. Evolution 43: 223–225.

Robertson, A., 1975. Remarks on the Lewontin-Krakauer test. Genetics 80: 396.

Ruzzante, D. E., S. Mariani, D. Bekkevold, C. Andre, H. Mosegaard, L. Clausen, T. Dahlgren, W. Hutchinson, E. Hatfield, E. Tortensen, J. Brigham, E. Simmonds, L. Laikre, L. Larsson, R. Stet, N. Ryman & G. Carvahlo, 2005. Biocomplexity in a highly migratory marine fish, Atlantic herring. Proceedings of the Royal Society of London B 273: 1459–1464.

Schlotterer, C., 2002. A microsatellite-based multilocus screen for the identification of local selective sweeps. Genetics 160: 753–763.

Shaw, P. W., C. Turan, J. M. Wright, M. O'Connell & G. R. Carvalho, 1999. Microsatellite DNA analysis of population structure in Atlantic herring (*Clupea harengus*), with direct comparison to allozyme and mtDNA RFLP analyses. Heredity 83: 490–499.

Vitalis, R., K. Dawson & P. Boursot, 2001. Interpretation of variation across marker loci as evidence for selection. Genetics 158: 1811–1823.

Walsh, P. S., D. A. Metzger & R. Higuchi, 1991. Chelex-100 as a medium for simple extraction of DNA for PCR-based typing from forensic material. Biotechniques 10: 506–513.

Ward, R. D., 2000. Genetics in fisheries management. Hydrobiologia 420: 191–201.

Watts, P. C. & J. P. Thorpe, 2006. Influence of contrasting larval developmental types upon the population-genetic structure of cheilostome bryozoans. Marine Biology 149: 1093–1101.

Weir, B. S. & C. C. Cockerham, 1984. Estimating *F*-statistics for the analysis of population structure. Evolution 38: 1358–1370.

Whitlock, M. C. & D. E. McCauley, 1999. Indirect measures of gene flow and migration: $F_{ST} \neq 1/(4Nm + 1)$. Heredity 82: 117–125.

Wright, S., 1951. The genetical structure of populations. Annals of Eugenics 15: 323–354.

Hydrobiologia (2008) 606:35–44
DOI 10.1007/s10750-008-9347-7

Marine reserves: the need for systems

W. J. Ballantine · T. J. Langlois

© Springer Science+Business Media B.V. 2008

Abstract Highly protected marine reserves are areas of the sea in which human disturbances are minimised so that the full natural biological diversity is maintained or, more often, allowed to recover to a more natural state. Europe has very few marine reserves; they are very small and almost all are in the Mediterranean. There are at present no official plans to create effective systems of marine reserves. Europe has many so-called Marine Protected Areas (MPAs). These are marine areas with some extra regulations or planning procedures. MPAs aim to make human activities more efficient and more sustainable. MPAs are user-orientated, knowledge-based, locality-dependent, problem-solving extensions of standard marine planning and management. Marine reserves are quite different. All extractive and potentially disturbing human activities are prohibited. The burden of proof is reversed; no evidence of damage or danger to particular species or habitats is required; all marine life is protected on principle. The concept of marine reserves is simple and practical, but because it is new, different and additional to existing marine management, the idea is seen by many as revolutionary. Basic biological principles and practical experience in many countries make it clear that marine reserves are important to science and education, essential for conservation and useful in resource management. These features apply in all regions and ecosystems. They are independent of climate, biogeography, current human activities and the present management. Representative and viable systems of marine reserves are needed in all regions. Fishing and other human disturbances have been widespread and intensive for so long that it is very difficult to predict the stages of recovery that occur in marine reserves. Furthermore, while some features change rapidly (e.g. numbers of previously targeted species), recovery continues for a long time (e.g. fourth- and fifth-order trophic and structural changes after >25 years). None of this alters the fact that, in scientific terms, marine reserves are controls not manipulations. Such controls are required if scientists are to understand the intrinsic processes and obtain data that are not confounded by human activities (e.g. separating natural variation from fishing effects). No significant progress will be made to establish marine reserves in Europe until scientists speak out strongly and clearly on the issue. We consider it is part of our professional duty as marine biologists to state publicly and frequently the need for a representative, replicated, networked and sustainable system of highly protected marine reserves. We doubt if our grandchildren will accept any excuses if we fail.

Guest editors: J. Davenport, G. Burnell, T. Cross, M. Emmerson, R. McAllen, R. Ramsay & E. Rogan
Challenges to Marine Ecosystems

W. J. Ballantine (✉) · T. J. Langlois
Leigh Marine Laboratory, University of Auckland, Box 349, Warkworth, New Zealand
e-mail: b.ballantine@auckland.ac.nz

Keywords Marine reserves · Systems · Scientific benefits · Professional duty

Introduction

Marine reserves are places in the sea that are left undisturbed so that they continue in their natural state or recover towards a more natural state. No fishing of any kind is permitted, no removal of anything (living, dead or mineral), no dumping, dredging, constructions or direct pollution. Within these restrictions, people are encouraged to appreciate and study the marine life that develops in the reserves.

Europe has very few marine reserves; they are all small and largely restricted to the Mediterranean. For instance, the UK has only one such reserve (3 km^2 off Lundy Island, in the Bristol Channel). There are no official plans to develop systems of such reserves anywhere in Europe.

Europe has many Marine Protected Areas (MPAs) under a very wide range of labels. MPAs are areas with some extra regulations or planning procedures. They may be organised at local, regional or national levels by a large number of authorities (e.g. fishing, transport, defence, conservation, mining, erosion or pollution control, etc.). MPAs are problem-orientated. They aim to reduce conflicts between human activities and make these activities more efficient, more effective or more sustainable. They are extensions of standard marine planning. MPAs concentrate on solving problems that have arisen. They depend on detailed information about the activities in each area, and what impact these have on particular species and habitats.

Marine reserves are completely different. All potentially disturbing activities are prohibited. The burden of proof is reversed; no evidence of damage or danger is needed. All marine life is protected. The concept is simple to state and can be understood easily by non-experts. Marine reserves are a practical idea, because most people (once they think about the idea at all) feel it is sensible and worthwhile to ensure the continued existence of the full range of marine life and its processes.

But it is a new idea; it requires new laws and new attitudes. It is different and additional to existing forms of marine planning and management. Because of these features, marine reserves are generally regarded by policy makers as revolutionary. Most professional marine planners and managers are restricted to specific sectors (fishing, shipping, mining, etc.), have little knowledge of wider aspects and are not asked to concern themselves with the total effect of human activities on marine life (Ballantine, 1999, 2001).

In this article we argue that marine reserves are essential for conservation, have important benefits to science and are necessary for effective management. These benefits are so basic that we believe it is the professional duty of marine scientists in general, and marine biologists in particular, to propose and support the establishment of systems of reserves.

Background

Although marine reserves have scarcely been considered in Europe, there is considerable practical experience in other regions (Roberts & Hawkins, 2000; Sobel & Dahlgren, 2004). In New Zealand the first marine reserve was established in 1975 adjacent to the laboratory where we work. There are now 32 reserves, covering a wide range of latitude, biogeography and habitat. Some are around remote oceanic islands, others are within the boundaries of metropolitan Auckland. They include fiords, harbours and open coasts. At least 12 of these are more than 10 years old.

While there are often considerable and lengthy discussions before a reserve is established (the first took 12 years!), within a few years of creation they prove to be socially popular, scientifically useful and important in conservation terms.

Action in New Zealand was continuous and has accelerated recently, but it was slow and included many unhelpful side tracks. The present scatter of marine reserves in New Zealand is still far from an adequate system. However the experience to date has allowed the recognition and development of the essential principles for such systems. It is now possible for other regions to move much more rapidly and effectively, provided, these principles are accepted.

This is already happening. The first representative system of marine reserves in the world was established in Victoria, Australia in 2002. A year later a representative and replicated system was created around the northern Channel Islands off the

California coast. In 2004 a major system, comprising at least 25% by area of 73 bioregions, was established in the Great Barrier Reef Marine Park (Australia). For more details see Chap. 11 in Sobel & Dahlgren (2004). In August 2006, the California Fish and Game Commission approved a set of no-take marine reserves in Central Region of the mainland coast comprising 8% of state waters (see www.dfg.ca.gov/news/news06/com06004.html).

The benefits of marine reserves to science

Marine reserves provide many benefits to science and these cover a wide range, from simple practicalities, through improved research opportunities, to advances in basic theory and understanding. Although, with hindsight, most of these benefits seem straightforward, indeed, rather obvious, few were planned or predicted, and each year more benefits are being discovered and demonstrated.

Practicalities

Scientists have no more rights in marine reserves than anyone else, but the managers of reserves will generally permit and licence scientific projects if these are carefully designed to involve only trivial and temporary disturbance and to provide interesting and useful information. This licensing is very important. It means that the protection of the reserve is extended to the scientific project, including apparatus, markers and actual experiments. It is difficult to explain to those who have not seen this happen how much it increases the scientific opportunities and freedom of action. In effect, the entire reserve becomes an extension of the scientist's laboratory bench and aquarium tanks.

Protection

This can cover:

(a) All forms of apparatus and equipment, from simple temperature sensors, through tide gauges, wave buoys, current metres, video cameras and microphones, to complex multi-level systems.
(b) Marker buoys, pegs, tagged individuals, transplanted specimens, permanent quadrats, acoustic trackers and fixed photographic sites.
(c) Experiments and manipulations on the shore and sub-tidally, including settlement plates, cages and fences for density manipulation, habitat alterations and in situ samplers.

In theory such protection can be arranged without a marine reserve, but this usually involves complex negotiations with a range of authorities and the problem of informing the public that special protection exists.

New levels of interest

Marine reserves automatically generate new levels of interest in scientists, policy makers and the general public. They also provide a practical focus and reason for action. Many types of study, while possible anywhere, are greatly encouraged and supported by the existence of a marine reserve.

These include (examples from the Leigh reserve):

(a) Detailed behavioural and taxonomic studies e.g. the separation of 13 species of Tripterygiidae each with three colour forms.
(b) Provision of identification guides (printed and photographic) and biological reviews. e.g. for fish (Thompson, 1981).
(c) Detailed surveys and habitat mapping e.g. a habitat map at 1:2,000 was made at the time of establishment (Ayling et al., 1981), and a second survey (using modern techniques) was made in 2006.
(d) Long-term physical monitoring e.g. Daily air and sea climate recording, e.g. Ballantine (1982).

More support and cooperation

The increased levels of interest and focus produce positive feedbacks (the actual examples are from the Leigh reserve).

(a) The greater number of scientists increases the chances of observing and recording events that are rare, local or sudden but ecologically important (e.g. free spawning in gastropods, disease mortality in echinoids, die-back in kelp and algal blooms).
(b) The greater range of scientific projects encourages interdisciplinary studies (e.g. the relation between climate anomalies and any biological phenomena, Rhodes et al. 1993).

(c) With more workers and projects, equipment and facilities can be shared and financial support is possible where no single project would warrant it [(e.g. aerial photography with helicopters and long-term monitoring of phytoplankton chlorophyll (Rees, 2003)].

Opportunities for improved understanding

Ecological changes within marine reserves

As soon as a high level of protection from human disturbance is provided by the establishment of a marine reserve, a wide range of changes begin to develop in the reserve. There are so many of these that, even in the best-studied and longest-established reserves, more changes are recognised every year.

The examples given in Table 1 are from NE New Zealand and include three replicate reserves. All the examples have been measured and those in bold type have been published. For more details see Babcock (2003) and Langlois & Ballantine (2005).

These changes are multiple, complex and often ecologically important. Some changes (e.g. Stages 1–3) form a trophic cascade, but others are the result of changes in structure. Few of the species in Stage 4 feed on kelp, but they require it as a substrate or shelter.

The changes continue to develop (for at least 30 years at Leigh) and there is no theoretical limit to this development. New changes are still being discovered and pre-existing ones recognised. The changes interact with large and small scale natural variations (e.g. El Nino events and local storms), and with each other.

Once the changes are observed, explanations are often clear, but the changes were not predicted or, in most cases, even thought about beforehand. For new reserves, the trends, amplitudes and timing of the changes are not predictable, because of the large number of variables involved. In the Te Angiangi reserve, snapper have not yet increased but the rock lobster increases were very large and rapid. The opposite occurred at the Poor Knights reserve.

An example of some of the complexities affecting the recovery of kelp forest (*Ecklonia radiata* (C.Ag)) is given in Table 2. The driving force is the reduction of grazing by the sea urchin *Evechinus chloroticus* (Valenciennes) due to predation by large crayfish *Jasus edwardsi* (Hutton) and snapper *Pagrus auratus* (Bloch and Schneider), but many factors can speed up or delay the recovery of the kelp.

New comparisons

In most of science, and especially in ecology, improved understanding develops mainly from comparisons (Table 3). The establishment of marine reserves allows a whole range of new comparisons, at least eight of which have already been used in experimental designs.

As systems of reserves are established, a further range of comparison will be possible, including changes outside reserves, which will vary with the proportion of the region in marine reserve (Ballantine, 2003b).

Table 1 Ecological changes in marine reserves (NE New Zealand)

Stage	Sign	Organisms involved
1st stage	Increases	**Snapper** (sparid) and **rock lobster** (palinurid)
		Numbers, sizes, biomass and fecundity
2nd stage	Decreases	**Sea urchin** (echinoid) **numbers and dominance**
		Large bivalves
		Cryptic fish
3rd stage	Increase	**Kelp forest** (laminarian) and coralline algal turf
4th stage	Increases	Mobile epifauna (small crustaceans) around kelp
		Sessile fauna on kelp fronds
		Canopy sheltering fish
		Kelp derived POC and DOC
5th stage	Increase	Juvenile spotties (labrid) feeding on mobile epifauna

Table 2 Factors affecting the recovery and of kelp forest (*Ecklonia radiata*)

Slower development	Faster development
(a) **Severe wave action** tearing off larger plants, especially in shallow water[a]	(a) Calm weather conditions increasing *Ecklonia* growth[b]
(b) **Lengthy algal blooms** reducing light, especially in deeper water[b]	(b) **Urchin mortality** due to toxic or smothering algal blooms[b]
(c) Development of coralline algal turf, inhibiting *Ecklonia* recruitment	(c) Urchin mortality due to disease
(d) Increased (compensatory) grazing by gastropods (e.g. *Cookia sulcata* (Gmelin))	(d) **Increased *Ecklonia* recruitment**
(e) **Increased sand movement** and/or deposition inhibiting *Ecklonia* recruitment[b]	(e) Decreased sand movement and/or deposition[a]
(f) Increased urchin recruitment	(f) Reduced urchin recruitment

References: Babcock et al. (1999); Parsons et al. (2004)

Bold type effects have been measured

[a] More likely in La Nina conditions

[b] More likely in El Nino conditions

Table 3 Ecological comparisons possible with marine reserves

Comparative study	No. of examples
Inside a single marine reserve to outside it	++
Inside a reserve to outside over time	++
Replicated marine reserves to outside	++
Different distances from marine reserve boundary	+
Different sizes of marine reserves	+
Inside and outside reactions to severe natural changes	+
Different dates of reserve establishment	++
Degree of protection (no-take reserves versus partial bans)	+

References: Edgar & Barrett (1999); Langlois et al. (2006); Shears et al. (2006); Willis et al. (2001), (2003)

+, 1 or 2; ++, >2

Increased scales of time and space

Most manipulative experiments in marine ecology have space scales measured in metres; marine reserves offer the opportunity to operate in kilometres. Scale effects can be very important and are not necessarily predictable. Small scale manipulations of sea urchins at Leigh did not suggest the changes that occurred at a reserve scale.

Most research projects are limited by the time scales of grants and individual careers. Marine reserves are permanent and do not have any direct cost to the scientists.

Time lags in trophic cascades at Leigh included 5–7 years for the main predators to recover, a further 10 years for them to abolish grazer control and at least three more years for kelp recovery. This picture was built up through several projects and multiple researchers (Babcock et al., 1999).

Provision of controls and baseline information

Controls are essential for scientific understanding and baselines are important for scientific measurements. Marine reserves provide controls for all direct human disturbances and more natural baselines for impact measurements.

In a standard experimental design, the scientist arranges the manipulation and merely selects the control areas. But marine reserves need to be set up, and manipulations continue to occur everywhere else. Normally, no changes are expected in the control areas (except natural variation), marine reserves continue to change in many ways, while the manipulated areas (outside) may not.

None of this alters the fact that, in scientific terms, marine reserves are controls not manipulations. Such controls are necessary if scientists are to obtain data and understanding that is not confounded by human activities (e.g. separating natural variation from fishing effects). Despite the shifting baseline, marine reserves provide the only objective measures of human-induced impacts and, hence, ecosystem health (Ballantine, 2003a, b).

The wider benefits of marine reserve

Politicians do not rush to make new arrangements simply because these offer special advantages to scientists, especially if such arrangements require new and, possibly, controversial legislation. Fortunately, marine reserves have many wider benefits, which have strong appeal to the general public, as well as scientific value.

Advanced training

This includes:

(i) Practical courses for tertiary students in many aspects of marine science.
(ii) Special courses for post-graduate students (e.g. underwater fish identification).
(iii) A range of professional courses (e.g. habitat classification and mapping).
(iv) Advanced diving courses (e.g. night diving and fish behaviour).

Education

Primary and secondary school trips to the marine reserve at Leigh are now common, and several educational tour operators specialise in arranging these. Whole classes as young as 11 years old go snorkelling in the reserve. Education kits have been published (e.g. Whitley & Ballantine, 1995; Walsby, 2003). An education centre with an aquarium and library has been set up nearby. The reserve is a favoured location for dive training.

Public awareness and recreation

Large numbers of the public visit marine reserves ($\sim$300,000 per year at Leigh). These numbers are at least 10 times greater than for nearby areas where fishing is permitted. Many programmes exist to increase public knowledge and enjoyment of the marine life they find there. These include displays on adjacent land, a glass-bottomed boat at Leigh, hire centres for snorkel and SCUBA gear, dive charter boats, videos (e.g. Ballantine, 1993), popular books (e.g. Enderby & Enderby, 2006), magazine articles (e.g. Warne, 2006), and pamphlets (e.g. Walsby, 2001).

Conservation

The primary aim of marine reserves is the conservation (or recovery) of the full range of marine life and its intrinsic processes. This primacy has four aspects: its importance, practicality, self-evidence and the inter-relationships with other benefits.

(i) It would be difficult to exaggerate the level of importance. Although our knowledge of biogeochemical cycles and climate control processes is still limited, it is already clear that marine life is an essential part of these. The future of the human race depends on maintaining these processes, regardless of our present levels of understanding.
(ii) Even in Europe, our knowledge of marine life is still at a low level. Many species remain to be described, many habitats are not yet mapped and we only know some examples of the ecological processes that are involved. While we must use the knowledge we have, it would very unwise to assume it is sufficient for all purposes. The only practical way to ensure full conservation is to keep representative areas free from all exploitation.
(iii) The fact that marine reserves are less disturbed and more natural than areas that are fished, dredged, mined, dumped on, etc., is self-evident and requires no detailed data.
(iv) All the other aims and benefits of marine reserves are largely dependent on the success of conserving a high level of undisturbed marine life.

Standard science—which involves the determination of detailed facts, their careful analysis and the 'reduction' of problems so that precise conclusions can be clearly stated—is not very helpful when we are trying to conserve the full range of marine life and its processes (including those that are little known or even undiscovered).

In particular, it is not helpful to reduce biodiversity to some simple indices of species or groups. As well as the major phylogenetic groups, their species richness and abundance, biodiversity includes many aspects, for example,

- Size ranges
- Metabolic diversity (including mutualisms and parasites)

- Genetic, physiological and developmental diversity within, as well as between, species
- Behaviour and movement
- Life histories
- Biogeography
- Ecosystems, habitats and communities
- Patterns in time, including large-scale biogeochemical cycles, ENSO, etc.

The general public is very ignorant about marine life and its diversity because it is so different from their terrestrial experience. However, for precisely the same reason, the public is very interested in any aspect of marine life that is brought to their attention and they intuitively recognise the range of diversity.

Resource management

Most discussion about marine reserves focuses on fishing. This is unfortunate for three reasons.

(i) We need marine reserves whether or not they do anything for fishing.
(ii) Although it is likely that marine reserves will assist fishing in various ways, it is not possible to predict these in any precise sense.
(iii) It is rarely possible to prove such effects even when they have occurred (because fishing adapts rapidly to every type of change, Ludwig et al. 1993).

The most important use of marine reserves for marine management is the provision of better information. In particular, marine reserves allow us (for the first time) to unequivocally separate natural changes from human-induced events (especially fishing). For example,

(i) The major decline in crayfish in the Leigh region in the mid 1990s was not a fishing effect since it occurred (proportionately) inside the reserve as well as outside (Kelly et al., 2000).
(ii) The 'urchin barrens' habitat at Leigh is a fishing effect, since it slowly disappears inside established reserves but remains (or increases) in fished areas. (Babcock et al., 1999).
(iii) Snapper abundance at Poor Knights Is. was severely limited by recreational fishing, since it recovered spectacularly when the area became a fully protected marine reserve (Denny et al., 2004).

Recently the phrases—'sustainable development', 'ecosystem health' and 'spatial planning'—have become popular in marine policy and planning circles (e.g. Defra, 2006). Without systems of marine reserves such slogans are likely to remain pious hopes, but marine reserves would provide the objective standards necessary to make them a practical reality. How could we know the state of health of an ecosystem unless some parts of it were free from human disturbance? How could we determine what was sustainable unless we had areas that were not exploited? The spatial planning of a region needs a standard zone which provides the unexploited baseline for the major habitats.

The need for principles and systems

The main benefits of marine reserves to science are largely independent of region, ecosystem or habitat, and are completely independent of political boundaries. It is important to science to establish a full system of such reserves in every region. These systems need to be based on principles which are acceptable to both professionals and the public. Such principles must be clear and either self-evident or demonstrable. It is the responsibility of scientists to provide and explain these principles. Fortunately, most of this work has already been done, but the results have not yet been given much publicity, even in scientific circles.

The wider benefits of marine reserves, especially those to conservation, are sufficiently obvious to warrant the support of any thoughtful citizen, but the ordinary public needs its intuitive views to be given formal and public backing by scientists, if the politicians are to resist the claims of those profiting from the status quo.

The United Nations technical expert group on the conservation of marine biodiversity produced its final report in 2004 (CBD, 2004). This includes the necessary principles for marine reserves (described there as Highly Protected Marine and Coastal Areas). These are summarised below. Statements in "quotation marks" are from pages 23–25 of CBD (2004).

For each reserve

High levels of protection

All species are protected "because ecological interactions are complex and mostly unknown. Allowing any fishing jeopardizes goals of maintaining ecological structure and function and confounds the scientific ability to achieve understanding."

Permanence

"The protection...should be permanent" because reversion to more natural conditions continues for a long time and the benefits accumulate over time. "Long term changes cannot be effectively measured if highly protected areas are temporary."

Viability

Each reserve should be "large enough so that most ecological processes will be able to operate within the area". Reserves "should also be legally and socially viable". "Boundaries should be simple to identify and enforce."

For the system

Representation

"All biogeographic regions should be represented. Within each region, all major habitats should be represented." We must resist attempts to restrict this representation by the application of any theory or assumption. We do not want just the well-studied areas, the biodiversity hotspots, the rare or beautiful, the places considered important with existing information, those known to be under threat or those under any particular jurisdiction.

Replication

"All habitats in each region should be replicated within the network, and these should be spatially separate, to safeguard against unexpected failures or collapse of populations."

A network arrangement

"A network design should be prepared for each national or regional area, including the exclusive economic zones and the High Seas." The reserves should be spread across the whole region to provide:

Maximal connections (including larval dispersal) between all areas.

An additional range of natural variation (both known and unknown).

Spread of benefits (and any inconveniences).

A sustainable amount

"The ultimate objective is to create a network of geographically dispersed sites that is self-sustaining, independent (as far as possible) of what happens in the surrounding area." Each reserve should be as ecologically viable as possible, but the whole system must be capable of sustaining itself. System size would be measured as the percentage of the area at each level—region, ecosystem and habitat.

Some guidelines:

At least 10% by area of all bioregions and habitats is needed for science and education.

To ensure conservation, we would need at least 20% of each region, ecosystem and habitat.

To maximise benefits to fisheries, this should rise to 30%.

In regions of very intensive use, it would be sensible to aim for at least 50%.

Conclusions

1. When marine reserves are established the range of benefits proves to be large, and many of these are important, but most of them are surprises and were not predicted even by the proponents.
2. When high levels of protection are provided by marine reserves the ecological changes that occur are multiple, complex and on-going. Again, few, if any, of these are predictable.
3. It is self-evident that marine reserves are a highly effective means for conserving (or recovering) the full range of marine biodiversity. Indeed, marine reserves are the only practical method for ensuring this.
4. Marine reserves are important to science in many ways. The most important is that by acting as controls for the effects of human disturbance, they provide for the first time a practical and objective method for determining the basic (natural) content and processes of marine life.

5. It is time for Europe to take action to establish systems of marine reserves that are representative, replicated, networked and sufficiently large to be sustainable.
6. It will be necessary for marine biologists to lead in this matter, and we have a professional duty to do so. At present no one in government has the job of promoting marine reserves. There is a policy vacuum. Unless we, as professionals capable of understanding the issues, speak out clearly and strongly, little will happen. The planners, politicians and the public need to be told that, while the interests of existing user groups are important, the key issue is what will be maintained for the next generations. The great majority of the general public, once the idea is presented to them, find this reasonable and worthwhile. The politicians take their cue from the public (voters). The existing policy makers and planners will tend to resist, but will take instructions from the politicians.

Acknowledgements This review covers more than 30 years of investigations by a large number of workers. We wish to thank them all—especially (in alphabetical order) Tony Ayling, Russ Babcock, Chris Battershill, Howard Choat, Bob Creese, Chris Denny, Debbie Freeman, Roger Grace, Dennis Gordon, Tim Haggitt, Geoff Jones, Shane Kelly, Vince Kerr, Mike Kingsford, Allie MacDiarmid, John Morton, Darren Parsons, Emma Rushcombe, Nick Shears, Sue Thompson, Natalie Usmar, Jarrod Walker, Kathy Walls, John Walsby, Kim Westerskov and Trevor Willis.

References

Ayling, A. M., A. Cumming & W. J. Ballantine, 1981. Map of Shore and Sub-tidal Habitats of Cape Rodney to Okakari Point Marine Reserve, North Island, New Zealand in 3 Sheets, Scale 1: 2000. Department of Lands and Survey, Wellington.

Babcock, R. C., 2003. The New Zealand marine reserve experience: the science behind the politics. In Hutchings. P. & S. D. Lunney (eds) Conserving Marine Environments: Out of Sight, out of mind. Royal Zoological Society of New South Wales, Mossman: 105–119.

Babcock, R. C., S. Kelly, N. T. Shears, J. W. Walker & T. J. Willis, 1999. Changes in community structure in temperate marine reserves. Marine Ecology Progress Series 189: 125–134.

Ballantine, W. J. (ed.), 1982. Sea Surface Temperatures. Leigh Laboratory Bulletin No 5. University of Auckland: 43 pp.

Ballantine, W. J., 1993. Marine reserves for New Zealand. Leigh Marine Laboratory, University of Auckland. 20 minute video.

Ballantine, W. J., 1999. Marine reserves in New Zealand: The development of the concept and the principles. Proceedings of an International Workshop on Marine Conservation for the New Millenium. Korean Ocean Research and Development Institute, Cheju Island, Korea: 3–38 [available at www.marine-reserves.org.nz].

Ballantine, W. J., 2001. Marine Reserves. Paper presented to the first meeting of UN experts on marine biodiversity (AHTEG). 10 pp [available at www.marine-reserves.org.nz].

Ballantine, W. J., 2003a. Scientific principles for marine reserve systems. MPA News 4(9): 5.

Ballantine, W. J., 2003b. Practical action with marine reserve systems. MPA News 4(10): 5.

CBD (Convention on Biological Diversity), 2004. Technical advice on the establishment and management of a national system of marine and coastal protected areas. Secretariat of the Convention of Biological Diversity. CBD Technical series 13: 40 pp.

Defra, 2006. A Marine Bill: A Consultation Document. Department for Environment, Food and Rural Affairs, London: 309 pp.

Denny, C. M., T. J. Willis & R. C. Babcock, 2004. Rapid recolonisation of snapper *Pagrus auratus* Sparidae within an offshore island marine reserve after implementation of no-take status. Marine Ecology Progress Series 273: 183–190.

Edgar, G. T. & N. S. Barrett, 1999. Effects of marine reserves on Tasmanian reef fish, invertebrates and plants. Journal of Experimental Marine Biology and Ecology 242: 107–144.

Enderby, J. & T. Enderby, 2006. A guide to New Zealand's marine reserves. New Holland Publishers, Auckland: 176 pp.

Kelly, S., D. Scott, A. B. MacDiarmid & R. C. Babcock, 2000. Spiny lobster, *Jasus edwardsi*, recovery in New Zealand marine reserves. Biological Conservation 92: 359–369.

Langlois, T. J., M. J. Anderson, R. C. Babcock & S. Kato, 2006. Marine reserves demonstrate trophic interactions across habitats. Oecologia 147: 134–140.

Langlois, T. J. & W. J. Ballantine, 2005. Marine ecological research in New Zealand: developing predictive models using no-take marine reserves. Conservation Biology 19: 1763–1770.

Ludwig, D., R. Hilbron & C. Walters, 1993. Uncertainty, resource exploitation and conservation: lessons from history. Science 260: 17, 35.

Parsons, D. M., N. T. Shears, R. C. Babcock & T. R. Haggitt, 2004. Fine-scale habitat change in a marine reserve, mapped using acoustically positioned video transects. Marine and Freshwater Research 55: 257–265.

Rees, T. A. V., 2003. Safety factors and nutrient uptake by seaweeds. Marine Ecology Progress Series 263: 29–42.

Rhodes, L. L., A. J. Haywood, W. J. Ballantine & A. L. MacKenzie, 1993. Algal blooms and climate anomalies in north-east New Zealand August-December 1992. New Zealand Journal of Marine and Freshwater Research 27: 419–430.

Roberts, C. M. & J. P. Hawkins, 2000. Fully Protected Marine Reserves: A Guide. WWF Endangered Sea Campaign, Washington D.C.

Shears, N. T., R. V. Grace, N. R. Usmar, V. Kerr & R. C. Babcock, 2006. Long-term trends in lobster populations in a partially protected vs. no-take Marine Park. Biological Conservation 132: 222–231.

Sobel, J. & C. Dahlgren, 2004. Marine Reserves: A guide to Science, Design and Use. Island Press, Washington D.C.: 383 pp.

Thompson, S., 1981. Fish of the Marine Reserve: A Guide to the Identification and Biology of the Common Coastal Fish of North-eastern New Zealand. Leigh Laboratory Bulletin No 3. University of Auckland.

Walsby, J. R., 2001. Cape Rodney—Okakari Point Marine Reserve: The Underwater World. A3 pamphlet. Department of Conservation, Auckland.

Walsby, J. R., 2003. Marine Reserves Education Kit, 3rd edn. Department of Conservation, Auckland: 158 pp.

Warne, K., 2006. Poor Knights—rich seas. New Zealand Geographic 78: 52–75.

Whitley, J. & W. J. Ballantine, 1995. Marine Reserves: An Education Resource Kit. Leigh Marine Laboratory, University of Auckland: 86 pp.

Willis, T. J., R. B. Millar & R. C. Babcock, 2003. Protection of exploited fish in temperate regions. Journal of Applied Ecology 40: 214–227.

Willis, T. J., D. M. Parsons & R. C. Babcock, 2001. Evidence for long-term site fidelity of snapper (*Pagrus auratus*) within a marine reserve. New Zealand Journal of Marine and Freshwater Research 35: 581–500.

Hydrobiologia (2008) 606:45–53
DOI 10.1007/s10750-008-9348-6

The challenge of assessing whether the OSPAR network of marine protected areas is ecologically coherent

Jeff A. Ardron

© Springer Science+Business Media B.V. 2008

Abstract As MPA networks continue to grow worldwide, decision makers and the public will need to be kept informed regarding progress made in meeting the underlying ecological goals of the networks. The twelve coastal European nations of the OSPAR Commission have agreed to establish within the waters of the Northeast Atlantic an "ecologically coherent" network of marine protected areas by 2010. An initial set of 81 MPAs in the Northeast Atlantic was accepted by the OSPAR Commission in June 2006, and as of 31 December 2006 six additional sites have been put forward. They amount to approximately 26,500 km^2, submitted by six of the twelve coastal nations. The job of assessing their ecological coherence has proven to be a difficult issue, and one that has not been attempted before. Assessment techniques that can cope with data-poor situations will prove most helpful, and with this in mind, three general approaches that can be used in data-poor situations and four specific examples of assessment techniques, are presented in this paper. Additionally, a brief summary of the current status of the OSPAR MPA network is provided.

Guest editors: J. Davenport, G. Burnell, T. Cross, M. Emmerson, R. McAllen, R. Ramsay & E. Rogan
Challenges to Marine Ecosystems

J. A. Ardron (✉)
German Federal Agency for Nature Conservation, Isle of Vilm, 18581 Putbus, Ruegen, Germany
e-mail: jeff.ardron@bfn-vilm.de

Keywords OSPAR · MPA · Ecological coherence

Introduction

In 2003, environmental ministers from European nations bordering on the Northeast Atlantic (OSPAR Convention) and the Baltic (HELCOM) gathered in Bremen, Germany. There they signed a joint-ministerial statement agreeing to identify a first set of marine protected areas (MPAs) by 2006, establish what gaps then remain, and complete by 2010 a joint network of well-managed marine protected areas that, together with the European Natura 2000 network, is "ecologically coherent" (JMM, 2003). However, it was not explicitly stated what *ecological coherence* meant, nor how it could be assessed.

As the 2010 deadline approaches, the urgency to assess ecological coherence increases; however, proper scientific assessment is hampered by a current lack of detailed ecological data. Approaches that can make do with what little information is available are being developed and are the focus of the latter sections of this paper. The first section of this paper provides background on the current MPA network.

Background

OSPAR is the mechanism by which fifteen Governments of the western coasts and catchments of

Europe, together with the European Community (known collectively as *Contracting Parties*), cooperate to protect the marine environment of the North-East Atlantic. It began in 1972 with the Oslo Convention against dumping. It was broadened to cover land-based sources and the offshore industry by the Paris Convention of 1974. These two conventions were unified, updated and extended by the OSPAR (*Os*lo-*Par*is) Convention of 1992. The new annex on biodiversity and ecosystems was adopted in 1998 to cover non-polluting human activities that can adversely affect the sea, which has allowed for the development of marine protected areas (OSPAR, 2006a, Sect. 1).

When OSPAR was established, the Contracting Parties agreed that questions of fisheries management are appropriately regulated under other international and regional agreements dealing specifically with such questions. In 2000, OSPAR identified the environmental impact of fisheries as one of the two most important issues for the future of the marine environment. Because it does not have direct regulatory authority, however, OSPAR is only able to draw attention to such fisheries' issues, in the hope that the competent international bodies and national authorities will take notice (OSPAR, 2006a, Sect. 7). Nonetheless, OSPAR is proceeding with the establishment of MPAs, and in 2003 adopted Recommendation 3 with that explicit purpose to:

a. protect, conserve, and restore species, habitats, and ecological processes which have been adversely affected by human activities;
b. prevent degradation of, and damage to, species, habitats, and ecological processes, following the precautionary principle;
c. protect and conserve areas that best represent the range of species, habitats, and ecological processes in the maritime area. (Sect. 2.1, OSPAR, 2003).

The first MPA status report was accepted by the OSPAR Commission in June 2006 (OSPAR, 2006b). Since then, a second annual report has also been drafted, though it is still under review (OSPAR, 2007a). Aspects of both reports, which were written by the author, will be drawn upon in the current paper when outlining MPA progress to date. In the ensuing discussion of ecological coherence, however, very little has yet been finalised by OSPAR or HELCOM.

OSPAR MPA Nominations

As of December 31, 2006, the following Contracting Parties have reported areas as components of the OSPAR network of MPAs: France, Germany, Norway, Portugal, Sweden, and UK. Contracting Parties that did not report any OSPAR MPAs are: Belgium, Denmark, Iceland, Ireland, the Netherlands, and Spain. A full listing can be found in Annex 2 of the OSPAR MPA status reports (OSPAR, 2006b, 2007a) and is summarized in Table 1. Germany has reported the largest total area, approx. 12,000 km^2. The UK has reported the most sites, which range in size. Although the total area of the UK sites has not yet been accurately determined, it is expected to exceed 10,000 km^2, making it the second largest contribution. Norway has nominated about 2,000 km^2, Portugal about 1,700, Sweden about 1,000, and France about 300.

Jurisdictions

The vast majority of sites nominated to date fall within territorial waters. Only five sites of the 87 nominated fall within an EEZ (3 Norway, 1 Germany,

Table 1 OSPAR MPAs as of 31 December 2006

	MPAs	Km^2
Belgium	0	0
Denmark	0	0
France	8	275
Iceland	0	0
Ireland	0	0
Germany	4	11,923
Netherlands	0	0
Norway	6	1,905
Portugal	7	1,686
Spain	0	0
Sweden	6	972
UK[a]	56	9,858
Total[a]	87	26,619
OSPAR area[b]		13,618,510

[a] Areas for the UK sites are estimates based on Natura 2000 marine area, which does not include the intertidal. Actual UK OSPAR numbers have not yet been received, but are expected to be somewhat higher, thereby also raising the overall total

[b] Not all of the OSPAR Maritime Area is navigable waters, with significant ice cover in the Arctic region

2 Sweden). No sites are in areas beyond national jurisdiction. One site (Portugal: Rainbow hydrothermal vents) is on an extended continental shelf. In cases where the continental shelf extends beyond 200 nautical miles, the seafloor may be considered under the jurisdiction of the concerned party (up to 350 nm, as provided by the UN Convention on the Law of the Sea), but the water column above it will already be an area beyond national jurisdiction (ABNJ). In such cases, coordination and cooperation between OSPAR and the relevant Contracting Party(ies), as well as the relevant organisations with management authority over ABNJ in the OSPAR maritime area, is required. This recent Portuguese MPA currently has no management plan in place.

Overall area

These initial nominations represent about 26,500 km^2. In relation to the overall OSPAR maritime area, this is still very small, even after ice-bound areas are removed. (It can be assumed that no more than 20% of the OSPAR Maritime Area is icebound.) However, as explained below, some biogeographic areas have better protection than others.

OSPAR biogeographic regions

The Dinter (2001) biogeographic classification has been accepted by OSPAR as one indication of broad-scale representativity for the purpose of establishing the OSPAR network. Dinter delineated the OSPAR maritime area into three large *biomes*. The first Biome considered the pelagic waters less than 1,000 m in depth, of which there were three *provinces*. The second biome considered the seafloor up to 1,000-m depth, of which there were 17 provinces and sub-provinces. Finally, waters and the seafloor deeper than 1,000 m were treated together with two broad provinces. Thus altogether, there are 22 biogeographic zones. Using biogeographic zones as a surrogate for describing different ecosystems is generally accepted as the only way forward in the absence of detailed biological information (Zacharias et al, 1998; Day & Roff, 2000; ICES, 2004; CSAS, 2005). If an MPA network contains representative portions of each bioregion, it can be considered more likely to meet broad-scale ecological objectives, than a network that does not. As can be seen in Table 2, representativity of the current network ranges up to 6.1% (due to the 2006 MPA nominations in the Azores) of each of the biogeographic zones. However, thirteen of the twenty-two biogeographic zones continue to have effectively no representation at all (<0.00%). This currently uneven distribution is in part because some Contracting Parties have not yet nominated sites. There is also a general tendency to nominate near-shore sites first, thus leaving the further offshore waters vacant for the time being.

Natura 2000

Of the 81 sites submitted by EU Member States, 78 of them are Natura 2000 sites (EC Birds Directive, 1979; EC Habitats Directive, 1992). Generally, the boundaries are the same as for the OSPAR sites; however, four Portuguese nominations have smaller Natura sites that are contained within much larger OSPAR designations; conversely, for the Swedish nominations, the OSPAR sites were sometimes smaller. Of the EU Member States, only Portugal has submitted three sites that are not also Natura 2000, all of them hydrothermal vents.

Given that the geographical scope of the OSPAR network is much larger than that of Natura 2000 (including waters of Iceland, Norway, and areas beyond national jurisdiction), and that the ecological criteria for MPA selection within OSPAR are broader (including a different list of threatened and endangered species and habitats), it is expected that if nominations by EU states are limited to existing Natura 2000 sites, exclusively, then it is unlikely that the OSPAR network's ecological goals can be met.

Ecological coherence

Given the very incomplete nature of the initial MPA nominations, as described above, it is self-evident to note that these do not yet represent an ecologically coherent network.

Neither OSPAR nor HELCOM has a formal definition for *ecological coherence*, though both have begun preliminary work on the topic (HELCOM, 2005; OSPAR, 2006c). Likewise, though "coherent", "coherence" (and one instance of "ecological coherence" in Art. 10) is used throughout the EC Habitats

Table 2 Biogeographic representation as of 31 December 2006

Biome	Region/sub-region	Province	MPAs	Km^2	Bioregional area (%)
Pelagic	Arctic	na	0	0.0	0.00
	E. Atlantic Temperate	Cool-temperate	75	24,780.5	0.37
	E. Atlantic Temperate	Warm-temperate	12	1,838.1	0.05
Shelf and slope	Arctic	Barents Sea	0	0.0	0.00
	Arctic	Barents Sea: White Sea	0	0.0	0.00
	Arctic	High Arctic Maritime	0	0.0	0.00
	Arctic	NE Greenland Shelf	0	0.0	0.00
	Arctic	NE Water Polynya	0	0.0	0.00
	Arctic	SE Greenland, N Iceland	0	0.0	0.00
	E. Atlantic Temperate	Boreal	43	19,530.4	2.72
	E. Atlantic Temperate	Boreal-Lusitanean	24	2,805.0	0.62
	E. Atlantic Temperate	Lusitanean-Boreal	4	130.4	0.09
	E. Atlantic Temperate	Lusitanean Cool	0	0.0	0.00
	E. Atlantic Temperate	Lusitanean Warm N	1	22.0	0.05
	E. Atlantic Temperate	Lusitanean Warm S	0	0.0	0.00
	E. Atlantic Temperate	Macaronesian: Azores	4	1,376.4	6.10
	E. Atlantic Temperate	Norway: Finnmark	0	0.0	0.00
	E. Atlantic Temperate	Norway: Skagerrak	4	543.4	2.23
	E. Atlantic Temperate	Norway: W. Norwegian	4	1,901.7	0.55
	E. Atlantic Temperate	S Iceland-Faero Shelf	0	0.0	0.00
Deep Sea	Arctic[a]	na	0	0.0	0.00
	Atlantic	na	3	309.3	0.00

[a] Not all of the OSPAR Maritime Area is navigable waters, with significant ice cover in the arctic

(1992) and EC Birds (1979) Directives, these terms are not explicitly defined either. Ecological coherence is not often applied as a term in the scientific literature, and its occasional usage is usually fairly different from what is intended here (e.g. genetic relatedness). The term does show up in the "grey literature" reports often in the context of Natura 2000, but is also not clearly defined. It is generally used to imply some sort of connective structure (e.g. corridors—Good, 1998) existing amongst, and binding together, ecological processes and functions (Bull et al., 2003; STRA-REP, 1998).

So that there is a general understanding of what is meant by this term, as applied in this paper, the following working definition is put forward based on previous work by OSPAR (2006c) and also by Laffoley et al. (2006):

a. An ecologically coherent network of MPAs:
   i. interacts and supports the wider environment (OSPAR, 2006c, Sects. 5.3, 6);
   ii. maintains the processes, functions, and structures of the intended protected features across their natural range (Laffoley et al., 2006); and
   iii. functions synergistically as a whole, such that the individual protected sites benefit from each other to achieve the above two objectives (based on OSPAR, 2006c, Sect. 5.2).
b. Additionally, an ecologically coherent network of MPAs *may*:
   i. Be designed to be resilient to changing conditions (OSPAR, 2006c, Sect. 5).

As the MPA network grows, so will the need to better assess it using agreed-upon criteria. Such criteria already exist in the literature on MPA *selection* and can be readily employed as *Assessment Criteria* as well. Four Assessment Criteria are currently under discussion in OSPAR and HELCOM: adequacy/viability, representativity, replication, and connectivity (OSPAR, 2006d, 2007b; BALANCE-HELCOM 2006). Initial assessments should cast light on how well these criteria are being met. They should not involve complex or detailed analyses, but instead techniques that can provide general indications where gaps may lie. Three different, non-exclusive approaches are being explored by OSPAR, each focussing on different sources of information:

1. *Self-assessment based on expert knowledge*. In this approach, those involved in the design of the particular reserves report subjectively on how well they feel certain criteria were considered in the MPA selection. This has the advantage of speed and simplicity, and draws upon expert knowledge and intuition. However, it lacks objective rigour, is limited in the questions that can be reasonably answered, and results can vary across experts. It comprises a checklist and an additive scoring system (OSPAR, 2007c).
2. *Species-habitat assessment based on reporting*. In this approach, the species and habitats reported to be contained within the reserves are cross-tabulated against biogeographic regions. The advantages include making use of data that are already being reported (thus easing additional reporting requirements), providing an overview of whether certain agreed-upon (or legislated) species and habitats are being protected, and can give some indication if different functional groups are being protected. The disadvantages include a reliance on reporting accuracy, a lack of clarity about what these occurrences of MPAs containing species and habitats are actually telling us about the overall network's ecological coherence, and a lack of spatial information to determine the distribution and sufficiency of protection. This involves a spreadsheet approach, where species and habitats may be further grouped according to their ecological characteristics (OPSAR, 2007d).
3. *Spatial assessment based on GIS data*. In this approach, the overall network is examined based on a battery of simple spatial tests to provide an indication if it is likely meeting the criteria of ecological coherence. The advantages of this approach include that it is less reliant on subjective opinions or reporting accuracy than the above two approaches; that it is not reliant on any single diagnostic but instead is fairly holistic, considering the overall picture that emerges; and, that it deals explicitly with spatial concerns. (This is important since unlike other marine management options such as emission limitations, fishing gear restrictions, or extraction quotas, MPAs are primarily a spatial tool.) The disadvantages include that it requires additional work above the minimum reporting requirements and requires the collection/collation of spatial data. Further, the use of simplified analyses inevitably raises scientific questions concerning whether these tests or "rules of thumb" are ultimately supportable. This approach involves GIS analyses (OSPAR, 2007b).

Each of the above approaches is an attempt to make use of available sources of information, balancing scientific rigour with political and administrative realities. They should be viewed as different tools in a toolbox. The last one is potentially the most sophisticated, but is also the least explored. Because of novelty of this last approach, four examples of such techniques under consideration by OSPAR, hereafter called *Assessment Guidelines*, are given below. These could belong to a suite of perhaps twenty or thirty to be used in an MPA network assessment. As that in most cases the available data are limited, such Assessment Guidelines must follow the philosophy of "making the most out of what little you've got;" i.e. they must be realistic in their data requirements and achievable in their execution. As such, any individual guideline should be seen as rather too simplistic to stand on its own, but, when considered altogether, the suite of guidelines can nevertheless produce a valid overall picture of various aspects of reserve design. The following four examples focus mainly the size and distribution of sites.

Example Assessment Guideline 1: *Owing to the multiplicity of its objectives, the MPA network can be expected to contain a variety of reserve sizes and corresponding spacing between sites.*

Corollary: If a region were dominated by only large widely spaced reserves, or only several small reserves, then it would suggest that the network in this region may not be adequately diverse to meet all its objectives.

Justification: Various site shapes and configurations have been found to be appropriate for different situations. The venerable *SLOSS* debate (single large or several small; e.g. Diamond, 1975; Game, 1980) is these days usually side-stepped, by saying, "it depends". For example, metapopulation theory initially appears to suggest that larger reserves may be more efficacious, but once several assumptions are removed, it may actually be that moderate sizes are better (Rampal & Heesterbeek, 2000). Empirical evidence suggests that larger reserves often have beneficial effects disproportionate to their size (Halpern, 2003). However, for coastal areas, modelling suggests that a number of smaller "stepping stones" may be more efficacious than single large reserves separated by greater distances (Gains et al., 2003).

Example Assessment Guideline 2: *MPAs should be checked to ensure that they are broadly distributed across biogeographic regions. In such an assessment, there is no need to restrict the assessment to just one biogeographic classification system. So long as they appear to be rigorous (e.g. peer-reviewed), the use of a variety of systems is encouraged.*

Justification: Different classification systems will emphasize different aspects of the marine ecosystems and are therefore valuable in highlighting possible gaps in the MPA network that another system might have overlooked. An extensive review of existing biogeographic classification systems within the OSPAR Area has been given in Dinter (2001). Out of that work emerged a new classification system that was accepted by the OSPAR working group on marine protected areas, species and habitats. Since then, at the request of the EC, ICES also developed a classification system for the NE Atlantic (ICES, 2004). Currently, The Nature Conservancy is developing global Marine Ecoregions Of the World (MEOW), which was presented at a side event to COP-8 of the Convention on Biological Diversity and includes shelf regions of the OSPAR Area (Spalding et al., 2006). To an outside observer, this plethora of various biogeographic systems can be confusing: Why is there still not universal agreement? This question is treated quite seriously in Dinter (2001), where it is pointed out that there are so many variables to consider, and so many different ways to put them together, that it seems almost inevitable that, "… there are as many methods as biogeographers". (Dunbar, 1979, quoted in Dinter, 2001). It should be added that data quality and availability can also influence these decisions.

Example Assessment Guideline 3: *Features that are very common and broadly distributed will likely require less proportional (percentage) protection than those features that are less common.*

Justification: An assessment of representative features does not necessarily imply that an equal proportion of all features should be included in a network. Indeed, differing representative features will likely require differing levels of protection, depending on factors such as rarity, ecological significance, and level of threat. On the other hand, if fixed constant proportions are applied uniformly across a network, then that network will become dominated by the protection of vast swaths of what are acknowledged to be very common features and which are probably not under particularly serious threat. For example, protecting 10% of a 10,000,000 ha feature is a considerably larger undertaking than protecting 10% of a 10,000 ha feature.

Example Assessment Guideline 4: *As a rule of thumb taking into consideration the assumptions listed below, within a given feature class, representative protection can be expected to be roughly proportional to the square root of the ratio of representative features' overall areas.*

Thus, within a given feature class (e.g. seabird species, benthic habitats, or marine biomes), the proportions of any two (x and y) should be such that: $(x_p/y_p) \approx (x_t/y_t)^{0.5}$ where the subscript "$_p$" represents the area protected of a given feature and the subscript "$_t$" represents the total area of a given feature in the network. That is, the rule of thumb is stating that the distribution of protection of multiple representative features of the same general kind should fall within a continuum roughly proportional to the square root of their respective total areas.

Another way to express this identity is to introduce an average ratio, which is labelled here as the "Protection Ratio" (PR): $\mathrm{PR} = (\sum((i_p)/(i_t)^{0.5}))/n$, summed for each protected feature occurrence "i" in a given representative network where "n" is the total number of features. If these feature occurrence data are unavailable, then individual sites that are said to protect a given feature can be used as very approximate proxies for "i". Once PR is calculated, then individual protected areas can be compared to it. To evaluate the representative network overall, a coefficient of variation (CV) for the ratios across the network can be calculated in the standard fashion. High CV values should trigger further investigation, as that it would indicate that protection in a given class varies widely across the network of representative sites, even after varying feature sizes have been accounted for.

Justification: The underlying assumptions that validate this Guideline are:

(a) The features can be reasonably grouped into the given feature class whereby the rationale of why representativity is important is more –or less the same, no matter which particular feature is being considered within that class;
(b) The delineation of the protected representative areas has not been unduly confounded by other overlapping features or other considerations; and,
(c) That mathematical distribution of the total areas of representative features is skewed, with a long right-hand tail, and that taking the square-root of these area measures helps to normalize the distribution.

Assumption (a) is pointing out that different representative feature classes could have differing protection requirements. If such information is available, then the representative features should be stratified according to protection requirements, and then assessed using this formula. However, in many situations, representativity may be treated as a "catch-all" category, when no other data are available. In these situations, assumption (a) can be seen as holding true across all representative sites.

In practice, so-called "representative" sites will likely be protected with other considerations also in mind, and thus assumption (b) should be seen as guidance, not a strict criterion, when selecting which sites in a network will be assessed using this rule –of thumb.

Assumption (c) is a specific example of the common assumption of normality that is usually not stated but nonetheless widely applied in statistics. Fortunately, standard deviation (and thus CV) is robust to this assumption (Zar, 1999). Bimodal or multimodal distributions, if they exist, should be separated out into separate distributions that can be assessed separately, as per assumption (a). This may be facilitated by identifying a determining variable (e.g. territorial vs. EEZ waters or two different nationalities), which has led to different selection methodologies and/or protection requirements.

Conclusions

The ministerial commitment to achieving ecological coherence within the OSPAR and HELCOM networks of MPAs is broad and ambitious. Yet, there is a pressing need to be able to assess it with a minimum of time and expense. Addressing this difficult question will likely involve a process of staged assessments, beginning with preliminary assessments that are straightforward and achievable.

Early assessments of ecological coherence should make the most of expertise and data available to provide guidance to Contracting Parties as to where the major gaps and deficiencies likely lie. However, as the network of MPAs develops and becomes more sophisticated, so will the need to assess it. This will require a greater depth of information and expertise. The need to coordinate the acquisition and compiling of data sets is universal.

The assessment of ecological coherence can be grouped under four general criteria, already generally accepted in the literature surrounding MPA selection: adequacy/viability, representativity, replication, and connectivity. Each of these criteria should then be broken down into several aspects, for which plausible assessment guidelines can be developed. It is not expected that all these guidelines will necessarily be applied in any given assessment. It is helpful to bear in mind that in certain places, there may be better data and better understanding of ecological processes, which could lead to the development of more locally appropriate and sophisticated assessment techniques. However, in many places data for even the most rudimentary assessments may be missing, and simple guidelines may be the only way forward.

Because ecological coherence is a holistic concept reliant on many constituent parts, it is much easier to develop tests that indicate when it has *not* been achieved (i.e. some of the parts are missing) than it is to test when it has been achieved (i.e. when all the parts are present *and* interacting as expected). Thus, achieving the goal of ecological coherence is one that ultimately cannot be measured exactly, but must rather be approached in a stepwise fashion, stated as a converging likelihood, based on a growing suite of approaches and indicators.

Acknowledgements The author would like to thank the two anonymous reviewers who provided helpful comments on the initial draft of this manuscript.

References

BALANCE-HELCOM, 2006. BALANCE-HELCOM Workshop on Ecologically Coherent Network of MPAs in the Baltic Sea and the North East Atlantic, 25th–27th of October 2006, Helsinki, Finland. Final minutes from the meeting.

Bull, K. S. E. & D. d'A. Laffoley, 2003. Networks of Protected Areas in the Maritime Environment. A Report for the Review of Marine Nature Conservation and the Marine Stewardship Process on a Stakeholder Workshop held in London on 19 June 2003. Peterborough: English Nature Research Reports, No 537, 35 pp. http://www.english-nature.org.uk/pubs/publication/PDF/537.pdf.

CSAS, 2005. Canadian Science Advisory Secretariat. Framework for Classification and Characterization of Scotia-Fundy Benthic Habitats. Science Advisory Report 2005/071. http://www.dfo-mpo.gc.ca/csas/Csas/status/2005/SAR-AS2005_071_E.pdf.

Day, J. C. & J. C. Roff, 2000. Planning for Representative Marine Protected Areas: A Framework for Canada's Oceans. Report Prepared for World Wildlife Fund Canada, Toronto.

Diamond, J. M., 1975. The island dilemma: lessons of modern biogeographic studies for the design of natural reserves. Biological Conservation 7: 129.

Dinter, W. P., 2001. Biogeography of the OSPAR Maritime Area. German Federal Agency for Nature Conservation, Bonn: 167 pp.

EC Birds Directive, 1979. European Commission Council Directive 79/409/EEC on the conservation of wild birds. http://europa.eu.int/eur-lex/en/consleg/pdf/1979/en_1979L0409_do_001.pdf.

EC Habitats Directive, 1992. European Commission Council Directive 92/43/EEC on the Conservation of natural habitats and of wild fauna and flora. http://ec.europa.eu/environment/nature/nature_conservation/eu_nature_legislation/habitats_directive/index_en.htm.

Gains, S. D., B. Gaylord & J. L. Largier, 2003. Avoiding current oversights in marine reserve design. Ecological Applications 13: S32–S46.

Game, M., 1980. Best shape for nature reserves. Nature 287: 630–632.

Good, J. A. 1998. The potential role of ecological corridors for habitat conservation in Ireland: a review. Irish Wildlife Manuals, No. 2, 72 pp. http://www.npws.ie/Publications Literature/AllPublications/file,2388,en.pdf.

Halpern, B., 2003. The impact of marine reserves: do reserves work and does size matter? Ecological Applications 13: S117–S137.

HELCOM, 2005. Helsinki Commission, Nature Protection and Biodiversity Group, seventh meeting Kalmar, Sweden, 10–14 October 2005. Status Report on Analysis of Ecological Coherence of the Network of Baltic Sea Protected Areas. HABITAT 5.2/7 http://sea.helcom.fi/dps/docs/documents/Nature%20Protection%20and%20Biodiversity%20Group%20(HABITAT)/HABITAT%207,%202005/5.2-7.pdf.

ICES, 2004. ICES response to EC request for information and advice about appropriate ecoregions for the implementation of an ecosystem approach in European waters. http://www.ices.dk/committe/acfm/comwork/report/2005/ace%20advice/ICES%20Eco-regions%20advice%20to%20EC%202004.pdf.

JMM, 2003. Joint Ministerial Meeting of the Helsinki and OSPAR Commissions (JMM). Declaration of the Joint Ministerial Meeting of the Helsinki and OSPAR Commissions. Convention on the Protection of the Marine Environment of the Baltic Sea Area (Helsinki Convention) and OSPAR Convention for the Protection of the Marine Environment of the North-East Atlantic, Bremen: 25–26 June 2003. JMM 2003/3. www.ospar.org.

Laffoley, D.d'A., S. Brockington & P.M Gilliland. Developing the concepts of Good Environmental Status and Marine Ecosystem Objectives: Some Important Considerations. English Nature, Peterborough, English Nature Research Report, 29 pp, in press.

OSPAR, 2003. OSPAR Convention for the Protection of the Marine Environment of the Northeast Atlantic, Meeting of the OSPAR Commission, Bremen, 23–27 June 2003. Recommendation 2003/3 on a Network of Marine Protected Areas. OSPAR 03/17/1-E; Annex 9.

OSPAR, 2005. OSPAR Convention for the Protection of the Marine Environment of the Northeast Atlantic, Meeting of the Working Group on Marine Protected Areas, Species and Habitats, Bristol, 3–7 Oct. 2005. MASH 05 Summary Record, Annex 7.

OSPAR, 2006a. OSPAR Convention for the Protection of the Marine Environment of the Northeast Atlantic, Meeting of the OSPAR Commission, Stockholm, 26–30 June 2006. OSPAR briefing note. OSPAR 06/3/4; Annex B.

OSPAR, 2006b. 2005/2006 Report on the Status of the OSPAR Network of Marine Protected Areas. Publication Number: 268/2006. www.ospar.org.

OSPAR, 2006c. OSPAR Convention for the Protection of the Marine Environment of the Northeast Atlantic, Meeting of the Biodiversity Committee, Trondheim, 13–17 March 2006. Guidance on Developing an Ecologically Coherent Network of OSPAR Marine Protected Areas. BDC 06/3/7.

OSPAR, 2006d. OSPAR Convention for the Protection of the Marine Environment of the Northeast Atlantic, Meeting of the Working Group on Marine Protected Areas, Species and Habitats, Horta, 2–5 Oct. 2006. Criteria and Guidelines for Assessing whether the OSPAR Network of Marine Protected Areas is Ecologically Coherent. MASH 06/5/3.

OSPAR, 2007a. OSPAR Convention for the Protection of the Marine Environment of the Northeast Atlantic, Meeting of the Intersessional Group on Marine Protected Areas, Secretariat, London, 5–7 Feb. 2007. DRAFT 2006/2007 Report on the status of the OSPAR Network of Marine Protected Areas. ICG-MPA 07/02/01-rev.

OSPAR, 2007b. OSPAR Convention for the Protection of the Marine Environment of the Northeast Atlantic, Meeting of the Intersessional Group on Marine Protected Areas, Secretariat, London, 5–7 Feb. 2007. Background Document to Support the Assessment of Whether the OSPAR Network of Marine Protected Areas is Ecologically coherent. ICG-MPA 07/03/01-rev.

OSPAR, 2007c. OSPAR Convention for the Protection of the Marine Environment of the Northeast Atlantic, Meeting of the Intersessional Group on Marine Protected Areas, Secretariat, London, 5–7 Feb. 2007. Rapid Self-assessment

Checklist for Ecological Coherence of the OSPAR Network of Marine Protected Areas. ICG-MPA 07/03/02-rev.

OSPAR, 2007d. OSPAR Convention for the Protection of the Marine Environment of the Northeast Atlantic, Meeting of the Intersessional Group on Marine Protected Areas, Secretariat, London, 5–7 Feb. 2007. A Matrix Approach to Assessing the Ecological Coherence of the OSPAR MPA Network. ICG-MPA 07/03/03.

Rampal, S. E. & J. A. Heesterbeek, 2000. On optimal size and number of reserves for metapopulation persistence. Journal of Theoretical Biology 203: 33–50.

Spalding, M., H. Fox, N. Davidson, Z. Ferdana, B. Halpbern, M. Jorge, A. Lombana, S. Lourie, K. Martin, E. McManus, J. Molnar, K. Newman, C. Recchi, J. Robertson, 2006.Global coastal and marine biogeographic regionalization as a support tool for implementation of CBD programmes of work. UNEP/CBD/COP/8/INF/34. http://www.biodiv.org/doc/meetings/cop/cop-08/information/cop-08-inf-34-en.pdf.

STRA-REP, 1998. National and Regional Approaches for Ecological Networks in Europe: Document Presented by Rob H. G. Jongman (Wageningen, The Netherlands) and I. B. Kristiansen (Roskilde, Denmark), and submitted by ECNC Committee of Experts for the Development of the Pan-European Ecological Network. Strasbourg, 6 May 1998. STRA-REP (18). http://www.strategyguide.org/docs/rep98e18.doc.

Zacharias, M. A., D. E. Howes, J. R. Harper & P. Wainwright, 1998. The British Columbia marine ecosystem classification: rationale, development, and verification. Coastal Management 26: 105–124.

Zar, J. H. 1999. Biostatistical Analysis, 4th edn. Prentice Hall, NJ, USA: 663 pp plus appendices.

Hydrobiologia (2008) 606:55–61
DOI 10.1007/s10750-008-9345-9

CHALLENGES TO MARINE ECOSYSTEMS

Individual-based movement behaviour in a simple marine reserve—fishery system: why predictive models should be handled with care

Edward A. Codling

© Springer Science+Business Media B.V. 2008

Abstract The problem of overexploitation and unsustainability is a major issue in global fisheries. Marine reserves or protected no-take zones have been suggested as a possible solution that would maintain yield and protect stocks indefinitely. A key factor in the effectiveness of a marine reserve—fishery system is the rate of exchange of biomass between reserve and fishery: if the rate of exchange is too low then the fishery is not viable, but if the rate of exchange is too high then stocks may be exploited unsustainably and the reserve is rendered ineffective. The rate of exchange is determined by both the physical design and shape of the reserve, and the movement and dispersal behaviour of both the adult and larval-stage fish. Previous models looking at optimal reserve design usually only consider a diffusive population scale movement and dispersal, even though most animal movement is more realistically modelled as being correlated at the individual level. In this article, a deliberately simple simulation of a theoretical marine reserve—fishery system is used to demonstrate the danger of making predictions using only a population-level simplistic diffusive movement model. Further predictions based on the population average of a more realistic correlated movement model are also shown to be inaccurate. This result is due to both the high levels of individual variability in movement behaviour, and the heterogeneity of the environment. This suggests that in future studies, individual-based (rather than population-level) simulations and models are likely to give more useful insights into the dynamics of the marine fishery environment.

Keywords Marine reserves · Marine protected areas · Individual-based model · Random walk · Predictive model · Environmental heterogeneity

Introduction

The problem of overexploitation and unsustainability is a major issue in global fisheries (e.g. Pauly et al., 2002). Marine reserves or protected no-take zones have been suggested as a possible solution that would maintain yield and protect stocks indefinitely (Sladek Nowlis & Roberts, 1997, 1999; Pezzy et al., 2001; Roberts et al., 2001; Neubert, 2003; Gerber et al., 2003; Rodwell & Roberts, 2004). Assuming there is a desire to maintain a viable fishery (rather than setting

Guest editors: J. Davenport, G. Burnell, T. Cross, M. Emmerson, R. McAllen, R. Ramsay & E. Rogan
Challenges to Marine Ecosystems

E. A. Codling (✉)
Department of Biological Sciences, University of Essex, Colchester, Essex CO4 3SQ, UK
e-mail: ecodling@essex.ac.uk

E. A. Codling
Department of Mathematical Sciences, University of Essex, Colchester, Essex CO4 3SQ, UK

up a reserve purely for conservation), then the key factor in the effectiveness of a marine reserve - fishery system is the rate of exchange of biomass between reserve and fishery: if the rate of exchange is too low then the fishery is not viable, but if the rate of exchange is too high then stocks may be exploited unsustainably and the reserve is rendered ineffective (Watson et al., 2000; Gerber et al., 2003). The rate of exchange is determined by both the physical design and shape of the reserve, and the movement and dispersal behaviour of eggs, juvenile (larval-stage) and adult fish (e.g. Gerber et al., 2003; Neubert, 2003; Pelletier & Mahevas, 2005). It is this latter point that is considered in more detail in this article.

Previous models looking at optimal reserve design usually only (either implicitly or explicitly) consider a diffusive population scale movement and dispersal, i.e. individual movement is effectively assumed to be random and uncorrelated (e.g. Watson et al 2000; Rodwell et al., 2002, 2003; Neubert, 2003; Pelletier & Mahevas, 2005), and advection is usually only due to localised oceanographic features such as current or eddies. This is reasonable when considering the dispersal of passive particles such as eggs, but most animals are motile and their movement has been more realistically modelled as an individual-based correlated random walk (Bovet & Benhamou, 1988; Okubo & Levin, 2001; Benhamou, 2004; Codling et al., 2004). (A correlated random walk is a process where, over small time scales, individuals are more likely to continue moving in the same direction and movement is not completely random.) In many situations described in the literature, the assumption of diffusive movement may be reasonable and model predictions may match well with reality (see examples in Okubo & Levin, 2001). However, this is not always the case (see discussion in Bovet & Benhamou, 1988), and in this article a simple example is given of a virtual marine reserve-fishery system where the assumption is not reasonable. The simulation used is deliberately simple and general—the aim is not to replicate all the complexities of a real fishery system, but instead to highlight how the modelling process can lead to the wrong conclusions being drawn if incorrect assumptions are made about key features of the system dynamics. In particular, the simulation results highlight the problem of using a bulk population average measure to make predictions when the environment is heterogeneous and highly variable. The results suggest that in future studies, individual-based (rather than population-level) simulations and models may give better insights into the key features and dynamics of complex systems like the marine fishery environment.

Methods

Simulation model

A simple virtual model of a marine reserve-fishery environment is used to illustrate how movement behaviour can affect the performance of a reserve to protect a fish stock from exploitation (measured in terms of individual mortality or survival rate) and how this may differ from theoretical predictions. The system is non-dimensionalised so that units have no quantitative meaning and distances/speeds/times are all on a relative scale. For this reason, any results should be considered in a qualitative sense only.

The simulation is run with 1,000 ‘individuals’ (where an ‘individual’ could be a single fish or a sub-group/school of the main population) for $n = 1000$ discrete time steps. All individuals are assumed to move independently and no interactions with either other individuals or the underlying environment are modelled. For simplicity, all individuals start at the origin point $O = (0,0)$ (the centre of the reserve) with randomly distributed directions and are then assumed to disperse away. The simulation is spatially heterogeneous with a marine reserve covering a circular area with radius 50 units centred on the origin point $O = (0,0)$. The area inside the reserve is closed to fishing and the area outside the reserve is open to fishing. At every time step, each individual may ‘die’ due to natural mortality or fishing pressure depending on spatial position. This mortality is modelled as a random process with the following probabilities of death at each time step:

P(death in reserve) $= 0.0002$ (natural mortality only),
P(death outside reserve) $= 0.002$ (natural and fishing mortality).

Thus due to fishing pressure, an individual suffers a tenfold increase in the probability of death at each time step when it moves outside the reserve and into the fishery.

At each time step, those individuals that are still alive will subsequently move a fixed distance of 1 unit in a new direction (i.e. a constant speed is assumed). The new direction is randomly chosen from a *circular normal distribution* (Mardia & Jupp, 1999) with variance ${\sigma_0}^2$ centred on the previous direction of movement so that individuals move as a *correlated random walk*. Simulations were run across a range of values of ${\sigma_0}^2$ corresponding to different levels of correlation in individual movement (${\sigma_0}^2 = 0$ corresponds to purely linear movement and as ${\sigma_0}^2$ increases correlation is lost and movement becomes more random). A static environment is assumed: there are no fixed oceanographic features or currents that could introduce bias into the movement. However, low-level turbulence and environmental fluctuations are implicitly modelled as part of the random choice of direction at each step. No global-directed movement, corresponding to, e.g. migration to or from a spawning or feeding ground, is considered, although this will not alter the key points made. In fact, including a global preferred direction in the model may actually highlight further the discrepancy between simple diffusion-advection movement and more realistic correlated and biased movement; see Codling et al. (2004) for details. Similarly, the model assumes that each individual maintains the same movement behaviour throughout the whole reserve-fishery system and no local-directed movements are modelled. Thus, possible changes in movement patterns due to group interactions or the spatial distribution of resources/habitat quality are not considered (i.e. only the simplest foraging behaviour is assumed).

A more detailed description of the correlated random walk movement algorithm is given in Codling et al. (2004) and further discussion of correlated random walks and their relation to diffusion models of movement can be found in Bovet & Benhamou (1988), Benhamou (2004) and Okubo & Levin (2001). The algorithm produces movement trajectories similar to those mapped out in Fig. 1.

Predicted dispersal and survival

The expected *mean squared dispersal distance*, E(MSD), of a group of individuals moving as a correlated random walk after n steps with unit step length is given by (Bovet & Benhamou, 1988; Benhamou, 2004):

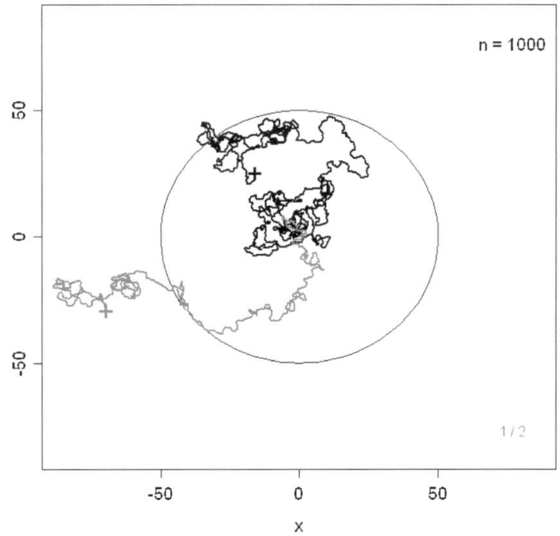

Fig. 1 Examples of simulated random walk trajectories. The trajectories are for two different individuals moving as a correlated random walk with ${\sigma_0}^2 = 1$ after $n = 1{,}000$ time steps. The grey trajectory shows an individual that has left the reserve area and is now 'dead'

$$E(\mathrm{MSD}) = \{n(1+r)/(1-r)\} - \left\{2r(1-r^n)/(1-r)^2\right\} \quad (1)$$

where $r = \exp(-{\sigma_0}^2/2)$ is the first moment of the circular normal distribution (Mardia & Jupp, 1999).

The corresponding *mean dispersal distance*, E(MDD), is given by (Bovet & Benhamou, 1988; Benhamou, 2004):

$$E(\mathrm{MDD}) = \sqrt{[\pi E(\mathrm{MSD})/4]}. \quad (2)$$

For purely diffusive (uncorrelated) movement, $r = 0$, and Eqs. 1 and 2 give $E(\mathrm{MDD}) = 28$ after $n = 1000$ time steps. Since in this case E(MDD) < 50 (reserve radius), individuals are expected to stay within the reserve for the entire time period ($n = 1{,}000$ time steps) if purely diffusive movement is assumed.

To study the dynamics of the system it is now assumed that the correlated movement behaviour in the system can be described by the population average given in Eqs. 1 and 2. For correlated movement ($r > 0$), Eqs. 1 and 2 are used to calculate the expected time spent in the reserve (t_{res}, the number of time steps when $E(\mathrm{MDD}) \leq 50$) and the expected time spent outside the reserve (t_{fish}, the number of time steps when $E(\mathrm{MDD}) > 50$), where $t_{res} + t_{fish} = n = 1{,}000$. Thus t_{res} and t_{fish} give the expected time spent in the reserve and fishery respectively, averaged across the whole population. Using the mortality rates given earlier, the expected survival rate for the 'average' fish is then:

$$P_s = 0.9998t_{res} \times 0.998t_{fish}. \quad (3)$$

From Eq. 3, the theoretical maximum survival rate is $P_s = 0.82$ when $t_{res} = 1{,}000$ and $E(\mathrm{MDD}) < 50$ for all time steps (i.e. an individual stays inside the reserve for the entire time period). Equation 3 gives the survival probability for the 'average individual' in the simulation and, although it represents correlated movement better than a simple diffusion model, it should be emphasised that it is not necessarily the best model to describe the dynamics of this simplified virtual system. Instead, it is used as an example to illustrate the point that, as with the diffusion model, a predictive model based on a simple population average may be oversimple and lead to the wrong conclusions about the system being drawn.

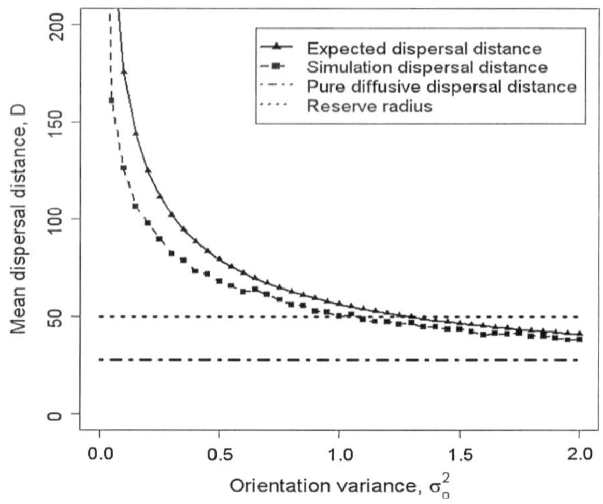

Fig. 2 Mean dispersal distance after $n = 1{,}000$ time steps for different values of orientation variance, σ_0^2. The theoretical expected dispersal distance [see Eqs. 1 and 2] is slightly higher than simulated values because trajectories are truncated for 'dead' individuals

Results

From Fig. 2, the mean dispersal distance after $n = 1{,}000$ time steps, $E(\mathrm{MDD})$, decreases as σ_0^2 increases (and correlation is lost) for both the simulation results and for the theoretical results based on Eq. 2. There is a slight difference between simulation and theoretical results. This is because simulated MDD values are calculated across the whole population, and since truncated trajectories corresponding to 'dead' individuals are included, the MDD is slightly less than expected (the results match if simulations are completed with zero mortality). As can be clearly seen in Fig. 2, predictions of $E(\mathrm{MDD})$ based on purely diffusive movement [i.e. $E(\mathrm{MDD}) = 28$] are highly inaccurate when compared to simulation results, even with low correlation in movement (high σ_0^2).

From Fig. 3, both the theoretical survival rate predicted from Eq. 3 and the simulation survival rate increase as σ_0^2 increases, corresponding to less correlated movement and a higher likelihood of staying within the reserve (and not being exposed to higher mortality). However, although both the theoretical survival rate from Eq. 3 and the simulated results have the same general qualitative trend, they are not a good match, with the theoretical survival rate being an overestimate for high σ_0^2 and an underestimate for low σ_0^2. As might be expected, if predictions on survival rate were made assuming purely diffusive movement then there is clearly a very poor match to the simulation results (where correlation is present), even for high σ_0^2 (low correlation).

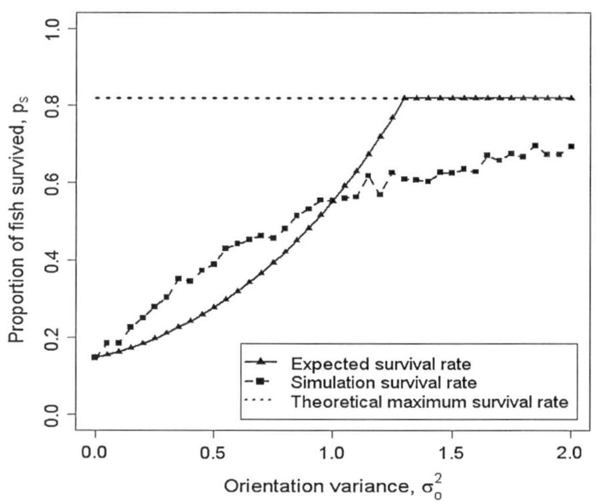

Fig. 3 Proportion of individuals that survive for $n = 1000$ steps for different values of orientation variance, σ_0^2. The theoretical maximum survival probability [see Eq. 3] corresponds to all individuals staying within the reserve for the entire time period (expected for diffusive movement and movement with low σ_0^2)

Discussion

The simple model used in this article is only realistic at a most basic level, but it is sufficient to illustrate the main points—(i) that movement behaviour can make a dramatic difference to the apparent effectiveness of a marine reserve and (ii) care should be taken when using theoretical models to evaluate and predict reserve effectiveness.

If predicted results for the virtual fishery in this example are based on a simple diffusion (uncorrelated) model of movement, then the population is expected to stay in the reserve [$E(\mathrm{MDD}) = 28$], and the expected survival rate is given by the maximum theoretical rate of $P_s = 0.82$ (Fig. 3). However, this is a very poor match to the actual simulation results (Fig. 3), and it is clear that for this simple example of a virtual fishery, the assumption of a diffusion model of movement is not appropriate. It is not uncommon in the literature for such simple assumptions about movement to be made for far more complex fishery systems than in this example and few authors question the validity of the diffusion movement model (see discussions in Bovet & Benhamou, 1988). If predictions about the effectiveness of marine reserves are made using analytic or simulation models that use only a simple diffusion-type movement (e.g. Watson et al., 2000; Rodwell et al., 2002, 2003; Neubert, 2003; Pelletier & Mahevas, 2005), then care should be taken in interpreting results before presenting conclusions and advice to fishery managers. The uncorrelated diffusion movement model may be appropriate in many situations [e.g. if individuals are mainly passive relative to the environment, or the spatial scale being observed is very large compared to individual movement scales (see discussions in Okubo & Levin, 2001)], but most fish populations are likely to demonstrate far more complex behaviour as discussed below.

Predicted results from Eq. 3 that include correlated movement behaviour are also shown to differ from simulated results for the virtual fishery. This difference between simulation and predicted results can be explained by the fact that Eqs. 1–3 are based on a group population average, namely $E(\mathrm{MDD})$. In the simulations some individuals move further and some shorter distances in the same time period [where the *average* distance moved will correspond to $E(\mathrm{MDD})$]. This individual variability is highly important when the environment is non-homogenous and different mortality is suffered depending on spatial position (i.e. inside or outside the reserve). The theoretical predicted survival rate from Eq. 3 is based on this population average and thus implicitly assumes that all of the population have moved the same distance and ignores possible re-crossing over the reserve-fishery boundary by individuals. In fact, for high σ_0^2 (low correlation or diffusion model) more individuals are exposed to the higher mortality outside the reserve than the bulk average prediction from Eq. 3 suggests. Similarly, for low σ_0^2 (high correlation) the opposite is true and the predicted survival rate from Eq. 3 is lower than simulation results. The key point is that in a system with environmental heterogeneity and high individual variability, the survival rate predicted for the 'average individual' (Eq. 3) is not the same as the average survival rate across the whole population.

The effectiveness of the reserve in this example is measured in terms of survival rate of individuals. This may not always be an appropriate measure, particularly as some reserves are designed for other reasons, e.g. to maximise spawning or to protect a special habitat, rather than to reduce mortality. Similarly, the virtual fishery in this example has adult fish moving out of a reserve and being exploited by the fishery, but many studies in the literature have assumed no adult movement and only a transfer of eggs and larvae between the reserve and fishery (e.g. Sladek Nowlis & Roberts, 1997, 1999; Pezzy, 2001). However, most of these studies have assumed passive larval dispersal through diffusion and/or advection by oceanographic features such as currents and eddies. The results in Fig. 2 clearly show that if correlation is included in movement behaviour (i.e. larvae are self-motile) then dispersal distances can be much greater than by simple diffusion (see also Codling et al., 2004). This may be important when considering the connectivity of populations at different locations and the subsequent design of marine reserves.

The underlying simulation model used for the virtual fishery is basic and could be made more realistic. The correlated random walk model that governs the movement behaviour of individuals in the virtual fishery has been used to model and describe the trajectories of fish movements in various different contexts and environments (e.g. settling reef fish larvae in Codling et al., 2004, and aggregating tuna in Girard et al., 2004). However, most fish populations are likely to exhibit far more complex behaviour than that described by a simple correlated random walk. For example, the same species of fish may behave differently in different areas and/or environments (Righton et al., 2001), while individual fish may also exhibit foraging behaviour where they adapt their movement relative to the location of resources or the habitat quality in their local environment (e.g. Ollason et al., 2006). Interactions between both individuals and groups in the population may also be important to the system dynamics and could be included in the behavioural model (Hoare et al., 2004; Codling et al., 2007). All these possible factors need to be at least considered when interpreting results from a simple theoretical predictive model.

Unfortunately, highly complex models are much harder to analyse (and hence it is harder to produce testable predictions) and are arguably less transparent (see discussions in Gerber et al. 2003; Pelletier & Mahevas, 2005). Adding this sort of complexity to the simulation model in this example will not alter the key point—the assumption and subsequent interpretation of a simple population average 'bulk measure' in a predictive model is not the best way to consider these systems. For simplicity and ease of understanding, many theoretical models in the literature involve several parameters that are (either explicitly or implicitly) defined as average quantities, usually based on bulk population measures. However, in a system where there is a high degree of individual and / or environmental variability (as in most real life fishery systems), these average measures may not truly reflect the real dynamics, particularly in a non-homogenous environment with different levels of mortality as in this example (see Pitchford & Brindley, 2001; Pitchford et al., 2005, for similar examples). Thus, to gain insights into such a highly dynamic and variable system as a marine fishery environment, an individual-based simulation model that includes all the key dynamical features of the system is likely to be more appropriate than a predictive model based on bulk population average measures.

Acknowledgement I am grateful to J. W. Pitchford for helpful comments on an earlier draft of this article.

References

Benhamou, S., 2004. On the expected net displacement of animals' random movements. Ecological Modelling 171: 207–208.

Bovet, P. & S. Benhamou, 1988. Spatial analysis of animals' movements using a correlated random walk model. Journal of Theoretical Biology 131: 419–433.

Codling, E. A., N. A. Hill, J. W. Pitchford & S. D. Simpson, 2004. Random walk models for the movement and recruitment of reef fish larvae. Marine Ecology Progress Series 279: 215–224.

Codling, E. A., J. W. Pitchford & S. D. Simpson, 2007. Group navigation and the 'many wrongs principle' in models of animal movement. Ecology 88: 1864–1870.

Gerber, L. R., L. W. Botsford, A. Hastings, H. P. Possingham, S. D. Gaines, S. R. Palumbi & S. Andelman, 2003. Population models for marine reserve design: a retrospective and prospective synthesis. Ecological Applications 13: S47–S64.

Girard, C., S. Benhamou & L. Dagorn, 2004. FAD: fish aggregating device or fish attracting device? A new analysis of yellowfin tuna movements around floating objects. Animal Behaviour 67: 319–326.

Hoare, D. J., I. D. Couzin, J.-G. Godin & J. Krause, 2004. Context-dependent group size choice in fish. Animal Behaviour 67: 155–164.

Mardia, K. V. & P. E. Jupp, 1999. Directional Statistics. Wiley, Chichester.

Neubert, M. G., 2003. Marine reserves and optimal harvesting. Ecology Letters 6: 843–849.

Okubo, A. & S. Levin, 2001. Diffusion and Ecological Problems: Modern Perspectives. Springer, New York.

Ollason, J. G., J. M. Yearsley, K. Liu & N. Ren, 2006. Modelling the behaviour of individuals and groups of animals foraging in heterogeneous environments. In Boyd, I., S. Wanless & C. J. Camphuysen (eds), Top Predators in Marine Ecosystems. Cambridge University Press, Cambridge: 294–309.

Pauly, D., V. Christensen, S. Guenette, T. J. Pitcher, U. R. Sumaila, C. J. Walters, R. Watson & D. Zeller, 2002. Towards sustainability in world fisheries. Nature 418: 689–695.

Pelletier, D. & S. Mahevas, 2005. Spatially explicit fisheries simulation models for policy evaluation. Fish and Fisheries 6: 307–349.

Pezzy, J. C. V., C. M. Roberts & B. T. Urdal, 2001. A simple bioeconomic model of a marine reserve. Ecological Economics 33: 77–91.

Pitchford, J. W. & J. Brindley, 2001. Prey patchiness, predator survival and fish recruitment. Bulletin of Mathematical Biology 63: 527–546.

Pitchford, J. W., A. James & J. Brindley, 2005. Quantifying the effects of individual and environmental variability in fish recruitment. Fisheries Oceanography 14: 156–160.

Righton, D., J. Metcalfe & P. Connolly, 2001. Electronic tags reveal different behaviour of North and Irish Sea cod. Nature 411: 156.

Roberts, C. M., J. A. Bohnsack, F. Gell, J. P. Hawkins & R. Goodridge, 2001. Effects of marine reserves on adjacent fisheries. Science 294: 1920–1923.

Rodwell, L. D. & C. M. Roberts, 2004. Fishing and the impact of marine reserves in a variable environment. Canadian Journal of Fisheries and Aquatic Sciences 61: 2053–2068.

Rodwell, L. D., E. B. Barbier, C. M. Roberts & T. R. McClanahan, 2002. A model of tropical marine reserve-fishery linkages. Natural Resource Modeling 15: 1–34.

Rodwell, L. D., E. B. Barbier, C. M. Roberts & T. R. McClanahan, 2003. The importance of habitat quality for marine reserve-fishery linkages. Canadian Journal of Fisheries and Aquatic Sciences 60: 171–181.

Sladek Nowlis, J. C. & C. M. Roberts, 1997. You can have your fish and eat it, too: theoretical approaches to marine reserve design. Proceedings of the 8th International Coral Reef Symposium 2: 1907–1910, Smithsonian Tropical Research Institute, Panama City, Panama.

Sladek Nowlis, J. C. & C. M. Roberts, 1999. Fisheries benefits and optimal design of marine reserves. Fishery Bulletin 97: 604–616.

Watson, R., J. Alder & C. Walters, 2000. A dynamic mass-balance model for marine protected areas. Fish and Fisheries 1: 94–98.

Hydrobiologia (2008) 606:63–68
DOI 10.1007/s10750-008-9346-8

Effects of marine reserve protection on spiny lobster (*Palinurus elephas* Fabr., 1787) in a central western Mediterranean area

Maria Cristina Follesa · Danila Cuccu · Rita Cannas · Serenella Cabiddu · Matteo Murenu · Andrea Sabatini · Angelo Cau

© Springer Science+Business Media B.V. 2008

Abstract The contribution that a small restocking area (central western Mediterranean) has made to the *Palinurus elephas* (Fabr., 1787) population was examined by comparing the abundances recorded inside and outside the area before (1997) and after the establishment of the reserve (1998–2005). From 1998 to 2002 a progressive percentage increase of *P. elephas* biomass values was recorded both inside the area and in the surrounding zones. The total mean abundance within the reserve (CPUE = 0.23 ± 0.10 kg/50 m/boat) was 7.5 times greater than that for the neighbouring zone (CPUE = 0.03 ± 0.07 kg/50 m/boat). The inter-annual analysis of lobster size inside the area also showed a progressive increase of adults and juveniles. The results highlighted the effectiveness of fishing restrictions in rebuilding the lobsters population and suggest that small MPAs should be set up.

Keywords Spiny lobster · *Palinurus elephas* · Marine reserve area · Central western Mediterranean

Guest editors: J. Davenport, G. Burnell, T. Cross, M. Emmerson, R. McAllen, R. Ramsay & E. Rogan
Challenges to Marine Ecosystems

M. C. Follesa (✉) · D. Cuccu · R. Cannas · S. Cabiddu · M. Murenu · A. Sabatini · A. Cau
Dipartimento di Biologia Animale ed Ecologia, Università di Cagliari, Viale Poetto, 1, 09126 Cagliari, Italy
e-mail: follesac@unica.it

Introduction

Marine reserves, areas where all forms of fishing and extractive use are banned, have been widely promoted as conservation and fishery management tools. In the past decade, studies on the conservation benefits of marine reserves have produced a burgeoning literature (Roberts & Polunin, 1991; Roberts & Hawkins, 2000; Goni et al., 2001; NCR, 2001; Gell & Roberts, 2002; Russ, 2002; Gerber et al., 2003) but the utility of no-take areas for fishery management remains controversial (NCR, 2001; Hilborn, 2002). The main expectations of marine reserves are that the cessation of fishing might promote an increase in abundance as well as in the mean size of protected populations and sustain fishing outside them by their becoming net exporters of adults (the spillover effect) and of recruits (the recruitment effect) (Russ, 2002). Unequivocal evidence to support these expectations has so far been lacking (Russ, 2002).

Many studies have highlighted that the species most likely to respond to the 'reserve effect' are long-living predators, organisms highly vulnerable to fishing, while other species may not be influenced by protection of marine protected areas (MPAs).

The main problem in setting up MPAs is that many marine reserves have often been created to meet unspecified conservation benefits rather than verifiable management objectives (Jennings, 2001). In many cases, the effectiveness of MPAs has been difficult to demonstrate empirically, partly because the location, size and habitats protected were chosen with no particular species in mind and partly due to the lack of fishery data before the reserve was set up (Palumbi, 2001; Willis et al., 2003). Moreover, clear demonstrations of spillover affecting local fishery catches are rare and results are often equivocal (Cole et al., 2000; McClanahan & Mangi, 2000; Davidson, 2001; Roberts et al., 2001; Galal et al., 2002; Kelly et al., 2002; Russ, 2002; Goni et al., 2006). In addition, effective tests proving the movements of the species from MPAs are hampered by the difficulty of achieving a good study design and also by the lack of a tagged-fish catch time series and fish recovery data (Roberts & Polunin, 1993; McClanahan & Kaunda-Arara, 1996).

To date, many studies have been conducted in the Mediterranean Sea to assess the effectiveness of marine protected areas (Sanchez Lizaso et al, 2000) but few authors have assessed the response to the protection of spiny lobster populations in MPAs (Marin, 1987; Goni et al., 2001, 2006; Follesa et al., 2007).

The aim of the present study is to examine the effect of 8 years of protection (1998–2005) on the population of *Palinurus elephas* in a restocking area and the neighbouring zone in the central western Mediterranean. Answers to two questions will be attempted: have spiny lobsters increased in abundance and mean size in the marine reserve in the 8 years since it was established? Are spiny lobsters more abundant inside the marine reserve than in the adjacent fished areas? The data of the effect of no-take areas will also be highlighted by comparing these data with those for lobster abundance before setting up the reserve.

Materials and methods

The study was carried out in a restocking area of central-western Sardinia (central-western Mediterranean) located at a bathymetric depth of from 50 to 100 m and the neighbouring zone (the zone around the restocking area within a radius <5 km from the centre of the area) (Fig. 1).

The area, first identified in 1997, was chosen as a restocking site on account of its geomorphological and bionomic characteristics. Directly investigated by scuba divers, it proved to be characterized by formations comparable to coastal precoralligenous and coralligenous detritus (Peres & Picard, 1964).

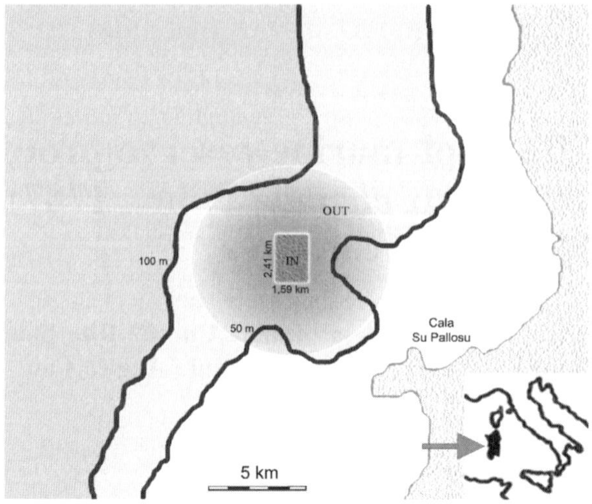

Fig. 1 Map of the protected area (IN) and the surrounding zone (OUT)

Since 1998, fishing was prohibited in the ca. 4-km^2 area (Regional Law No. 776 of 6-5-1998) (Secci et al., 1999). The neighbouring zone consists of coralligenous habitats with patches of sand and mud at a depth of from 50 to 80 m.

In 1997 (before the reserve was set up), ten fishing samplings were carried out inside and ten outside the restocking area, respectively, all in the period when commercial fishing allowed (March–September). The samples were conducted using trammel nets of 1,000 m length (nominal mesh from 50 to 73 mm), those inside the area following a sampling plan with transects set up in such a way as to enable the whole no-take area to be investigated. The first transect was placed from west to east in the centre of the area, with the remaining nine parallel to it at a distance of ca. 250 m from each other. Fishing outside the area was with 1,000-m-long trammel nets (nominal mesh from 50 to 73 mm) arranged along transects positioned in such a way as to monitor the whole protected area (diameter ca. 3 km). All these data provided the mean abundance of the lobsters present in the study area and in the neighbouring zone before the establishment of the reserve. In this year no specimens-size data were collected.

From 1998 to 2005 (in every year except for 2002 and 2003), a total of 5044 individuals (mean Carapace Length (CL) of 65.17 mm ± 7.31 SD for females and 66.22 ± 9.04 SD for males, respectively) were tagged with Plastic T-bar-type tags (inserted dorso-laterally, using an appropriate pistol, between the first and second abdominal segments) and released inside the

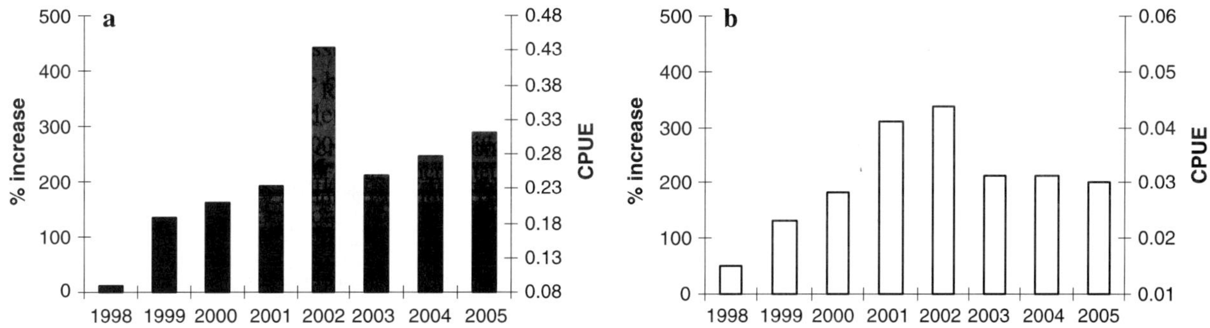

Fig. 2 Rate of increase (%) of *Palinurus elephas* abundance inside (**a**) the reserve area and in the surrounding zone (**b**)

area. During the same years, in the period from May to September, an average of ten sampling sessions were carried out each year (except for 2004, when there were only four). Every year, the same sampling plan was adopted inside the area as that for 1997.

Between May and August each year, catch data (in kg) were recorded for lobsters fished commercially (an average of 30 samplings/year) in the surrounding zone, in which the commercial catch was mainly made by the same boats that belonged to the Su Pallosu fishing cooperative. The data are taken from the delivery notes supplied daily by the cooperative. The fishermen were generally the same as those employed in the experimental sampling inside the reserve.

The number, weight and sex of all lobsters caught inside the reserve were recorded. Carapace Length (CL $\pm$ 1 mm) was measured mid-dorsally from the posterior edge of the carapace to the tip of the rostral spine. CPUE (catch per unit effort) was calculated by abundance (kg) of species caught per 50 m of trammel net and boat (kg/50 m/boat). The catch data for the period from 1998 to 2005, from both inside and outside the area, were compared with the 1997 values (% increase).

The mean trends in the catch of lobsters were analysed statistically by means of Kendall's τ test (Zar, 1996).

Data on the size of lobsters from the protected area were presented graphically and analysed by means of the Kolmogorov–Smirnov test (Zar, 1996).

Results

From 1998 to 2002, a progressive percentage increase of *Palinurus elephas* biomass values was recorded both inside the area and in the surrounding zones (Fig. 2). From 2003 to 2005, the increase ranged from 210% to 290% inside the observed area (IN) and was about 200% in the surrounding zones (OUT) (Fig. 2). The absence of restocking with tagged specimens in 2002–2003 could be one of the causes of the drop in biomass percentage increase; however, in 2005, the values rose above the mean (ca. 290%).

Overall, the whole mean abundance within the reserve (mean = 0.23 $\pm$ 0.10 kg/50 m/boat) was 7.5 times greater than that for the neighbouring zone (mean = 0.03 $\pm$ 0.07 kg/50 m/boat).

Inside the area, together with the increase of spiny lobster biomass, a steady and significant increase of carapace length in the captured specimens, both male and female, was recorded from 1998 to 2003 (τ Kendall: male $P < 0.01$, female $P < 0.05$), leading then to a decrease in 2004–2005 (Fig. 3).

A size-frequency analysis of the captured specimens highlighted a gradual increase over time of large-sized specimens (i.e. $\geq$100 mm CL) as well as of those with CL $\leq$ 65 mm (Fig. 4). In fact, the statistical comparison, made using the Kolmogorov–Smirnov test, indicated highly significant differences between the histograms of the relative size frequency for the 3 years (KS $P < 0.01$) (cf. Fig. 4). In particular, for 2005 (in which two modes can be distinguished, corresponding to 55 and 90 mm CL), the statistical differences from the histograms of 1999 and 2003 are attributed to the appearance of a large number of juveniles (109 out of the total 182 lobsters caught).

Discussion

Our study showed evident effects on abundance and size of *Palinurus elephas* inside the MPA compared

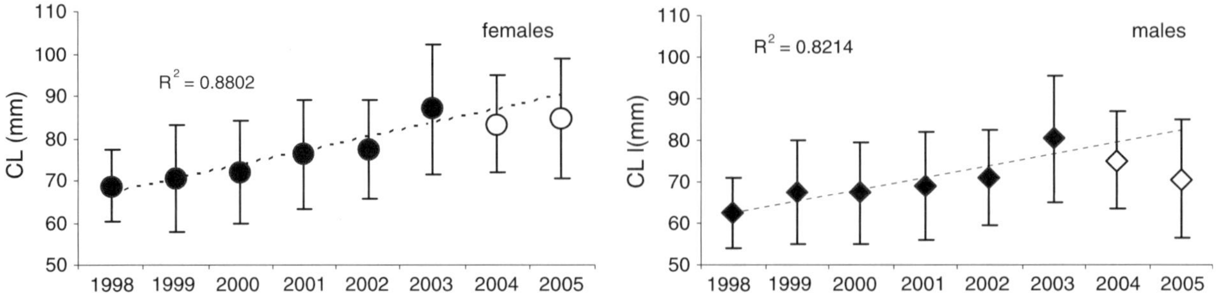

Fig. 3 Analysis of mean carapace length, CL (±SD) in female and male of *Palinurus elephas* caught inside the MPA. A positive trend (dashed line) was highlighted in the period 1998–2003

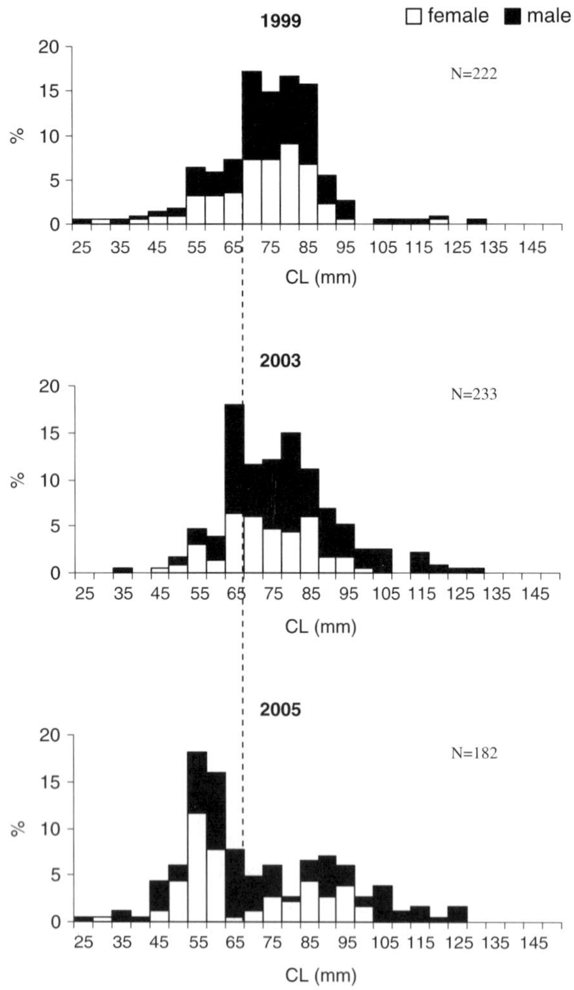

Fig. 4 Annual size-frequency histograms of *Palinurus elephas* caught inside the reserve area. The dashed line denotes the division between the juveniles (CL < 65 mm) and the rest of the population

with the adjacent fishing zones. The overall higher mean abundance in the protected area than in the surrounding areas strongly suggests that the cessation of fishing led to an increase in the lobster population in the marine reserve.

It is known that the spillover level from a protected area is closely connected with the size of the area, and to the mobility level of the species in question (Kramer & Chapman, 1999; Sanchez Lizaso et al., 2000; Goni et al., 2001, 2006; Kelly & MacDiarmid, 2003). Species with a medium-low mobility, like the lobster, need small areas for an immediate beneficial effect to be felt in the surrounding fishing areas (Sanchez Lizaso et al., 2000; Kelly et al., 2002).

The small size of our study area (4 km^2) seems to have ensured, even in the first years of experimentation, the movement of lobsters from the IN zone to the OUT ones, at a rate that appeared constant throughout the 8-year period (Follesa et al., 2007). The similarity in the trends for the percentage increase in biomass inside and outside the area offers further strong support for the hypothesis of constant movement of *Palinurus elephas* between the two areas (even in the first year of experimentation), also confirmed by applying the Cormack–Jolly–Seber (CJS) model to the specimens tagged inside the area (Follesa et al., 2007).

Considering the fact that all the lobsters recaptured had covered a mean distance of ca. 1.8 km/year and that the distance a specimen would have to cover in order to move outside the area (calculated as a straight line from the centre of the area) would be between 1.4 and 1.6 km, it appears clear that the spillover of specimens outside the area proved to be consistent after 1 year of the area being set up.

Moreover, recent interviews with local fishermen have revealed that even from the first year of experimentation, spillover was apparent because of the increased capture of medium- to large-sized adults and, in the last 2 years, by the pronounced increase in the number of smaller-sized specimens (CL $\leq$ 65 mm). Undoubtedly, the large-sized lobsters could represent a portion of those introduced inside the area, which had grown in size with time and moved across the boundary of MPA. The analysis of size-frequency histograms also indicated a reasonably good recruitment over the last years, with a marked increase occurring in 2005. A portion of these juveniles (aged ca. 3–4 years, Marin, 1987; Follesa et al., 2003) were probably responsible for the decrease in the mean CL values recorded in the last 2 years. These animals could also have migrated to the surrounding areas.

The results of the present study are encouraging for the establishment of small protected areas. From the literature, small areas (<6 km^2) are known to generally make a moderate contribution to a reduced commercial fishing area, while at the same time offering a greater and more immediate opportunity for increasing the commercial catch (McClanahan & Kaunda-Arara, 1996). Moreover, numerous studies have pointed out how, due to the reduced size of an area, movements of lobsters can be observed outside the Marine Protected Area (MPA) as a result of nocturnal foraging or simply from seasonal migration (Hunt et al., 1991; MacDiarmid & Breen, 1993; Goni et al., 2001). The relatively small size of our study area, compared with those reported in other studies (Cole et al., 1990; Hunt et al., 1991; Goni et al., 2001), has produced a two-fold benefit: the area has served as a protection and collection basin for the species, but it has also contributed to a local restoration of stock, increasing the number of specimens in the neighbouring zones, where commercial fishing is normally practised. Hence, even if our results cannot be extended as a generalization to apply to other protected areas, undoubtedly in some circumstances, the setting up of small MPAs could offer benefits at a local level, and with appropriate encouragement, could lead to an increase in the stocks of this commercially important species.

References

Cole, R., T. M. Ayling & R. G. Creese, 1990. Effects of marine reserve protection at Goat Island, northern New Zealand. New Zealand Journal of Marine and Freshwaters Research 24: 197–210.

Cole, R. G., E. Villouta & R. J. Davidson, 2000. Direct evidence of limited dispersal of the reef fish *Parapercis colias* (Pinguipedidae) within a marine reserve and adjacent fished areas. Aquatic Conservation Marine Freshwater Ecosystem 10: 421–436.

Davidson, R. J., 2001. Changes in population parameters and behaviour of blue cod, (*Parapercis colias*: Pinguipedidae) in Long Island-Kokomohua Marine Reserve, Marlborough Sounds, New Zealand. Aquatic Conservation: Marine and Freshwater Ecosystems 11: 417–435.

Follesa, M. C., D. Cuccu, F. Damele, A. Sabatini & A. Cau, 2003. Valutazioni sull'accrescimento di *Palinurus elephas* (Fabr., 1877) tramite marcatura e ricattura nei mari sardi. Biologia Marina Mediterranea 10(2): 253–256.

Follesa, M. C., D. Cuccu, R. Cannas, A. Sabatini & A. Cau, 2007. Emigration and retention of *Palinurus elephas* (Fabricius, 1787) in a region of the Central Western Mediterranean MPA. Scientia Marina 71(2): 279–285.

Galal, N., R. F. G. Ormond & O. Hassan, 2002. Effect of a network of no-take reserves in increasing catch per unit effort and stocks of exploited reef fish at Nabq. South Sinai, Egypt. Marine and Freshwater Research 53, 199–205.

Gell, F. R. & C. M. Roberts, 2002. The fishery effects of marine reserves and fishery closures. World Wildlife Fund-United States, Washington, D.C., USA.

Gerber, L. R., L. W. Botsfrod, A. Hastings, H. P. Possingham, S. D. Gaines, S. R. Palumbi & S. Andelman, 2003. Populations models and marine reserve design: a retrospective and prospective synthesis. Ecological Applications 13: S47–S64.

Goni, R., O. Renones & A. Quetglas, 2001. Dynamics of a protected Western Mediterranean population of the European spiny lobster *Palinurus elephas* (Fabricius, 1787) assessed by trap surveys. Marine Freshwater Resources 52: 1577–1587.

Goni, R., A. Quetglas & O. Renones, 2006. Spillover of spiny lobsters *Palinurus elephas* from a marine reserve to an adjoining fishery. Marine Ecology Progress Series 306: 207–219.

Hilborn, R., 2002. Marine reserves and fisheries management-Reply to C.M. Roberts. Science 295: 1233–1234.

Hunt, J. H., T. R. Matthews, D. Forcucci, B. S. Hedin & R. D. Bertelsen, 1991. Management implications of the trends in the population dynamics of the Caribbean spiny lobster, *Palinurus argus*, at Looe Key National Marine Sanctuary. NOAA Contract N. 50-DGNC-6-0093, Final Report.

Jennings, S., 2001. Pattern and predictions of population recovery in marine reserves. Review in Fish Biology and Fisheries 10: 209–231.

Kelly, S., D. Scott & A. B. MacDiarmid, 2002. The value of a spillover fishery for spiny lobsters around a marine reserve in New Zealand. Coastal Management 30: 153–166.

Kelly, S. & A. B. MacDiarmid, 2003. Movements patterns of mature spiny lobsters, *Jasus edwardsii*, from a marine reserve. New Zealand Journal of Marine and Freshwaters Research 37: 149–158.

Kramer, D. L. & M. R. Chapman, 1999. Implication of fish home range size and relocation for marine reserve function. Environmental biology of Fishes 55: 65–79.

Marin, J., 1987. Exploitation, biologie et dynamique du stock de langouste rouge de Corse. *Palinurus elephas* Fabricius. Thèse Univ. Aix-Marseille. Facultè Sciences Luminy: 328 pp.

MacDiarmid, A. B. & P. A. Breen, 1993. Spiny lobster population change in a marine reserve. Proceedings of the Second International Temperate Reef Symposium: 7–10.

McClanahan, T. R. & B. Kaunda-Arara, 1996. Fishery recovery in a coral reef marine park and its effects on the adjacent fishery. Conservation Biology 10(4): 1187–1199.

McClanahan, T. R. & S. Mangi, 2000. Spillover of exploitable fishes from a marine park and its effect on the adjacent fishery. Ecological Applications 10: 1792–1805.

NCR [National Research Council (USA)], 2001. Marine Protected Areas: Tools for Sustaining Ocean Ecosystems. National Academy Press, Washington, D.C., USA.

Palumbi, S. R., 2001. The ecology of marine protected areas. In Bertness, M. D., S. D. Gaines & M. E. Hay (eds). Marine Community Ecology. Sinauer, Sunderland, Massachusetts, USA: 509–530.

Peres, J. M. & J. Picard, 1964. Nouveau manuel de bionomie benthique der la Mediterranee. Recueil des Travaux de la Station Marine d'Endoume 31(47): 1–37.

Roberts, C. M. & N. V. C. Polunin, 1991. Are marine reserve effective in management of reef fisheries? Reviews in Fish Biology and Fisheries 1: 65–91.

Roberts, C. M. & N. V. C. Polunin, 1993. Effects of marine reserve protection on Northern Red Sea fish populations. Proceedings of the 7th International Coral Reef Symposium 2: 979–987.

Roberts, C. M. & J. P. Hawkins, 2000. Fully-protected Marine Reserve: A Guide. Endangered Seas Campaign, World Wildlife Fund-United States, Washington, D.C., USA.

Roberts, C. M., J. A. Bohnsack, F. Gell, J. P. Hawkins & R. Goodridge, 2001. Effects of marine reserve on adjacent fisheries. Science 294: 1920–1923.

Russ, G. R., 2002. Yet another review of marine reserves as reef fisheries management tools. In P. F. Sale (ed.), Coral Reef Fishes: Dynamics and Diversity in a Complex Ecosystem. Academic Press, San Diego, California: 421–443.

Sanchez Lisazo, J. L., R. Goni, O. Renones, J. A. Garcia Charton, R. Galzin, J. T. Bayle, P. Sanchez Jerez, A. Perez Ruzafa & A. A. Ramos, 2000. Density dependence in marine protected populations: a review. Environmental conservation 27(2): 144–158.

Secci, E., D. Cuccu, M. C. Follesa & A. Cau, 1999. Fishery and tagging of *Palinurus elephas* in Sardinian seas. The biodiversity and crisis and crustacean: 665–672.

Willis, T. J., R. B. Millar, R. C. Babcock & N. Tolimieri, 2003. Burdens of evidence and the benefits of marine reserves: putting Descartes before des horse? Environ Conservation 30: 97–103.

Zar, J. H., 1996. Biostatistical analysis, 3rd edn. Prentice-Hall, Englewood Cliffs-NJ.

Hydrobiologia (2008) 606:69–79
DOI 10.1007/s10750-008-9343-y

Incorporating ecological functioning into the designation and management of marine protected areas

C. L. J. Frid · O. A. L. Paramor · S. Brockington · J. Bremner

© Springer Science+Business Media B.V. 2008

Abstract Marine protected areas are generally designed and managed on the basis of the presence and extent of specific habitat types or the habitats of important species. However, it has become clear that in addition to including these 'structural' elements of marine systems, management strategies should incorporate a consideration of the functional aspects of the ecosystems. Biological traits analysis (BTA) has been successfully used to describe ecological functioning in marine benthic systems. BTA uses a number of biological characteristics expressed by the taxa present as indicators of key ecosystem functions. Two expert workshops were used to examine the potential for the application of BTA in the designation and management of MPAs. They concluded that BTA represented the best tool currently available for quantifying ecological functioning and agreed on 10-key ecological functions delivered by marine benthic communities. Twenty-four biological traits were also identified by the workshops as indices of these ten functions. In order to demonstrate the practical utility of the approach, BTA using these traits, was applied to a dataset covering benthos from within and around the proposed Eddystone Special Area of Conservation (SW England). The case study demonstrated that with the type of data normally available from conservation assessment type surveys, and a knowledge of the relevant biological traits, it is possible to use a consideration of ecological functioning to set boundaries for the MPA and to inform the site management objectives. The use of structure and function information to inform the designation process and subsequent management of marine protected areas is discussed.

Keywords Ecological services · Biological traits analysis · Conservation objectives · Benthos

Guest editors: J. Davenport, G. Burnell, T. Cross, M. Emmerson, R. McAllen, R. Ramsay & E. Rogan
Challenges to Marine Ecosystems

C. L. J. Frid (✉) · O. A. L. Paramor · J. Bremner
School of Biological Sciences, Biosciences Building, University of Liverpool, Crown Street, Liverpool L69 7ZB, UK
e-mail: C.L.J.Frid@liverpool.ac.uk

S. Brockington
English Nature, Northminster House, Peterborough PE1 1UA, UK

Present Address:
J. Bremner
Section of Marine Biology and Limnology, Department of Biology, University of Oslo, P.B. 1064, Blindern, Oslo 0316, Norway

Introduction

A variety of international conventions (e.g. RAMSAR, ASCOBANS), as well national legislation (EC,

1992; Australian Fish Resources Management Act, 1994; US Magnuson-Stevens Fishery Conservation and Management Act, 1996; Canadian Ocean's Act, 1997; EC, 2004), require the protection of sites for nature conservation purposes. It is now recognised that protection of the habitat, and its associated functional processes, is a key element of ensuring ecological sustainability and is therefore a key element of the application of the 'ecosystem approach' (Frid et al., 2005, 2006). Until recently, sites have been selected for protection based on the *presence* of specific habitats or species. However, a growing number of legislative agreements, including the Convention on Biological Diversity (United Nations, 1992), the European Marine Strategy Directive (EC, 2005) and the Habitats Directive (EC, 1992), require management schemes to address the *functioning* of ecosystems.

Consideration of functioning in the designation process has been difficult in the past due to the lack of science to support such an approach. A number of studies have recently sought to describe the delivery of 'ecological goods and services' from the marine ecosystem (Chapin et al., 1997; Snelgrove et al., 1997; Lasiak, 1998; Beaumont & Tinch, 2003; Frid & Paramor, 2006). Most of these have been descriptive accounts, or focussed on only one aspect (say fisheries resources or nutrient regeneration), and presented their analysis in units appropriate to that function. However, application of an ecosystem approach requires the integration of ecosystem components, and so requires a multivariate, multi-function, approach. Recent advances in data handling have made this possible and one approach that has been applied with some success is biological traits analysis (BTA). It uses the biological traits of taxa as indicators of key aspects of functioning. BTA was initiated in lentic systems (Statzner et al., 1994), and developed for application to marine ecosystems (Frid et al., 2000; Bremner et al., 2003, 2005, 2006a). It focuses on the behaviour and attributes of biological entities that contribute to the maintenance of ecosystem processes and differs from previous trait-based approaches (e.g. trophic or functional groups (Pearson & Rosenberg, 1986; Snelgrove & Butman, 1994; Grall & Glemarec, 1997; Whitlatch et al., 1997; Clarke & Warwick, 1998; Dauwe et al., 1998; Telesh et al., 1999; Herrando-Pérez & Frid, 2001)) as it utilises a wider range of information on organism functional traits. It can be applied to any taxonomic level and can incorporate indicators of several different aspects of functioning. Furthermore, BTA can be used in a 'fuzzy coding' mode where a species is not allocated to a single code for each trait but can be dispersed over a number of code classes to reflect its biology, or our uncertainty of its biology. For example in classic functional group analysis an organism might be scored as either a predator or a scavenger. However, many such taxa exhibit both feeding modes. Under a fuzzy coding approach the organism can be scored 0.5 to each, or if predominantly a carnivore then, may be, 0.8 carnivore and 0.2 scavenger. The use of fuzzy coding complicates the subsequent analysis but allows for a much more realistic representation on the functional biology of the assemblage.

The initial stages of BTA involve the identification of key aspects of functioning (e.g. the processes involved in energy/carbon/nutrient cycling) and the functional traits that can be used as indicators of these (e.g. feeding traits as indicators of carbon transport between the pelagos and benthos).

In this article, we investigate how BTA might be used to aid designation and management of MPAs by providing a wider ecosystem approach than traditional models of MPA management. In order to do this we: (i) establish what are the most important, or key, ecological functions delivered by marine benthic systems, and consider the biological traits that organisms possess that underpin delivery of these functions and (ii) consider how this information might be used in MPA management. As a case study to assess the feasibility of this, we apply our approach, includes consideration of species' identities and ecological functioning, to a UK-proposed Special Area of Conservation (SAC).

How can ecological functioning be used in MPA management

Delivery of the ecological functioning of a healthy ecosystem can be incorporated in the designation of MPA management in two ways. First, in setting the boundary of the MPA to ensure key functions are protected by the MPA designation, and second, in setting the management objectives for the site, so that performance of the management regime can be assessed against them.

Setting boundaries

As MPAs do not exist in ecological isolation, but exchange resources with a much larger area, there is an argument to extend the boundaries of the protected area beyond the physical extent of the habitat/feature being targeted. However, since the spatial and temporal scales of processes that need to be considered in order to protect the structure and functioning of the habitat/feature are considerable, a more pragmatic approach is to protect those areas close to the feature that exhibit similar ecological functioning. Thus, regions of rapid change in the multivariate descriptor of the biological traits (and hence ecological functions performed) might be considered as possible boundaries for the MPA.

Conservation objectives

Designating an area on the basis of the presence of a feature/habitat or even the spatial distribution of functions, is only the beginning of effective management. Current management approaches focus on the delivery of defined objectives (objective-based management). Traditionally, conservation objectives have either been set in broad terms, e.g. maintain habitat *x* in a good condition, or have focussed on key or indicator taxa, e.g. provide habitat for 5% of the regional population of species *y*. With the availability of tools that provide quantitative metrics of ecosystem functioning, either in total/aggregate form or by individual functions, it is possible to set conservation objectives for function delivery. BTA allows the species delivering any given function to be identified and so changes in functions can be traced back to changes in the biota and hence back to impacting activities that can be controlled by management measures.

Methods

Identification of the ecological functions and the traits that map them

The selection of the key functions and the traits to that can be used to measure them was supported by two international workshops, one in London and one in Plymouth. At each workshop around 10 experts in various aspects of marine benthic ecology and ecosystem functioning, were asked to develop a list of key ecosystem functions delivered by marine benthic systems. They were then asked to consider whether BTA was a suitable means of measuring the functioning, and if so, which traits of the biota would be the most useful proxies. This included a consideration of practicality, which included information availability.

This was achieved in a workshop format using a semi-structured, round-table, discussion approach. The results of the discussions were consolidated and, after a period for reflection, were re-presented to the groups for review. The two workshops were run independently with the aim of using the degree of congruency of the outputs as an indication of the robustness of the conclusions.

The workshops identified 10 aspects of benthic marine system functioning which were seen as their key functions (Table 1). There was complete agreement in the composition of this list between the two workshops. The two workshops considered that 24 biological traits that could be used as indicators of the contribution of the biota to delivery of the key functions (Table 2). The two workshops independently came up with the same 20 traits and the remaining three were identified by only one of the workshops, this difference being a reflection of the specialisms of the experts present

A full explanation of the links between traits and key aspects of functioning are given in Bremner et al. (2006b).

Some of the traits identified may be of greater practical use than others because they are indicators of more than one aspect of functioning. However, the workshop participants decided not to rank traits by importance, as this was considered too subjective and difficult to apply to different sites. Participants concluded that all the listed traits should be included in the analysis, with the condition that particular traits could be further considered in isolation, if this was considered appropriate for a particular site, for example because a particular function was seen to be of great importance (i.e. food resources for birds, breeding habitat for species of high conservation importance).

Table 1 Key aspects of functioning identified during two international workshops (Bremner et al. 2006a)

Process, property or activity	
1.	Energy and elemental cycling (carbon, nitrogen, phosphorus, sulphur)
2.	Silicon cycling
3.	Calcium carbonate cycling
4.	Food supply/export
5.	Productivity
6.	Habitat/refugia provision
7.	Temporal pattern (population variability, community resistance and resilience)
8.	Propagule supply/export
9.	Adult immigration/emigration
10.	Modification of physical processes

A case study: the Eddystone Reef, Cornwall

Data source

As no contemporary data were available we used a historical dataset for the site to demonstrate the feasibility of the approach. The Holme's 'scoop sample' dataset from soft sediment habitats in the area of Eddystone Reef (Holme, 1953) considers all macrofauna obtained from sieving a sample of sediment retrieved from the seabed in a 'scoop' sampler. Holme's sampler was similar to what might now be known as a pipe or anchor dredge, thus samples are not strictly quantitative but as the same method was applied at each site they are comparable. Holme sampled stations arranged on a series of transect running north–south and data used here cover three transects and an additional station located close to the Eddystone Reef (see station locations in Fig. 4). All Holmes transects lie to the east of the Eddystone and so cover only the eastern portion of the proposed SAC which is centred on the reef itself.

Data analysis

Ecological structure

Centred (covariance) PCA (Gower & Hand, 1996) was used to examine differences in ecological structure at stations on and around the potential SAC features. Unlike non-parametric procedures, like MDS, PCA produces scores for each station and these scores can be used as a measure of how different the communities are in terms of their ecological structure (stations with similar scores contain communities with similar ecological structure). The first and second set of scores produced by the analysis (axes 1 and 2 scores) can, therefore, be used to generate an ordination plot that allows the differences between the communities to be visualised and quantified.

Where the resulting ordinations contained outlying stations that compromised the ability of an analysis to describe emergent patterns in the data, the abundance dataset was transformed (double root) and the

Table 2 Biological traits identified as indicators of key aspects of functioning in potential MPAs (Bremner et al. 2006a)

1.	Maximum size	15.	Resource capture method
2.	Maximum growth rate	16.	Food type
3.	Longevity	17.	Energy transfer efficiency
4.	Time to maturity	18.	Tissue components
5.	Reproductive method	19.	Defence strategy
6.	Fecundity	20.	Movement method
7.	Propagule dispersal	21.	Mobility
8.	Body design	22.	Water column migration
9.	Living habit	23.	Horizontal migration
10.	Living location/environmental position	24.	Intra-specific sociability
11.	Exposure potential	25.	Predictability of dynamics
12.	Degree of flexibility	26.	Recruitment variability/success
13.	Degree of attachment	27.	Biogenic habitat provision
14.	Strength of attachment	28.	Scale of habitat provision

analysis repeated. If transformation of the abundance data did not prove sufficient to reduce the influence of outliers, these stations were removed and the dataset re-analysed.

Ecological functioning

Co-inertia analysis (Doledec & Chessel, 1994) was utilised to examine differences in ecological functioning over the sampling stations. Co-inertia analysis assesses the co-structure between two data tables by simultaneously ordinating them, maximising both the variance from the individual tables and the correlation between them (Doledec & Chessel, 1994). This analysis produces scores for the stations that can be used as a measure of how different the communities are, in the same way as PCA, but because they incorporate information on both the abundance of taxa at a station and the biological traits they exhibit (Doledec et al., 1999), these scores describe how different the communities are in terms of their ecological functioning. The co-inertia scores can also be plotted on an ordination map, with each point on the map representing the abundance-weighted 'biological trait composition' of an individual station. Interpretation of the contoured structure and functioning maps required reference back to the results of the biological ordinations.

The analysis utilised both the dataset of taxon abundance used previously to examine ecological structure over the sampling stations, and the biological traits tables prepared for each of the four datasets examined. First, separate ordinations of the individual data tables were carried out. As before, centred PCA was used to investigate the ecological structure of the stations. However, for this analysis the table was transposed so that the taxa were in rows. Fuzzy correspondence analysis was used to assess the biological traits table. This is a form of correspondence analysis used when the categories of variables are fuzzy coded (Chevenet et al., 1994).

Co-inertia analysis was then carried out using both ordinations and the significance of the resulting co-structure was examined with a random permutation test (Doledec & Chessel, 1994). This test randomly permuted the rows of the co-inertia table and recalculated the inertia statistics 100 times. The observed co-inertia value was then compared to the frequency distribution of the randomly permuted values to assess if it was significantly larger.

The co-inertia analysis was initially applied to the four datasets in-full, irrespective of whether outlying stations had been removed from any of these datasets for the purposes of describing ecological structure. This is because several taxa can exhibit the same traits; therefore extreme differences in abundance of particular taxa at a station may not necessarily translate into extreme differences in trait composition. However, if outlying stations were noted on the resulting co-inertia ordination plots, the datasets were transformed and stations removed as appropriate.

Boundary mapping

As a result of the way they are calculated, the first set of station scores (axis 1 scores) generated by both the PCA and co-inertia analysis contain the greatest amount of information on the variability in ecological structure or functioning among the sampling stations, making them useful variables for summarising differences among the communities. These axis 1 PCA and co-inertia analysis scores were plotted over maps of the Eddystone Reef survey area, to provide a visualisation of how the structure and functioning varied over the region and to identify the areas of greatest change.

The ecological structure and functioning scores were grouped into appropriate categories using the mapping package ArcGIS 9 (ESRI, California, USA) and coloured labels assigned to the categories to ease interpretation of the maps. The co-ordinates of the sample stations were plotted and overlaid with the ecological structure and functioning scores. The structure and functioning scores information was then contoured using a Triangular Irregular Network (TIN), see the caption for Fig. 4 for a full explanation. TINs were utilised, in this case, as they were found to best reflect the patterns in the data from the Holmes' (1953) Eddystone Reef surveys.

This allowed species and functions causing data clusters and regions of rapid change on the map to be identified.

Table 3 Multivariate analysis of the ecological structure (PCA) and functioning (co-inertia analysis) of benthic communities sampled by Holme (1953) in the area of Eddystone Reef

Ordination axis	Eigenvalue	Relative inertia (%)	Cumulative inertia (%)
Ecological structure (PCA)			
1	11.600	29.11	29.11
2	5.773	14.49	43.59
Ecological functioning (co-inertia analysis)			
1	0.202	50.01	50.01
2	0.069	17.14	67.15

Results

Benthic structure and function at Eddystone

Ecological structure

Analysis of untransformed data resulted in station A2 (a 'very fine muddy sand' area to the far north of Eddystone) as an extreme outlier on the ordination plot. Square-root transformation of the abundance data did not help to reveal the patterns in ecological structure over the remaining stations, so A2 was removed from the analysis. PCA of the remaining stations described approximately 43% of the variability in the dataset (Table 3).

Samples from 'clean sand' habitats (also the unclassified stations C5 and C6 from the sandy grounds) showed some degree of grouping on the plot, however, there was insufficient replication of the other habitat types to determine how strongly stations grouped together based on habitat type (Table 4, Fig. 1).

The variable distribution of *Phaxas pellucidus* (Pennant) and *Lumberineris* sp. (de Blainville) caused the separation of stations along the first axis. Station C6 (to the south-east of Eddystone) and, to a lesser extent the 'clean sand' stations (Fig. 1), had relatively high abundance of both taxa, while they were not recorded (except for four *Lumbrineris* sp. found in A4) in stations C1, B2, A4 and A5. These stations contained a variety of substrate types (Table 4). Station C1 was differentiated from the latter stations by markedly higher abundance of *Echinocyamus pusillus* (OF Müller) and indeterminate polychaetes.

Table 4 Substrate types at stations sampled by Holme (1953) in the area of Eddystone Reef

Substrate type	Stations
Very fine muddy sand	A2
Clean medium grade sand/mixed muddy sand and gravel	A3
Muddy sand with small stones and shell fragments	A4
Shell gravel	A5
Clean sand	A6, A7, B5, B6, B7, C4
Fine muddy sand with small stones	B2
Slightly muddy fine sand	B4
Fine gravel of shell fragments and small stones	C1
Muddy sand with a few stones	C2, C3
Unclassified (sandy grounds)	C5, C6

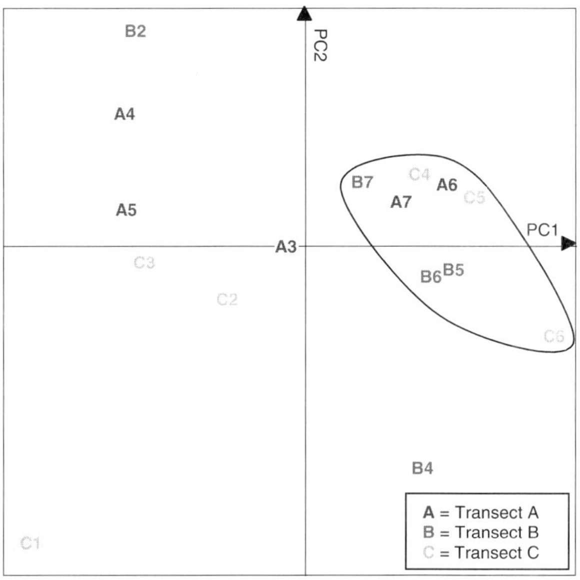

Fig. 1 PCA ordination of the stations analysed during investigation of ecological structure of benthic communities sampled by Holme (1953) in the area of Eddystone Reef. Information on substrate type for each station is given in Table 4. 'Clean sand' and 'unclassified' stations sampled from the sandy grounds are highlighted

Ecological functioning

Analysis of ecological functioning was based on a square-root transformation of the abundance data. Removal of station A2 was not necessary for this analysis, as it was not portrayed as an outlier in the ordination.

However, interpretation of the analysis was impaired by one-trait category, 'regular-seasonal/reproductive water column migration'. This trait category was so separated from all the others on the trait ordination plot that further interpretation of the results was not possible. This trait was only expressed by one taxon in the dataset (indeterminate polychaetes), and then only to a low degree (only a very small number of polychaetes undertake regular water column migrations, for reproductive purposes). For this reason, the analysis was repeated after exclusion of the category 'regular-seasonal/reproductive water column migration'.

The co-inertia analysis accounted for 67% of the variability in ecological functioning over the stations (Table 3). The 'clean sand' stations separated out from the other stations in terms of their trait compositions (Table 4, Fig. 2), although they were not as tightly grouped in general as they had been in the ordination of ecological structure (Fig. 1).

The traits most important for determining differences between stations were body design, living habit, exposure potential, degree of flexibility,

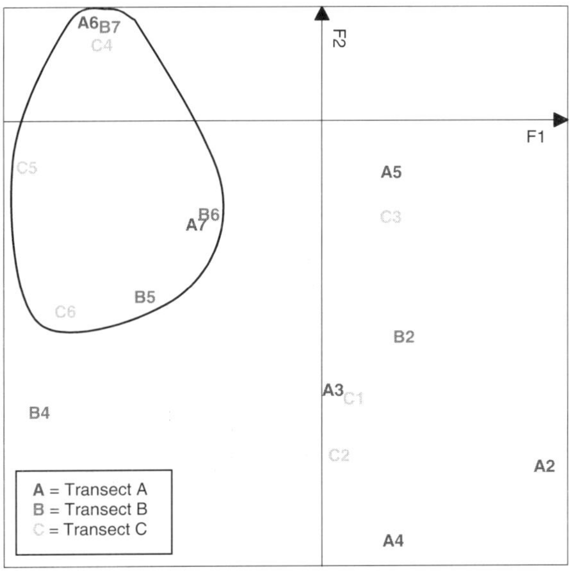

Fig. 2 Co-inertia ordination of the stations analysed during investigation of ecological functioning of benthic communities sampled by Holme (1953) in the area of Eddystone Reef. Information on substrate type for each station is given in Table 4. 'Clean sand' and 'unclassified' stations sampled from the sandy grounds are highlighted

horizontal and water column migrations and habitat provision (Table 5). Stations A4 and A2 ('muddy sand with small stones and shell fragments' and 'very fine muddy sand' stations to the north of Eddystone) were characterised by tube-dwelling organisms, those with high exposure potential and taxa forming habitat by the accretion of sediments (Fig. 2). Stations A6, B7 and C4, all 'clean sand' stations to the south (A6, B7) and east (C4) of Eddystone, were characterised by shelled organisms and taxa undertaking irregular or single horizontal migrations.

Stations in the lower left quadrant of the ordination plot, principally B4 but including the remaining 'clean sand' stations and the unclassified stations from the sandy grounds (C5 and C6), were characterised by flexible taxa (>45°), organisms inhabiting temporary burrows and those making irregular or seasonal water-column migrations (Fig. 2). The random permutation test indicated a lack of co-structure between the taxon abundance and biological traits tables ($P = 0.15$).

Boundary mapping

The depth contours around Eddystone Reef are shown in Fig. 3. The physical feature of the reef itself is clearly demarcated. The stations sampled by Holmes (1953) were all to the east of the Eddystone complex, on a well-dispersed grid. A few rapid changes in ecological structure were observed among the data points (Fig. 4), although this may have been due to the dispersed pattern of the sampling stations.

Table 5 Trait categories contributing most to differences in ecological functioning over stations sampled by Holme (1953) in the area of Eddystone Reef

Trait	Trait category
Body design	Hard-shell
Living habit	Tube
	Temporary burrow
Exposure potential	High (erect surface/interface dwelling)
Degree of flexibility	>45°
Horizontal migration	Irregular/single
Water column migration	Irregular/single
Habitat provision	Action—sediment accretion

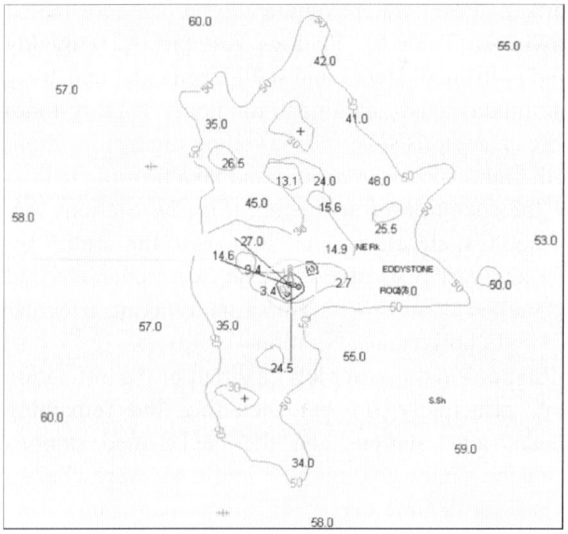

Fig. 3 The depth contours around the Eddystone Reef complex (adapted from Axelsson et al. 2006)

Figure 5 shows the functional changes across the survey area. The ecological functioning data shows a pattern distinct to that of the species composition dataset and, therefore, provides an additional information to the ecological structure analysis. However, it too is affected by edge effects due to the restricted nature of the sampling programme.

Delineation of the Eddystone Reef SAC would currently be limited to the physical extent of the reef feature as data were not available, at the time of analysis, on which to undertake a comprehensive assessment of gradients of change in the ecological structure and functioning of the reef and surrounding areas.

Discussion

This study has demonstrated a means by which ecological survey data can be used in the delimitation of boundaries for marine protected areas through the explicit consideration of the ecological structure (biological diversity) and functioning of the systems. Until now, delimiting nature conservation sites has either been done with reference to the spatial

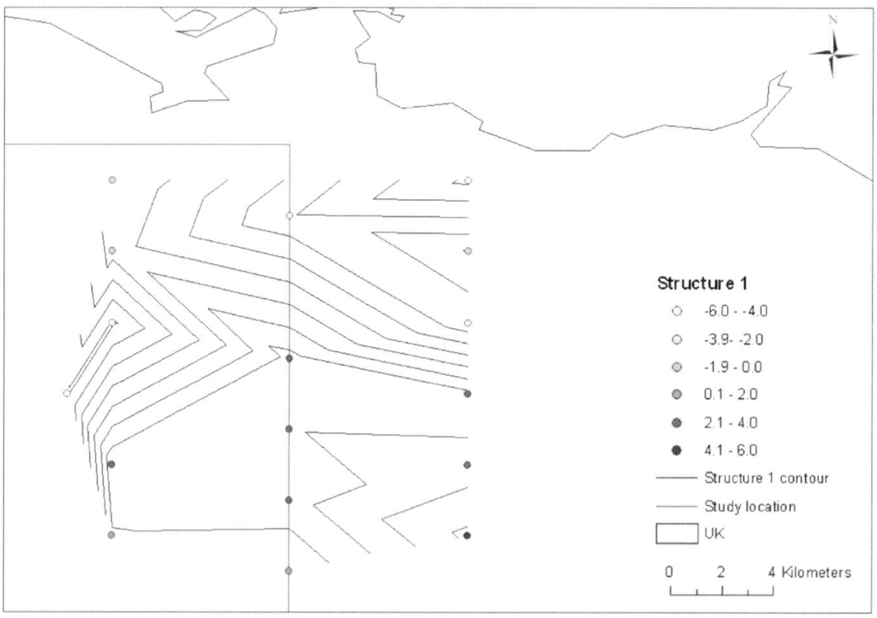

Fig. 4 The ecological structure to the east of Eddystone Reef, based on PCA scores of the biological communities sampled by Holme (1953) (see Fig. 1). These scores explain 29.1% of the variability in the data. A triangulated irregular network (TIN) was used to spatially contour the data. TINs are based on a set of adjacent, non-overlapping triangles with *x*, *y* coordinates and *z* vertical elevations for their vertices, with topological relationships between the triangles and their adjacent neighbours. The contour lines produced can therefore be read as a landscape map, with lines close together indicating areas of greatest change and lines further apart indicating areas of similarity. Each data point was categorised and labelled with a different colour to allow points of similarity to be identified. The box enclosing some of the sites represents the boundaries of the sampling area selected by the survey contractors (Axelsson et al. 2006)

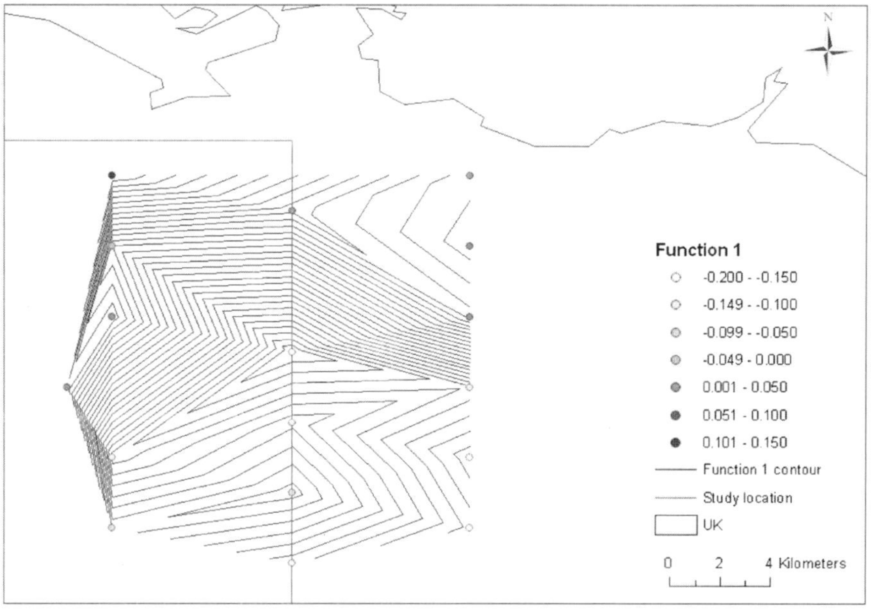

Fig. 5 The ecological functioning to the east of Eddystone Reef, based on co-inertia scores of the biological communities sampled by Holme (1953) (see Fig. 2). These scores explain 50% of the variability in the data. A triangulated irregular network (TIN) was used to spatially contour the data. See Fig. 4 caption for an explanation of TINs. The box enclosing some of the sites represents the boundaries of the sampling area selected by the survey contractors (Axelsson et al. 2006)

distribution of 'key' species or the extent of specific types of physical environment, i.e. the physical habitat. Marine systems are dynamic and, ecologically, open. This means that ecological processes extend across physical habitat boundaries and the health of the biological assemblage in a habitat may be dependent on processes occurring elsewhere. The approach developed and demonstrated here addresses these issues by explicit consideration of biological diversity and ecological functioning using BTA.

BTA has several potential uses with regards to SAC designation and management. First, BTA can be used as a tool to assist in boundary setting. A strict interpretation of the Habitats Directive requires SAC boundaries to follow the edge of the distribution of the species or habitat of concern. However, the Convention on Biological Diversity (United Nations, 1992) and subsequent treaty undertakings (e.g. World Summit on Sustainable Development, Johannesburg, 2002) require explicit conservation of ecological functioning. One can, therefore, envisage the application of BTA to delimit areas that function as Annex 1 habitats, even if our perception of them is that they are a different type of habitat. For example, conservation of offshore sandbanks less than 20-m deep is required under the Habitats Directive (EC, 1992). These areas deliver a range of ecological functions, so this delivery must, logically, also be protected. If an area adjacent to the banks is also delivering the same functions, but is, say, 24-m deep, then given the open nature of marine systems, it seems reasonable to include this as part of the bank system and incorporate it within the SAC boundaries.

Second, BTA allows the identification of both the ecological functions strongly expressed in a habitat (or unique to it), and the species delivering them. This information can then be used in the setting of conservation objectives for the site. For example, at Eddystone, body design, living habit, exposure potential, flexibility, migrations and habitat provision were the ecological traits that were most important in distinguishing among the stations.

In addition to its use for describing ecological functioning across a potential SAC site, the ability of the approach to identify whether communities in similar habitats but different geographic locations function in the same way (Bremner et al., 2006a), means it can provide information on differences or similarities in functioning between sites proposed for inclusion in SAC series. This information will be useful in the process of identifying sites for protection, because two sites with similar Annex I habitat types may not necessarily function in the same way.

The expert workshops reviewed the approaches available for providing information on ecological functioning and concluded that BTA was the most practical approach available at this time. In order to apply BTA requires two things, a knowledge of what are the important functions to include in consideration and secondly what traits can be used to index those functions in the biota. The workshops, with special reference to offshore sandbanks and subtidal reefs, identified a list of 10-key ecological functions (Table 1) and 24 biological indicator traits that could be used to index them (Table 2). It is reassuring that the combined expertise available to the two workshops should firstly agree on the list of functions and secondly that that lists of functions, and the linked traits, are not so long as to be impractical to apply using BTA. In order to examine the practical application of this conclusion we used these traits for the basis of the analysis presented here for the Eddystone Reef area.

Our analysis has shown how the approach can be applied to both characterize the ecological functioning of the assemblages present in an area and to delimit areas of different ecological functioning. Furthermore, the output from the fuzzy coded BTA could be used as input to a GIS system. The GIS then allowed spatial contouring of ecological functioning and this could assist in the selection of boundary points for an MPA.

Given the complex interactions at each stage of the ecological chain that links environmental conditions to biological assemblage composition to ecological function delivery managing the system to deliver tight targets for functioning is unlikely to succeed. However, it is clear that changes in functioning could be an important element of any failure to deliver a healthy ecosystem. It therefore follows that, in addition to a role in delimiting MPA boundaries, monitoring programmes should be designed to allow changes in functioning, in space or time, to be detected with a high degree of confidence. This will aid managers in determining the effects of natural change and/or human activities in MPAs. BTA allows the links between function delivery and the taxa responsible to be explicitly linked. It is impossible to manage the marine ecosystem and it is equally impossible to manage ecological functioning of the system, however, by being able to link functions back to taxa, our knowledge of the vulnerability of specific taxa to various human activities will allow management schemes to be advanced that do provide explicit consideration for, and protection of, ecosystem functioning.

Acknowledgements This manuscript derives from a study on designation of offshore SACs commissioned by English Nature and we acknowledge their financial and material support. We would like to thank Dan Laffoley and Leigh Jones of English Nature for their input to the project. We also thank Silvana Birchenough, Erik Bonsdorff, Tasman Crowe, Jan Hiddink, Keith Hiscock, Kerry Howell, Stuart Jenkins, Charlie Marshall, Paul Somerfield, Harvey Tyler-Walters and Jack Sewell for participating in the Plymouth and London workshops. We are indebted to Jean-Luc Solandt of the Marine Conservation Society and Paul Robinson of the Joint Nature Conservation Committee for providing the Seasearch diver survey data and access to the Marine Recorder software (www.jncc.org.uk), Chris Cesar and Diane Jones for help with collating the Eddystone Reef datasets and Sarah King of the Zoological Society of London for providing assistance with the ArcGIS mapping programme.

References

Australian Fish Resources Management Act, 1994. An Act relating to the management of fish resources, to repeal and amend certain Acts, and for related purposes. http://www.austlii.edu.au/au/legis/wa/consol_act/frma1994256/.

Axelsson, M., S. Dewey, S. Chaddock & S. Duke, 2006. Survey of the reef habitat around Eddystone Reef, Plymouth. Report to English Nature. SeaStar Survey, Southampton. p601.

Beaumont, N. J. & R. Tinch, 2003. Goods and services related to the marine benthic environment. CSERGE Working Paper ECM 03-14:12.

Bremner, J., C. L. J. Frid & S. I. Rogers, 2005. Biological traits of the North Sea benthos: does fishing affect benthic ecosystem function? American Fisheries Society Symposium 41: 477–489.

Bremner, J., S. I. Rogers & C. L. J. Frid, 2006a. Matching biological traits to environmental conditions in marine benthic ecosystems. Journal of Marine Systems 60: 302–316.

Bremner, J., O. A. L. Paramor & C. L. J. Frid, 2006b. Developing a methodology for incorporating ecological structure and functioning into designation of Special Areas of Conservation (SAC) in the 0-12 nautical mile zone. A report to English Nature from University of Liverpool, Liverpool.

Bremner, J., S. I. Rogers & C. L. J. Frid, 2003. Assessing functional diversity in marine benthic ecosystems: a comparison of approaches. Marine Ecology Progress Series 254: 11–25.

Canadian Oceans Act 1997. Oceans Act 1996, c. 31O-2.4 [Assented to December 18th, 1996]. http://www.pac.dfo-mpo.gc.ca/oceans/default_e.htm.

Chapin, F., B. Walker, R. Hobbs, D. Hooper, J. Lawton & O. Saia, D. Tilman, 1997. Biotic control over the functioning of ecosystems. Science 277: 500–504.

Chevenet, F., S. Doledec & D. Chessel, 1994. A fuzzy coding approach for the analysis of long-term ecological data. Freshwater Biology 31: 295–309.

Clarke, K. R. & R. M. Warwick, 1998. Quantifying structural redundancy in ecological communities. Oecologia 113: 278–289.

Dauwe, B., P. M. J. Herman, C. H. R. Heip, 1998. Community structure and bioturbation potential of macrofauna at four North Sea stations with contrasting food supply. Marine Ecology-Progress Series 173: 67–83.

Doledec, S. & D. Chessel, 1994. Co-inertia analysis: an alternative method for studying species-environment relationships. Freshwater Biology 31: 277–294.

Doledec, S., B. Statzner & M. Bournard, 1999. Species traits for future biomonitoring across ecoregions: patterns along a human-impacted river. Freshwater Biology 42: 737–758.

EC, 1992. COUNCIL DIRECTIVE 92/43/EEC of 21 May 1992 on the conservation of natural habitats and of wild fauna and flora. European Commission, Brussels.

EC, 2004. Council regulation no. 612/2004 of 22 March 2004 amending council regulation (EC) no. 850/98 as regards the protection of deep water coral reefs from the effects of trawling in an area north west of Scotland. European Commission, Brussels.

EC, 2005. Proposal for a Directive of the European Parliament and the Council establishing a Framework of Community Action in the field of Marine Environmental Policy (Marine Strategy Directive) Brussels, 24.10.2005 COM (2005) 505 final 2005/0211 (COD).

Frid, C. L. J., S. I. Rogers, M. Nicholson, J. R. Ellis & S. Freeman, 2000. Using biological characteristics to develop new indices of ecosystem health. 88th Annual ICES Science Conference; Mini-symposium on defining the role of ICES in supporting biodiversity conservation ICES CM Mini:02. ICES, Copenhagen.

Frid, C. L. J., O. A. L. Paramor & C. L. Scott, 2005. Ecosystem-based fisheries management: progress in the NE Atlantic. Marine Policy 29: 461–469.

Frid, C. L. J. & O. A. L. Paramor, 2006. Marine Biodiversity: The Rationale for Intervention. A Report to DEFRA. University of Liverpool, Liverpool.

Frid, C. J. L., O. A. L. Paramor & C. L. Scott, 2006. Ecosystem-based management of fisheries: is science limiting? ICES Journal of Marine Science 63: 1567–1572.

Gower, J. C. & D. J. Hand, 1996. Biplots, Vol. 54. Chapman & Hall, London.

Grall, J. & M. Glemarec, 1997. Assessment of benthic structure in the Bay of Brest. Annales De L Institut Oceanographique 73: 7–16.

Herrando-Pérez, S. & C. L. J. Frid, 2001. Recovery patterns of macrobenthos and sediment at a closed fly-ash dumpsite. Sarsia 86: 389–400.

Holme, N. A., 1953. The biomass of the bottom fauna in the English Channel off Plymouth. Journal of the Marine Biological Association of the UK 32: 1–49.

Lasiak, T., 1998. Multivariate comparisons of rocky infratidal macrofaunal assemblages from replicate exploited and non-exploited localities on the Transkei coast of South Africa. Marine Ecology-Progress Series 167: 15–23.

Pearson, T. H. & R. Rosenberg, 1986. Feast and famine: structuring factors in marine benthic communities. In: Gee, J. H. R. & P. S. Giller (eds), Organization of Communities: Past and Present. Blackwell Scientific: 373–398.

Snelgrove, P., T. Blackburn, P. Hutchings, D. Alongi, J. Grassle, H. Hummel, G. King, I. Koike, P. Larnbshead, N. Ramsing & V. Solis-Weiss, 1997. The importance of marine sediment biodiversity in ecosystem processes. Ambio 26: 578–583.

Snelgrove, P. V. R. & C. A. Butman, 1994. Animal sediment relationships revisited—cause versus effect. Oceanography and Marine Biology 32: 111–177.

Statzner, B., V. H. Resh & L. A. Roux, 1994. The synthesis of long-term ecological research in the context of concurrently developed ecological theory: design of a research strategy for the Upper Rhone River and it's floodplain. Freshwater Biology 31: 253–263.

Telesh, I. V., A. F. Alimov, S. M. Golubkov, V. N. Nikulina & V. E. Panov, 1999. Response of aquatic communities to anthropogenic stress: a comparative study of Neva Bay and the eastern Gulf of Finland. Hydrobiologia 393: 95–105.

United Nations, 1992. Convention on Biological Diversity. UN, New York.

US Magnuson-Stevens Fishery Conservation and Management Act, 1996. An Act to provide for the conservation and management of the fisheries, and for other purposes. As amended through October 11, 1996. Public Law 94–265.

Whitlatch, R. B., A. H. Hines, S. F. Thrush, J. E. Hewitt & V. Cummings, 1997. Benthic faunal responses to variations in patch density and patch size of a suspension-feeding bivalve. Journal of Experimental Marine Biology and Ecology 216: 171–189.

Hydrobiologia (2008) 606:81–92
DOI 10.1007/s10750-008-9344-x

CHALLENGES TO MARINE ECOSYSTEMS

Go with the flow: tidal import and export of larvae from semi-enclosed bays

M. J. Jessopp · R. J. McAllen

© Springer Science+Business Media B.V. 2008

Abstract There is a growing body of evidence that suggests the effective functioning of marine reserves is dependent on the dispersal and recruitment of larvae. Enhanced production inside reserves is predicted to lead to a net larval export and increased settlement and recruitment outside reserve boundaries. However, larval retention in bays is also well documented. Since bays are increasingly being used as reserve areas, planktonic larvae of benthic marine invertebrates were sampled from two semi-enclosed marine reserves during flood and ebb tides to determine whether these bays are acting as net exporters of larvae. Neither reserve was a net importer or exporter of species richness, larval abundance or diversity, although one reserve showed a small export of species richness during the hours of darkness. Both reserves balanced the net import of some species with a net export of others, which was generally related to adult or larval abundance, although exceptions were found in one reserve. Significant effects of light were found, with the net import or export of some species occurring exclusively during either the hours of daylight or darkness. An increased understanding of larval sink-source dynamics in bays is essential for ensuring their effective use as marine reserves to meet specific conservation needs.

Keywords Marine reserve · Larval dispersal · Larval retention · Bays

Introduction

Most benthic marine invertebrates are sedentary or sessile as adults, so their pelagic larval phase is the primary means of dispersal (Thorson, 1950). Marine invertebrate larvae are generally highly dispersive with some, but not all, behaving as relatively passive particles with respect to tidal advection (Banse, 1986; Shanks et al., 2003a), particularly where currents are greater than the swimming speed of larvae. Large-scale transport processes such as upwellings (Roughgarden et al., 1988) may aid in the wide dispersal of larvae, and while many coasts are not subject to upwelling conditions, coastal tidal currents can contribute to large-scale transport, linking populations over broad spatial scales. However, the presence of nearshore fronts (McCulloch & Shanks, 2003; Shanks et al., 2003b) may play a contrasting role in limiting dispersal, and the local recruitment of

Guest editors: J. Davenport, G. Burnell, T. Cross, M. Emmerson, R. McAllen, R. Ramsay & E. Rogan
Challenges to Marine Ecosystems

M. J. Jessopp (✉) · R. J. McAllen
Department of Zoology, Ecology and Plant Science, University College Cork, Enterprise Centre, Distillery Fields, North Mall, Cork, Ireland
e-mail: mjessopp@gmail.com

species with long-lived larvae (Palumbi, 1999, 2003) certainly supports this. The ability of zooplankton to influence horizontal transport by adjusting their vertical position in the water column (Kimmerer & McKinnon, 1987), can enhance their dispersal or retention and may lead to a selective advantage (Hill, 1991). While the active dispersal of larvae away from adult populations and to suitable nursery grounds can be accomplished by this process (DiBacco et al., 2001; Garland et al., 2002), the same process can be used to limit transport and prevent offshore advection (Shanks & Brink, 2005).

In terms of current efforts to implement an effective network of marine reserves, there is evidence that reserves can replenish exploited populations outside protected areas via larval export (Roa & Bahamonde, 1993; Murawski et al., 2000). This has important implications for future reserve designation as bays are increasingly being used in the management of marine biodiversity and populations of exploited species. Bays are a common component of many coastlines, and water retention (Sponaugle et al., 2002), and resulting increased aggregations of zooplankton and other passive particles (Black et al., 1990; Archambault & Bourget, 1999) in these areas suggest local retention of larvae. Assessing whether bays act as a net sink or source of larvae, and thus their potential effectiveness as reserve areas, requires accurate quantification of the tidal import and export of larvae.

In Ireland, two semi-enclosed bays (Strangford Lough in the north, and Lough Hyne in the south) provide an excellent opportunity to study the exchange of larvae between bays and their adjacent coastline. While these bays are not conventional in the sense that they are highly enclosed systems, similar coastal configurations are commonly found along eastern Canada and Norway, and other examples of marine reserves in semi-enclosed bays, lagoons and sea loughs can be found (e.g. Port Phillip Heads Marine national Park, Australia and Morro Natural Preserve, USA). Unlike estuaries, water movement in these sea loughs is almost entirely due to tidal currents, and the fully marine conditions remove any confounding effects of the need for some larvae to disperse from estuarine conditions to complete development in open coastal waters before returning to recruit to local populations (Christy, 1982). Tidal exchange between these bays and the adjacent ocean occurs through narrow constrictions making reliable quantifying of larval densities in flood and ebb tides possible. Small but significant export of phytoplankton (Johnson et al., 1995), increased local zooplankton abundance (Ballard & Myers, 2000), and a trend for the tidal import of zooplankton, particularly copepod species (Rawlinson et al., 2005) have been found at Lough Hyne, along with high local abundance of zooplankton at Strangford Lough (Boyd, 1973). However, the import and export of larvae from benthic marine invertebrates has not been studied in detail, and separating the confounding influences of light and dark, and ebb and flood tide on the exchange of larval abundance and assemblage structure has thus far proved difficult. It was therefore the purpose of this study to find out whether these bays were acting as net importers or exporters of marine invertebrate larvae. Second, if Lough Hyne and Strangford Lough are conserving diversity, and thus fulfilling one of the primary roles of reserves, samples from ebb tides should contain greater species richness and diversity than flood tides, and will determine the suitability of semi-enclosed bays for use as future reserve areas.

Materials and methods

Plankton samples were collected from surface waters in two Irish marine reserves, Strangford Lough, Co. Down, and Lough Hyne, Co. Cork (Fig. 1). Lough Hyne Marine Nature Reserve (51°30′ N 9°17′ W) is a small semi-enclosed fully marine sea lough, approximately 1 km long and 0.5 km wide. The lough consists of north and south basins, both approximately 18 m deep, connected by a deeper (50 m) trough in the western part of the reserve. The reserve is connected to the ocean through Barloge Creek, via a shallow (maximum depth 5 m at high water), narrow (25 m wide) channel known as the 'rapids'. The presence of a rock sill in the rapids modifies the usual semi-diurnal tidal cycle, resulting in an asymmetric tide with ebb flow lasting approximately twice as long as flood. Warmer summer water temperatures occur inside the reserve compared to the adjacent coastline (Rawlinson et al., 2004). Strangford Lough (54°27′ N 005°36′ W) is a much larger, semi-enclosed, fully marine sea lough, approximately 30 km long and 8 km wide. For the most part it is less than 10 m in depth, but a Y-shaped channel up to

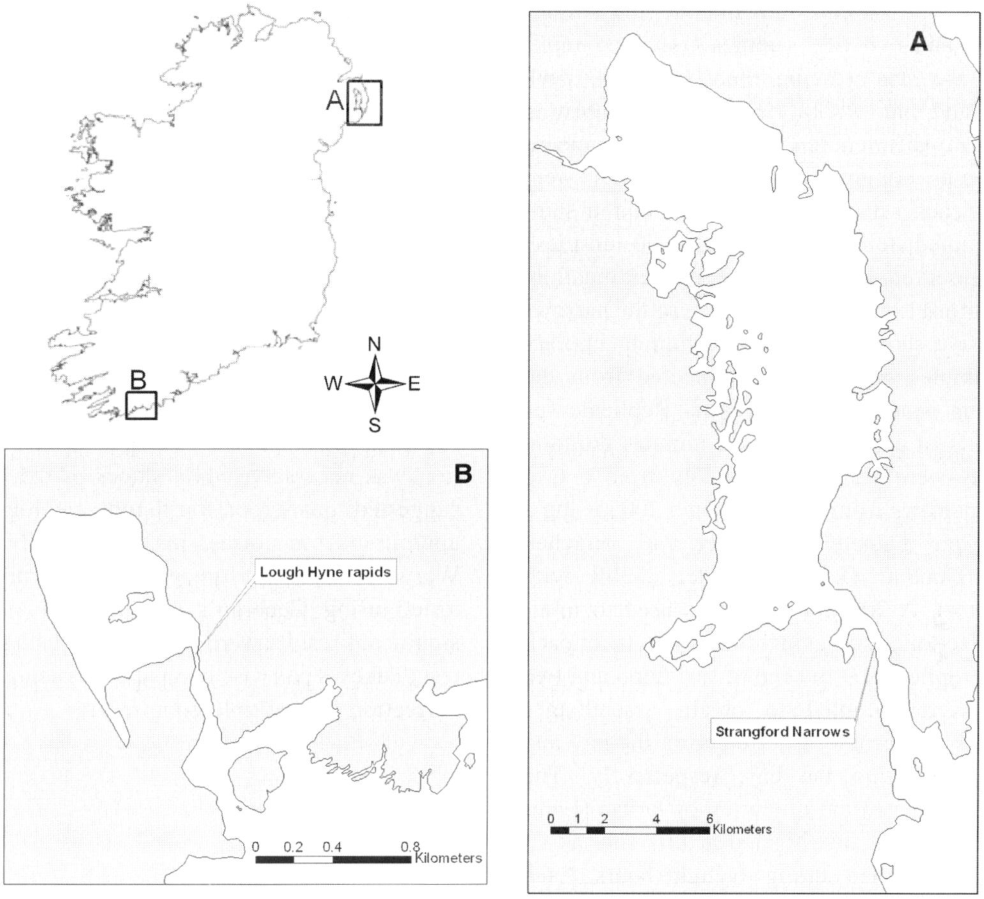

Fig. 1 Location of Strangford Lough (inset **A**), N. Ireland, and Lough Hyne (inset **B**), Republic of Ireland

66 m deep extends from the entrance up the central portion of the Lough. The reserve is connected to the open sea via a narrow, 8 km long channel (the 'narrows').

Due to differences in scale, and logistic constraints at each of the reserves, separate sampling protocols were used to independently measure larval abundances in flood and ebb tide waters. Direct comparisons between reserves can therefore not be made, and are outside the remit of this study. However, both protocols independently quantify the tidal import and export of larvae to assess whether these reserves are acting as a sink or source of larvae.

At Lough Hyne, reserve versus oceanic water could be differentiated by differences in water temperature, which was monitored over a number of tides using an optic tidbit temperature logger (Onset Computer Corporation, Massachusetts, USA), attached to a weighted buoy in the rapids. Plankton tows could not be conducted in the rapids due to the inability of the boat to sample in the low water level during ebb tides. Instead, a shore-based 'hose-on-suction-side' plankton pumping system (Miller & Judkins, 1981) was used to pump 200 litres of water through a mesh net using a portable generator, bilge pump and a 40 mm diameter hose suspended approximately 0.5 m below the surface of the rapids. As pump samples collect limited volumes of water and net clogging is negligible, a 50 μm mesh net was used to ensure collection of small larvae. Due to low volume of water sampled using the pump technique, months of high larval abundance were selected (following the spring plankton bloom) to ensure high larval abundance in samples, and sampling dates were chosen such that there were an equal number of flood and ebb tides during periods of daylight and darkness to ensure a balanced design. Samples were collected during the hours of daylight and darkness over 16 flood and 16 ebb tides in June and August 2004. Flood and ebb tide samples contained replicates from the start (during the first hour), peak (midpoint), and end (within the last hour) of flow to account for any differences in timing of larval release into the water column, and two replicates were collected at each sampling occasion. Plankton samples were immediately sieved through a 50 μm mesh sieve and preserved in 70% ethanol for identification.

At Strangford Lough, the larger size of the narrows made intensive shore-based pump sampling impractical, as samples could not be collected from the faster-flowing waters near the centre. Replicate sea surface trawls of approximately 10 minutes duration at 1–2 knots were conducted monthly from a 6 m boat in the narrows from March to June 2005 using a buoyed 40 cm diameter mesh net with attached flowmeter (General Oceanics model 2030R with standard rotor). A 200 μm mesh was used to minimise the effect of net clogging during trawls. In each month, two replicates at the end of two flood and two ebb tides were sampled to obtain assemblages representative of oceanic water entering the bay and reserve water exiting the bay, respectively. The effects of diurnal vertical migration of larvae (Lampert, 1989) were not investigated at this site, as all trawls were performed during daylight hours. Each sample was immediately taken to the Portaferry Marine Laboratory, sieved and sample volume adjusted to 50 ml, using 70% ethanol for preservation. Repeat subsamples in aliquots of 1 to 5 ml to yield approximately 300 individuals for identification (Venrick, 1971; Omori & Ikeda, 1992) were taken using a bulb pipette.

For both locations, planktonic larvae of benthic marine invertebrates were identified to species level (wherever possible) using the available literature, and abundances are given as the number of individuals per cubic metre of sampled water (ind. m^{-3}).

Larval species richness (S), abundance (N) and Shannon-Wiener diversity (H') were calculated and compared between flood and ebb tides across months and daylight/darkness using univariate ANOVA. It was envisaged that the relative abundance of species inside and outside the reserves would influence their import and export, so species occurring commonly in the plankton at each location were also investigated using ANOVA to test for differential larval import or export. For Lough Hyne, month was considered a fixed factor as periods of high larval abundance and an even number of flood and ebb tides in daylight and darknesss were selected. However, at Strangford Lough month was considered a random factor as no particular hypothesis about the effect of month on larval abundance or diversity was considered. Direction of flow (flood and ebb tides) and in the case of Lough Hyne, Light (daylight and darkness) were considered fixed, orthogonal factors. The most appropriate transformation of abundance data to remove large variances prior to ANOVA was determined by plotting log (standard deviation) against log (mean), and obtaining the value for the slope of the relationship. A slope of 0 suggested no transformation was necessary, while slopes of 0.5, 0.75 and 1 suggested square root, fourth root, and log as the most appropriate transformations, respectively (Clarke & Warwick, 2001). Homogeneity of variances was tested using Cochran's test prior to analysis, and significant results were investigated using a post-hoc test (Tukey's pairwise comparisons) using Bonferroni correction for multiple comparisons.

Results

Lough Hyne

Water temperature in the rapids (Fig. 2) shows a clear distinction between oceanic and the warmer reserve water. For approximately the first 50 minutes of flood tide, inflowing water was the same temperature as the last hour of ebb, suggesting that water from the end of ebb tide was retained in Barloge creek and transported back into the reserve at the start of the following flood tide. During ebb tide, water temperature steadily increased, suggesting mixed water with a progressively diminishing proportion of recently entered ocean water being swept back out of the reserve. This temperature profile was also apparent throughout the winter months (Jessopp, unpublished data).

Larval species richness, abundance and diversity were all significantly greater in darkness than daylight samples taken in August but not in June. Post-hoc pairwise comparisons showed that ebb tides had significantly greater species richness than flood tides during the hours of darkness, but no significant difference was found during the day. A similar

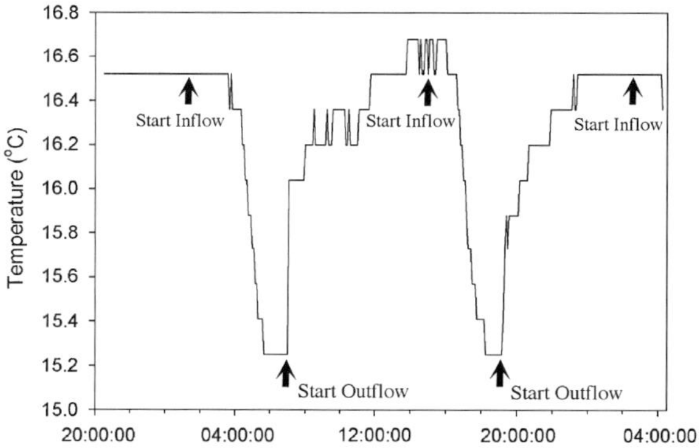

Fig. 2 Temperature (°C) taken from the rapids showing temperature difference between flood and ebb tide water. Arrows indicate timing of high and low slack water in the rapids, and consequent beginning of outflowing and inflowing tides

light*direction interaction for larval abundance highlighted no significant differences between flood and ebb tides within months, only between months. A month*light*direction interaction was noted for diversity, and post hoc pairwise comparisons again only highlighted significant differences in flood and ebb tides and daylight and darkness between month treatments (Table 1, Fig. 3).

ANOVA showed significant net tidal movement across the rapids in many of the commonly occurring species tested, and post-hoc tests showed under which months and light conditions this occurred (Table 2).

D-shaped bivalve veligers were more abundant in flood than ebb tides (net import) in daylight, as were veligers of the bivalve *Anomia ephippium*, Linnaeus. There was an export of veligers of the bivalve *Mytilus* spp. and an unidentified gastropod (spp 3, possibly young *Littorina* spp.) during darkness. A net export of spionid worm trochophores occurred during the hours of darkness in both months, and in addition during daylight hours in August. Bivalve veligers of *Hiatella* spp., Linnaeus, were more abundant in flood tides (net import) during daylight, and were also imported during both daylight and darkness in August, while gastropod veligers of *Turitella communis*, Risso, and *Balcis devians*, Monterosato, were more abundant in ebb tides (net export) during darkness in August only. Nauplii larvae of the barnacle *Chthamalus montagui*, Southward, were exported during August, while those of *Chthamalus stellatus*, Poli, were imported in the same month. Zoae larvae of the decapod *Jaxea nocturna*, Nardo, were exported during darkness in June, and low numbers of *Necora pube*, Linnaeus, zoea were imported in both months. The bryozoan *Membranipora membranacea*, Linnaeus, was imported into Lough Hyne in June but not August. Many species showed an effect of light, generally indicating higher larval abundance (due to vertical migration to surface waters) during the hours of darkness.

Table 1 Results of ANOVA for species richness (S), abundance (N) and Shannon-Wiener diversity (H') in the Lough Hyne rapids. Month, direction and light are all fixed, orthogonal factors

Source of variation	df	S			N (log transform)[a]			H′		
		MS	*F*	*P*	MS	*F*	*P*	MS	*F*	*P*
Month	1	145.04	18.63	***	7.011	20.42	***	0.52	13.63	***
Light	1	294	37.76	***	6.42	18.72	***	0.511	13.4	***
Direction	1	26.04	3.34	–	0.05	0.14	–	0.02	0.53	–
Month * light	1	40.04	5.14	*	1.76	5.13	*	0.182	4.78	*
Month * dir	1	8.16	1.05	–	0.19	0.56	–	0.149	3.91	–
Light * dir	1	135.38	17.39	***	3.55	10.37	**	0.108	2.82	–
Month * light * dir	1	0	0	–	0.17	0.5	–	0.234	6.13	*
Error	88	7.786			0.34			0.038		
Total	95									
		Aug: Dark > Day			Aug: Dark > Day			Aug: Dark > Day		
		Dark: Ebb > Flood								

[a] Indicates heterogeneous variances (Cochran's Test $P < 0.05$)

*$P < 0.05$, **$P < 0.01$, ***$P < 0.001$, – n/s

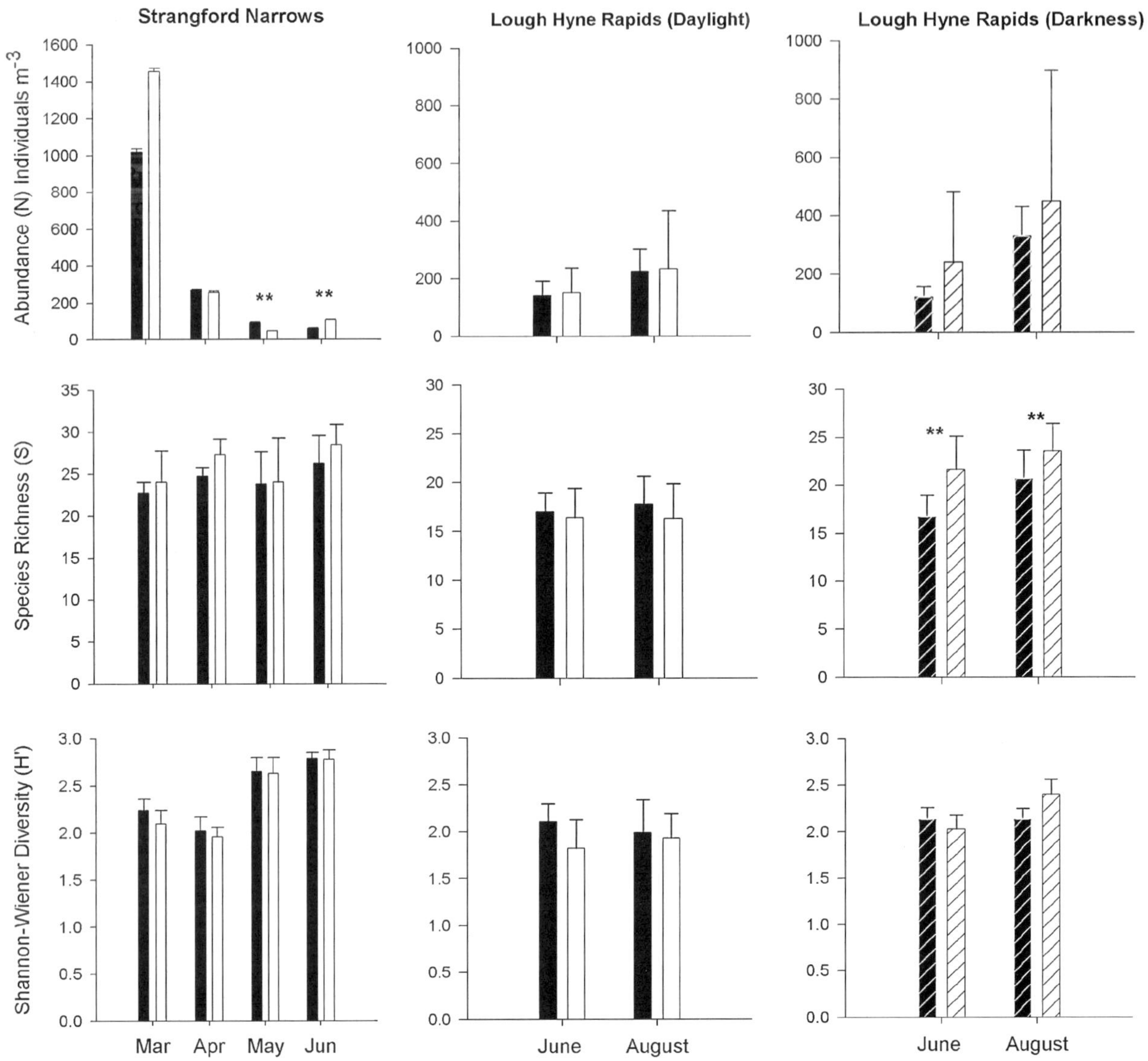

Fig. 3 Abundance, Species Richness and Shannon-Wiener Diversity (±s.e.) for samples collected at Lough Hyne (June, August 2004) and Strangford Lough (March–June 2005). ** denotes where significant differences were found between flood (dark bars) and ebb (white bars) tides. Hatched bars indicate samples collected during darkness (Lough Hyne only)

Strangford Lough

A significant effect of sampling month on abundance, diversity and species richness was observed, most likely due to different species releasing variable amounts of larvae into the plankton between months. No net import or export of abundance, species richness or diversity occurred, although there was a significant interaction between month and tidal direction for larval abundance. Since this interaction includes a random factor, post hoc tests are not possible. However, a test of the effect of direction within each level of month is possible (Underwood, 2002) and this showed no net unidirectional movement of larvae in March and April, a net import in May, and an export of larval abundance in June (Table 3, Fig. 3).

ANOVA on commonly occurring species (Table 4) showed significant differences in abundance between months, as well as interactions

Table 2 Summary of ANOVAs on common species found in plankton samples during daylight and darkness at Lough Hyne

Species	Transform	P Month	P Light	P Dir	P M * L	P M * D	P L * D	P M * L * D	Pairwise comparisons
Barnacles									
E. modestus	4th root	***	**	-	*	*	–	–	Jun: Dark < Day,
C. montagui[a]	4th root	***	–	*	–	***	–	–	Aug: Export
C. stelllatus[a]	ln(x + 1)	*	–	–	*	*	–	–	Aug: Import, June: Dark < Day
Decapods									
N. puber	square root	–	***	*	–	–	–	–	Import, Dark > Light
J. nocturna[a]	ln(x + 1)	**	**	**	**	**	**	**	Jun: Dark Export
Bivalves									
Bivalve D[a]	ln(x + 1)	–	***	*	*	–	***	–	Day: Import
A. eppiphium[a]	4th root	***	***	***	–	–	**	–	Day: Import
Hiatella spp.[a]	4th root	***	**	–	**	***	***	–	Day: Import, Aug: Import
Mytilus spp.[a]	ln(x + 1)	***	*	–	**	–	***	–	Dark: Export
Gastropods									
Gastropod spp1	square root	***	**	–	–	–	*	–	
A. pespelicani	4th root	***	–	***	–	**	–	–	Aug: Export
B. devians	square root	***	–	**	*	–	*	**	Aug: Dark Export
L. littorea[a]	ln(x + 1)	***	***	–	–	–	–	–	Dark > Day
Gastropod spp3[a]	square root	***	***	–	*	–	***	–	Dark: Export
C. imperforatum[a]	ln(x + 1)	***	*	*	*	*	–	–	Aug: Export, Aug: Dark > Day
T. communis	ln(x + 1)	–	***	*	–	–	*	*	Aug: Dark Export
Polychaetes									
Spionidae	ln(x + 1)	**	***	***	–	–	*	*	Dark: Export, Aug: Day Export
Bryozoans									
E. pilosa	4th root	***	–	–	–	–	–	*	–
M. membranacea	ln(x + 1)	***	–	–	–	**	–	–	Jun: Import

Month, Light and direction are all fixed, orthogonal factors. Results of pair-wise comparisons on significant interactions show where a net import or export of larvae occur

[a] Indicates heterogeneous variances (Cochran's Test $P < 0.05$), $*P < 0.05$, $**P < 0.01$, $***P < 0.001$, – n/s

between month and direction for many of these species. Most interactions were the result of species being present or highly abundant in only one of the four sampling months, with either a net import or export of larvae at this time. Two species of barnacle (Semibalanus balanoides, Linnaeus and Balanus crenatus, Brugiere,), one species of bryozoan (Electra pilosa, Linnaeus), a bivalve (Hitella spp., Linnaeus), two species of decapod zoea (Pisidia longicornis, Linnaeus, and Pilumnus hirtellus, Linnaeus) and barnacle cyprids were exported from the reserve in one of the four sampling months. One barnacle (Verruca stroemia, Müller), a decapod (Necora puber, Linnaeus), and a bryozoan (Membranipora membranacea, Linnaeus) were imported during one of the sampling months. Nauplii of the barnacles Verruca stroemia, Müller, and Elminius modestus, Darwin were imported in one month and exported in another, as were veligers of the bivalve Mytilus spp., Linnaeus.

the export of others, and it is suggested that this is related to the distribution of adult populations. At Lough Hyne, a net import of saddle oyster *Anomia ephippium* veligers was noted during daylight hours. *A. ephippium* is a common member of fouling communities inside the reserve, but this species has been found to be more representative of communities in adjacent Atlantic coastal sites (Watson, 2004), so its import probably reflects adult distribution. However, other sink-source dynamics appear to be unrelated to adult populations at Lough Hyne. Larvae of the invasive barnacle *Elminius modestus* are consistently more abundant inside Lough Hyne than in areas outside the reserve (Jessopp & McAllen, 2007), but no export of this species was detected. Conversely, increased settlement of the bryozoan *Electra pilosa* immediately outside compared to inside the reserve (D. Watson pers. comm.) suggests larger populations along the Atlantic coast, yet no import of these larvae was observed. Such discrepancies may partly be explained by the limited water exchange across the rapids at Lough Hyne preventing overall larval transport. It is apparent from the rapids temperature profile (Fig. 2), that flood tides contain a proportion of lough water from the previous ebb tide, and ebb tides contain oceanic water from the previous flood tide. Many larvae in these respective water bodies transported in or out of the reserve on one tide may simply be returned in the other direction at the beginning of the following tide, negating any net dispersal.

At Strangford Lough, the barnacles *Semibalanus balanoides* and *Elminius modestus* are certainly common and widespread inside the reserve (Erwin et al., 1986; Wilkinson et al., 1988), so the net export of their nauplii larvae is not unexpected. Zoea larvae of the velvet swimming crab (*Necora puber*) were imported, highlighting the possible effects of the commercial fishery for this species in Strangford lough. Potting for *N. puber* takes place in the Narrows and the southern half of the lough, reflecting the distribution records of this species (Roberts et al., 2004).

The contribution of imported or exported larvae to resulting marine communities either inside or outside reserve boundaries may depend on the abundance and developmental stage of larvae. While a large number of early stage larvae may initially seem important, a small number of late stage larvae may contribute more to the pool of settling larvae due to high levels of larval mortality in early stages. While decapod zoea were found in enough numbers to test for the effect of direction, megalopae were rare, and a test for net import or export of these later stages was not possible to compare to the observed import/export of the earlier stages. Barnacle nauplii tended to be exported from Strangford Lough, as did cyprid stages, further increasing the effect of larval export to areas outside the reserve. If conditions inside bays (such as the elevated water temperature in Lough Hyne) are more conducive to successful larval development, the potential for a net export of late stage larvae is considerable, and may override the effect of a net import of early stage larvae.

To date, investigating the interaction of tide and light on the exchange of larvae has proved difficult due to low replication (Fernandes et al., 2002; Rawlinson et al., 2005). While the lower sampling frequency at Strangford Lough made investigating such interactions impossible, the sampling regime adopted at Lough Hyne addressed this by studying sink-source dynamics at a fine temporal resolution, allowing separation of the confounding effects of dark and light, and ebb and flood tides on larval transport. This showed that the import and export of larvae was often dependent on light levels. Larvae of three gastropod species (*Turitella communis, Balcis devians* and an unidentified gastropod species 3), a bivalve (*Mytilus* spp.), and a decapod (*Jaxea nocturna*) were exported only during the hours of darkness. The import of bivalve veligers (*Anomia ephippium, Hiatella spp.* and early stage D-shaped veligers) occurred primarily during daylight hours. This suggests differential vertical migration strategies are being used to aid or limit dispersal. Active control of vertical position in the water column has been shown to influence dispersal in bivalve veligers (Shanks & Brink, 2005), and *Callinectes* post larvae utilise selective tidal stream transport to migrate into estuaries (Forward Jr. et al., 2003).

It is clear from the results that both bays are balancing the net import of some species with the net export of others. However, limited water exchange at Lough Hyne may be limiting larval dispersal, creating an isolated, self-seeding reserve area; illustrated by its long flushing time of 41 days (Johnson et al., 1995). This can be beneficial in retaining larvae to increase populations locally, but may lead to long-term genetic consequences such as inbreeding and reduced genetic diversity. The ability of such a reserve to act as a source of larvae to adjacent non-reserve areas, or recover from local catastrophes is also doubtful. Strangford Lough showed import and export of larvae reflecting known adult distributions, and it is much shorter flushing time of just over three days (Jessopp, unpublished data) suggests adequate connectivity to act as a source of larvae to adjacent areas.

There is a need for local recruitment in reserves to sustain populations, as well as dispersal of larvae to seed adjacent, non-protected areas. Finding the optimal balance between the two will be a function of reserve size and positioning. Small reserves, and those in bays with some degree of water and larval retention, may favour local recruitment of short dispersing larvae (Botsford et al., 2001). Future studies should utilise genetic techniques to determine the relatedness of short and long-duration larval species to local adult populations, illustrating the degree to which dispersal and recruitment processes are separated from local production in these contrasting strategies. If this is done across a gradient of exposure / flushing times, an optimal reserve configuration may be found for bays that can balance both the retention and export of larvae. Ultimately, understanding the larval sink-source dynamics of bays will contribute to their effective use as marine reserves in meeting specific conservation needs.

Acknowledgements This study would not have been possible without the assistance of staff at Portaferry Marine Laboratory and University College Cork; in particular, P. Johnston, G. Savidge, R. Thomas, and A. Dale. Samples were collected from Lough Hyne under research permit 08/04 granted by D. O'Donnell of the Department of Environment, Heritage and Local Government. This research was conducted in collaboration with M. Johnston and L. Allcock at Queens University Belfast and T. Crowe at University College Dublin. Funding was provided by the Irish Higher Education Authority, under strand 1 of the North-South Programme for Collaborative Research, which was supported by the National Development Plan 2000–2006.

References

Archambault, P. & E. Bourget, 1999. Influence of shoreline configuration on spatial variation of meroplanktonic larvae, recruitment and diversity of benthic subtidal communities. Journal of Experimental Marine Biology and Ecology 241: 309–333.

Ballard, L. & A. Myers, 2000. Observations on the seasonal occurrence and abundance of gelatinous zooplankton in Lough Hyne, Co. Cork. south-west Ireland. Biology and Environment-Proceedings of the Royal Irish Academy 100B: 75–83.

Banse, K., 1986. Vertical-distribution and horizontal transport of planktonic larvae of echinoderms and benthic polychaetes in an open coastal sea. Bulletin of Marine Science 39: 162–175.

Bassindale, R., F. J. Ebling, J. A. Kitching & R. D. Purchon, 1948. The ecology of the Lough Ine Rapids with special reference to water currents. Journal of Ecology 36: 305–322.

Black, K. P., S. L. Gay & J. C. Andrews, 1990. Residence times of neutrally-buoyant matter such as larvae, sewage or nutrients on coral reefs. Coral Reefs 9: 105–114.

Botsford, L. W., A. Hastings & S. D. Gaines, 2001. Dependence of sustainability on the configuration of marine reserves and larval dispersal distance. Ecology Letters 4: 144–150.

Boyd, R. J., 1973. A survey of the plankton of Strangford Lough, Co. Down. Proceedings of the Royal Irish Academy 73: 231–267.

Christy, J. H., 1982. Adaptive significance of semilunar cycles of larval release in fiddler crabs (genus Uca): test of a hypothesis. Biological Bulletin 163: 251–263.

Clarke, K. R. & R. M. Warwick, 2001. Change in Marine Communities: An Approach to Statistical Analysis and Interpretation. PRIMER-E, Plymouth.

DiBacco, C., D. Sutton & L. McConnico, 2001. Vertical migration behavior and horizontal distribution of brachyuran larvae in a low-inflow estuary: implications for bay-ocean exchange. Marine Ecology-Progress Series 217: 191–206.

Erwin, D. G., B. E. Picton, D. W. Connor, C. M. Howson, P. Gilleece & M. J. Bogues, 1986. The Northern Ireland Sublittoral Survey. Ulster Museum, Belfast.

Fernandes, L. D. A., S. L. C. Bonecker & J. L. Valentin, 2002. Dynamic of decapod crustacean larvae on the entrance of Guanabara Bay. Brazilian Archives of Biology and Technology 45: 491–498.

Fleminger, A. & R. I. Clutter, 1965. Avoidance of towed nets by zooplankton. Limnology and Oceanography 10: 96–104.

Forward R. B. Jr., R. A. Tankersley & J. M. Welch, 2003. Selective tidal-stream transport of the blue crab *Callinectes sapidus*: an overview. Bulletin of Marine Science 72: 347–365.

Fretter, V. & M. C. Pilkington, 1970. Veliger larvae of Taenioglossa and Stenoglossa. Fiches d'identification du zooplancton: 1–26.

Garland, E. D., C. A. Zimmer & S. J. Lentz, 2002. Larval distributions in inner-shelf waters: the roles of wind-driven cross-shelf currents and diel vertical migrations. Limnology and Oceanography 47: 803–817.

Heath, M. R., 1995. Size spectrum dynamics and the planktonic ecosystem of Loch-Linnhe. ICES Journal of Marine Science 52: 627–642.

Hill, A. E., 1991. Vertical migration in tidal currents. Marine Ecology-Progress Series 75: 39–54.

Jessopp, M. & R. McAllen, 2007. Water retention and limited larval dispersal: implications for short and long distance

dispersers in marine reserves. Marine Ecology Progress Series 333: 27–36.
Johnson, M. P., M. J. Costello & D. Odonnell, 1995. The nutrient economy of a marine inlet Lough-Hyne, South-West Ireland. Ophelia 41: 137–151.
Kimmerer, W. J. & A. D. McKinnon, 1987. Zooplankton in a marine bay II. Vertical migration to maintain horizontal distributions. Marine Ecology-Progress Series 41: 53–60.
Lampert, W., 1989, The adaptive significance of diel vertical migration of zooplankton. Functional Ecology 3: 21–27.
Mackas, D. L. & C. M. Boyd, 1979. Spectral analysis of zooplankton spatial heterogeneity. Science 204: 62–64.
McCulloch, A. & A. L. Shanks, 2003. Topographically generated fronts, very nearshore oceanography and the distribution and settlement of mussel larvae and barnacle cyprids. Journal of Plankton Research 25: 1427–1439.
Miller, C. B. & D. C. Judkins, 1981. Design of pumping systems for sampling zooplankton, with descriptions of two high capacity samplers for coastal studies. Biological Oceanography 1: 29–56.
Minchin, D., 1992. Multiple species, mass spawning events in an Irish sea lough – the effect of temperatures on spawning and recruitment of invertebrates. Invertebrate Reproduction & Development 22: 229–238.
Murawski, S. A., R. Brown, H. L. Lai, P. J. Rago & L. Hendrickson, 2000. Large-scale closed areas as a fishery-management tool in temperate marine systems: the Georges Bank experience. Bulletin of Marine Science 66: 775–798.
Omori, M. & W. M. Hamner, 1983. Patchy distribution of zooplankton: behaviour, population assessment and sampling problems. Marine Biology 72: 193–200.
Omori, M. & T. Ikeda, 1992. Methods in Marine Zooplankton Ecology. Krieger Publishing Company, Malabar, Florida.
Palumbi, S. R., 1999. Population biology – the prodigal fish. Nature 402: 734–735.
Palumbi, S. R., 2003. Population genetics, demographic connectivity, and the design of marine reserves. Ecological Applications 13: S146–S158.
Powlik, J. J., M. A. Stjohn & R. W. Blake, 1991. A retrospective of plankton pumping systems, with notes on the comparative efficiency of towed nets. Journal of Plankton Research 13: 901–912.
Rainbow, P. S., 1984. An introduction to the biology of British littoral barnacles. Field Studies 6: 1–51.
Rawlinson, K. A., J. Davenport & D. K. A. Barnes, 2004. Vertical migration strategies with respect to advection and stratification in a semi-enclosed lough: a comparison of mero- and holozooplankton. Marine Biology 144: 935–946.
Rawlinson, K. A., J. D. Davenport & D. K. A. Barnes, 2005. Tidal exchange of zooplankton between Lough Hyne and the adjacent coast. Estuarine Coastal and Shelf Science 62: 205–215.
Roa, R. & R. Bahamonde, 1993. Growth and expansion of an exploited population of the squat lobster (*Pleuroncodes monodon*) after 3 years without harvesting. Fisheries Research 18: 305–319.
Roberts, D., C. Davies, A. Mitchell, H. Moore, B. Picton, A. Portig, J. Preston, M. Service, D. Smyth, D. Strong & S. Vize, 2004. Strangford Lough Ecological Change Investigation (SLECI). Report to Environment and Heritage Service Queen's University, Belfast.
Roughgarden, J., S. Gaines & H. Possingham, 1988. Recruitment dynamics in complex life-cycles. Science 241: 1460–1466.
Shanks, A. & L. Brink, 2005. Upwelling, downwelling, and cross-shelf transport of bivalve larvae: test of a hypothesis. Marine Ecology-Progress Series 302: 1–12.
Shanks, A., J. L. Largier & J. Brubaker, 2003a. Observations on the distribution of meroplankton during an upwelling event. Journal of Plankton Research 25: 645–667.
Shanks, A., A. McCulloch & J. Miller, 2003b. Topographically generated fronts, very nearshore oceanography and the distribution of larval invertebrates and holoplankters. Journal of Plankton Research 25: 1251–1277.
Sponaugle, S., R. K. Cowen, A. Shanks, S. G. Morgan, J. M. Leis, J. S. Pineda, G. W. Boehlert, M. J. Kingsford, K. C. Lindeman, C. Grimes & J. L. Munro, 2002. Predicting self-recruitment in marine populations: biophysical correlates and mechanisms. Bulletin of Marine Science 70: 341–375.
Thorson, G., 1950. Reproductive and larval ecology of marine bottom invertebrates. Biological Review 25: 1–45.
Underwood, A., 2002. Experiments in Ecology. Their Logical Design and Interpretation Using Analysis of Variance. Cambridge University Press, Cambridge.
Venrick, E. L., 1971. Statistics of subsampling. Limnology and Oceanography 16: 811–818.
Watson, D. I., 2004. Recruitment variability in space and time by benthic invertebrates on Irish shores. PhD Thesis, University College Cork.
Wiebe, P. H., 1970. Small-scale spatial distributions in oceanic zooplankton. Limnology and Oceanography 15: 205–217.
Wiebe, P. H. & W. R. Holland, 1968. Plankton patchiness: effects on repeated net tows. Limnology and Oceanography 13: 315–321.
Wilkinson, M., I. A. Fuller, T. C. Telfer, C. G. Moore & P. F. Kingston, 1988. Northern Ireland Littoral Survey. A Conservation-Orientated Survey of the Intertidal Seashore of Northern Ireland. Institute of Offshore Engineering, Heriot-Watt University.

Hydrobiologia (2008) 606:93–103
DOI 10.1007/s10750-008-9341-0

CHALLENGES TO MARINE ECOSYSTEMS

Seabed mapping in the southern Irish Sea: predicting benthic biological communities based on sediment characteristics

Fionnuala McBreen · James G. Wilson · Andrew S. Y. Mackie · Caitríona Nic Aonghusa

© Springer Science+Business Media B.V. 2008

Abstract Marine habitat mapping is necessary to comply with European legislation (92/43/EEC, 79/409/EEC and 2001/60/EEC), with international obligations to organizations such as ICES and with agreements such as the OSPAR Convention. Clearly defined habitats are needed before conservation and management practices can be implemented. The BIOMÔR, SWISS and HABMAP projects have used sediment particle size, organic matter, organic carbon and nitrogen along with benthic macrofaunal species and abundance to define habitats in the southern Irish Sea. The sea floor of the southern Irish Sea is predominantly sandy gravel, grading down through sand to mud in the deeper parts of the Celtic Sea and similarly in the shallower northern basin beyond Anglesey. The physical gradient in particle size is correlated with organic content and also with biological communities. Nevertheless, there are marked discrepancies between boundaries defined by the Folk sediment characterization trigon, by the chemical characteristics of the sediments and by biological communities. The validity of using continuous physical and chemical sediment characteristics to more accurately predict categorical biological assemblages was tested with use of stepwise backward elimination Binary Logistic Regression (BLR). This method could be used as a tool to predict biological assemblages where there is a paucity of biological data. It lends support to the idea that benthic habitat mapping will have to take more account of biological structuring and system function.

Keywords Irish Sea · Benthos · Community · Habitat mapping · Sediment · Particle size

Guest editors: J. Davenport, G. Burnell, T. Cross, M. Emmerson, R. McAllen, R. Ramsay & E. Rogan
Challenges to Marine Ecosystems

F. McBreen (✉) · J. G. Wilson
Zoology Department, Trinity College Dublin, College Green, Dublin 2, Ireland
e-mail: mcbreenf@tcd.ie

A. S. Y. Mackie
Department of Biodiversity & Systematic Biology, Amgueddfa Cymru—National Museum Wales, Cathays Park, Cardiff CF10 3NP, Wales, UK

C. Nic Aonghusa
Marine Institute, Rinville, Oranmore, Co. Galway, Ireland

Introduction

Today marine benthic communities are being much more rigorously defined, mostly in response to conventions and EU legislation. The 2 main biological classification systems generally applied to the Irish Sea are the Marine Nature Conservation Review (MNCR) (Connor et al., 2004) and the European Nature Information System (EUNIS) classification systems (EUNIS: http://eunis.eea.europa.eu/). The MNCR system was originally designed with Britain & Ireland in mind, while EUNIS was designed to apply to the whole of north-east Europe and aimed to produce a network of protected sites across Europe known as Natura 2000. Both systems are converging under the 'Mapping European Seabed Habitats' programme (MESH: www.searchmesh.net). The EUNIS classification system classifies habitats on a hierarchical scale on a 7-point hierarchical scale, in which level 2 merely distinguishes between depth, sediment and rock down to level 6 which incorporates elements of depth (e.g. sublittoral), sediment type (e.g. muddy sand), exposure, current speed and biological communities, which themselves are subdivided depending on the contributing species.

Macrofaunal community assemblages have long been linked to habitat type. The idea of using a characteristic species dominant in either numbers or biomass to define sublittoral community assemblages was first put into practice by Peterson (1913). Peterson (1913) was also the first to separate macrofaunal communities into epifaunal and infaunal communities, distinguishing on the basis of the bottom type animals living in the sediment and animals living on substrates such as rocks and cobbles. Jones (1950) reviewed the literature and concluded that communities were dependent on a range of environmental conditions and that physical factors were more influential in determining community type than biological ones. His community classification system was based mainly on temperature, salinity, bottom type and depth. Jones also recognized the fact that communities did not have distinct boundaries, but graded into one another. Thorson (1957) expanded on these ideas by applying the community theory globally. Thorson (1957) further remarked that communities in different regions often had community assemblages containing the same dominant genera (but different species). Thorson's global parallel communities were characterized not only by taxonomic affinities but also by functional (particularly feeding) characteristics, thus emphasizing the biological aspects of the ecological niche (Thorson, 1957).

As many macrobenthic species are sessile as adults, the communities are strongly influenced by the habitat selection preferences of the larvae (Gray, 1974). The settlement of the larvae is governed by a combination of (often interrelated) physical, chemical and biological factors, such as the structure and contours of the surface, sediment particle size, the presence of organic and inorganic compounds, biofilms and populations of the same species (Gray, 1974). In addition, the presence of the organisms themselves can lead to changes in the composition of sediments (Gray, 1974; Rhoads, 1974). Snelgrove & Butman (1994) have reviewed the relationship between sediments and infauna and have concluded that particle size alone was not sufficient to explain community structure. However, they did emphasize that particle size may be correlated to other factors, such as the passive transport of larvae with sediment particles or the organic content of the sediments as food for deposivores (Snelgrove & Butman, 1994).

To conserve both species and spaces through enforcement and legislation, habitats need to be identified and mapped in detail. Benthic habitat mapping has tended to use physical sediment characteristics and, in more recent times, geophysical data to produce broad-scale habitat maps (Kenny et al., 2003; Roff et al., 2003) but with little basis on biological data due to a paucity of information (Roff & Taylor, 2000). Therefore by combining geophysical and sediment data and employing some biological ground-truthing, assumptions can be made about the location, biological boundaries and type of communities present. Physical sediment and biological boundaries are often extremely arbitrary (e.g. the BGS-modified Folk Trigon in which percentages are used to classify sediments (Folk, 1954)) and do not take into account biological structuring or ecosystem functioning. This is problematic when physical sediment maps are used as the basis for biological habitat maps. As a result more emphasis is being put on a biotope approach which incorporates both habitats and communities (Olenin & Ducrotoy, 2006).

Much work has been conducted on habitats and communities in the Irish Sea. The Irish Sea Study Group (1990) conducted an environmental review of the Irish Sea focusing on four main areas: nature conservation; waste inputs and pollution; exploitation of living marine resources; and planning, development and management. Within this, Mackie (1990) produced a benthic faunal map derived from earlier studies in the Irish Sea and delimited 9 communities. In 2002, the Irish Sea Pilot project was set up by the Joint Nature Conservation Committee (JNCC) (Vincent et al., 2004). This project recommended approaching marine conservation relative to marine landscapes. Geophysical and hydrographical data were used to define 18 different types of coastal and marine landscapes. Despite their different approaches, in practice both assessments of the Irish Sea benthic environment rely heavily on the distributions of the surficial sediments.

Mackie et al. (1995) found that gravel and silt content were the two main sediment variables that, together with depth, 'best explained' the benthic macrofaunal distributions in the southeast Irish Sea. A subsequent Canonical Correspondence Analysis (CCA) similarly found that the sediment composition (gravel, silt and sand) and depth were the main environmental variables influencing the polychaete worms (Mackie et al., 1997). A cluster analysis study of benthic variability (Mackie, 2004; corrected in Mackie et al., 2006, p. 209) showed that both macrofaunal replicates at 49 quantitative stations were paired. Despite the lack of high precision in station positioning, the replicates of only 2 stations were separated in the dendrogram of 102 samples. Therefore, in terms of broad-scale mapping, there appear to be large areas of the Irish Sea with broadly similar species and the gradients of change within sediment types may be low.

There have been a number of approaches using techniques such as logistic regression to examine the relationship between biological and environmental variables (Ysebaert et al., 2002; Thrush et al., 2003; Ellis et al., 2006). Ysebaert et al. (2002) used the abiotic variables of depth, salinity, current velocities, particle grain size and mud content to predict the presence or absence of macrobenthic species in the Schelde estuary while Thrush et al. (2003) looked at the species-specific model for 13 macrobenthic species to examine their relationship to sediment mud content. Ellis et al. (2006) examined the relationship between individual species and variable environmental conditions such as depth, silt/clay content, tidal currents and wind/wave disturbance in five estuaries in New Zealand. Ysebaert et al. (2002) and Ellis et al. (2006) found they could predict some species better than others, with correct predictions for the presence of species ranging from 59 to 97% in the 2006 study (Ellis et al., 2006).

This paper will further explore the sediment and biological data from the three EU funded projects in the southern Irish Sea, BIOMÔR (1989–91) (Mackie et al., 1995), the South West Irish Sea Survey (SWISS) (Wilson et al., 2001) and HABMAP (2005–2007: sediment data only). The present study will focus on the sediment characteristics and their potential for predicting benthic community distributions using binary logistic regression.

Material and methods

The study area is taken to be the area south from 54°50′ N to 51°06′ N between Ireland and Wales (Fig. 1). This area stretches from the southern Irish Sea and across the boundary of the Irish Sea in to the Celtic Sea. This area is particularly important as it crosses the Celtic Front which represents the boundary between southern Lusitanian species and the northern Boreal species. The sea floor of the southern Irish Sea is predominantly sandy gravel, grading down through sand to mud in the deeper parts of the Celtic Sea and also in the shallower northern basin beyond Anglesey (Dobson et al., 1971; Wilson et al., 2001). This article examines sediment and biological data from three surveys, 51 stations from BIOMÔR (Mackie et al., 1995), 55 from SWISS (Wilson et al., 2001) and 77 stations from HABMAP. During each survey sediment and replicate biological samples were taken at each station with a 0.1 m^2 Van Veen grab.

Sediment samples for each of the three surveys were taken from the surface sediment in a 0.1 m^2 Van Veen grab. The samples were frozen initially and subsequently dried in the oven at 105°C before chemical analysis for organic content, total organic carbon (TOC), calcium carbonate and particle size.

Organic content of the sediment was determined through loss-on-ignition at 550°C in a muffle furnace (SWISS & HABMAP) or from sediment pre-digested in concentrated hydrochloric acid (to remove calcium carbonate) at 600°C for two hours (BIOMÔR) (Buchanan, 1984; Mackie et al., 1995; Wilson et al., 2001). As carbonate minerals tend to breakdown at temperatures over 650°C, calcium carbonate should not have been combusted during the loss-on-ignition procedure at 600°C (Boyle, 2004); thus results from both procedures have been examined. Sediment samples used to estimate total organic carbon content were digested in sulphurous acid (SWISS & HABMAP) or hydrochloric acid (BIOMÔR) to remove any inorganic carbon (Bale & Kenny, 2005). The samples were then analysed using an elemental analyser (Verardo et al., 1990; Mackie et al., 1995; Wilson et al., 2001; Bale & Kenny, 2005). The calcium carbonate content of the sediment was determined through digestion with HCl (Buchanan, 1984; Mackie et al., 1995).

Particle size analysis was used to classify the sediment using the BGS-modified Folk classification system (Folk, 1954). Sediment samples were pre-treated with hydrogen peroxide to remove organic matter before analysis (Buchanan, 1984; Mackie et al., 1995). Sediment was sieved through 8-mm, 4-mm, 2-mm, 1-mm, and 500-μm sieves. Fractions less than 500-μm in size were analysed in the Malvern Mastersizer laser particle size machine (HABMAP) (Bale & Kenny, 2005). BIOMÔR and SWISS sediment samples were also dry sieved through 250-μm, 125-μm and 63-μm sieves. The silt/clay fraction for the BIOMÔR and SWISS sediments were analysed using pipette analysis (Buchanan, 1984; Mackie et al., 1995; Wilson et al., 2001).

Biological samples were also taken with a 0.1 m^2 Van Veen grab with two grabs at each quantitative station. The macrofauna were sieved through a 0.5-mm sieve. All macrofauna were identified and counted at each station. As some polychaete records were still incomplete they were excluded from the overall biological community analysis. The molluscs, arthropods and other animals included are estimated to comprise about 55% of the species and 45% of the total abundance (see Mackie et al., 1995). Biological data from the BIOMÔR and SWISS projects only were considered as no faunal data were available for HABMAP.

The results were analysed using two different approaches, one method examining categorized data and the second method using continuous data to predict biological communities. The biological data were analysed using cluster analysis and the SIMPER routine in PRIMER v.6 (Clark & Gorley, 2001). Species abundance data were $\log(x + 1)$ transformed before constructing a Bray–Curtis Similarity matrix. A dendrogram, which grouped individual sites based on their similarities and group average clustering, was created from the similarity matrix. Biological assemblages were then identified from groupings of sites at three different levels of similarity (15%, 23% and 29%). Assemblages at greater similarities than

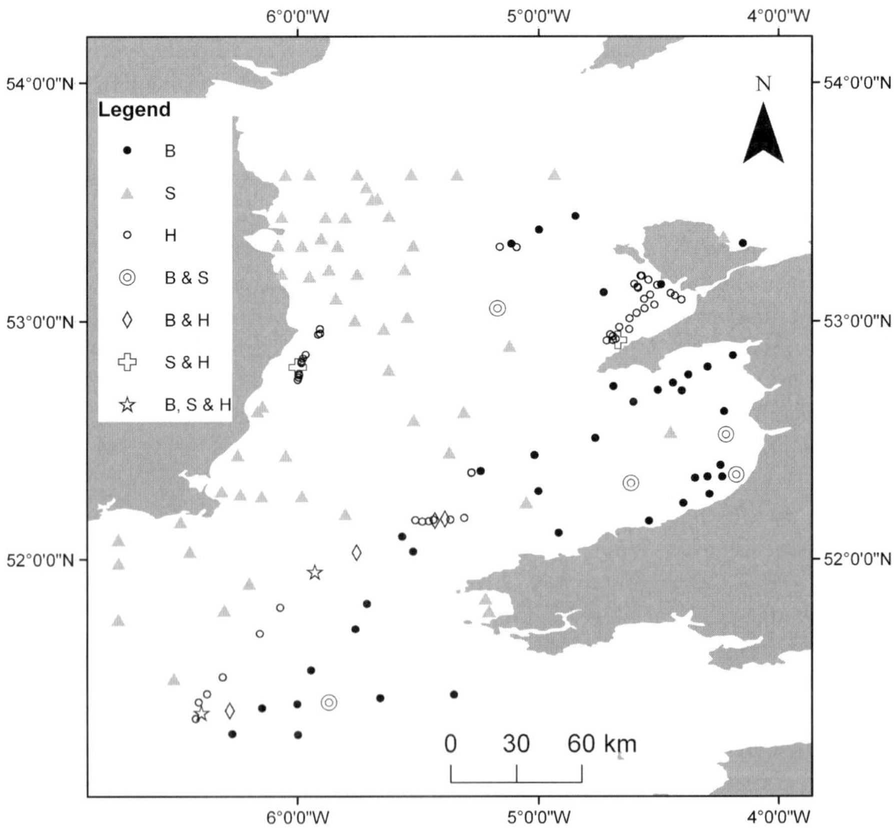

Fig. 1 Location of BIOMÔR (B), SWISS (S) and HABMAP (H) sites. Sites which were surveyed in more than one survey were combined into one symbol

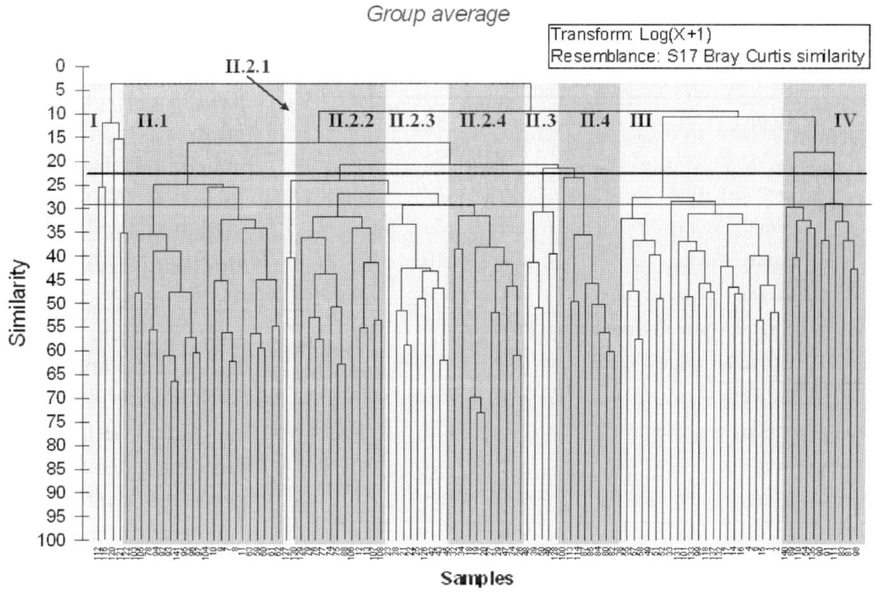

Fig. 2 Dendrogram of log $(x + 1)$ transformed species abundance data from the BIOMÔR and SWISS projects (excluding polychaetes). The dendrogram was constructed from a Bray-Curtis similarity matrix in PRIMER v6

29% were not examined as the number of sites per assemblage was deemed to be too low. The levels of similarity used were determined by the cluster analysis (Fig. 2). A two-crossed SIMPER test in PRIMER v.6 was used to distinguish the species which had the greatest effect on the assemblages identified in the dendrogram.

Binary logistic regression (SPSS 12.0.1) was used to predict the presence or absence of a categorical biological variable from a number of continuous environmental variables (depth, gravel, sand, mud, organic content, organic carbon and calcium carbonate concentrations). The biological assemblages designated from the cluster analysis were used as the categorical variables. A stepwise backward elimination approach was adopted in which the analysis started with all the continuous variables and eliminated the least significant variable in each progressive step. The optimum number of variables was taken to be the lowest number of variables which showed the highest percentages for correctly predicting the presence or the absence of a biological assemblage.

Results

The physical and chemical sediment characteristics of 183 quantitative stations from the three projects were analysed. Firstly, the sites were divided by categorizing data (using the BGS-modified Folk sediment classification) to give 11 different types of Folk habitat from a possible 15 (Table 1).

These habitats were further subdivided by organic matter concentration. Organic matter was divided into three categories low (<1%), medium (1–7.5%) and high (> 7.5%). In the absence of a marine model, these categories were derived from the estuarine Pollution Load Index (PLI) (Jeffrey & Wilson, 1985), where the baseline level for organic matter as a pollutant is 1% and the threshold is 7.5%. The addition of a second categorized environmental variable increased the number of habitats from 11 to 19 (Table 2).

Through this inclusion of chemical sediment characteristics, the physical sediment habitats could be further subdivided into categories more highly linked to biological assemblages (Table 3) (complete species abundance data, BIOMÔR project) (Mackie et al., 1995) In Table 3 differences can be seen between physical sediment habitat types and biological assemblages where levels of organic matter differ, e.g. the slightly gravelly muddy sand category divides into low and medium organic matter associated with biological assemblages B1 and B3, respectively. It should also be noted that the same biological assemblages occur over different physical sediment types, e.g. B1 occurs in

Table 1 Quantitative samples classified using the BGS-modified Folk sediment classification

	Sediment type	Number of sites
1	Sand	57
2	Sandy gravel	35
3	Gravelly sand	32
4	Muddy Sand	24
5	Sandy Mud	10
6	Slightly gravelly sand	9
7	Slightly gravelly muddy sand	6
8	Gravelly muddy sand	5
9	Gravel	2
10	Muddy sandy gravel	2
11	Mud	1
	Total	183

slightly gravelly muddy sand, gravelly muddy sand and muddy sand. This shows that categorical physical sediment classifications alone cannot predict biological assemblages. Also different biological assemblages such as A2, B2, B3, B4 and C1 were found in the same physical sediment category (Table 3).

In view of the limitations of categorized sediment data, a second approach was taken to test whether biological assemblages in the southern Irish Sea could be predicted from the continuous environmental variables. Using species abundance data (BIOMÔR and SWISS projects, polychaetes not included) a dendrogram was constructed from a Bray-Curtis similarity matrix of the $\log(x + 1)$ transformed data using PRIMER v6 (Fig. 2). From the dendrogram different assemblages were identified at different levels of similarity; level 1 at 15% similarity (I, II, III and IV), level 2 at 23% similarity (e.g. II.1, II.2, I.3, II.4, IV.1 and IV.2) and level 3 at 29% similarity (e.g. II.2.1, II.2.2, II.2.3 and II.2.4).

Stepwise backward elimination binary logistic regression (BLR) was used to test to what extent biological assemblages (Fig. 2) could be correctly predicted from the environmental variables; depth (m), gravel (%), sand (%), mud (%), organic content (%), organic carbon (%) and calcium carbonate (Table 4). From Table 4, it can be seen that Level 1 assemblages were not predicted with the same consistency from the environmental variables. The absence of assemblage I was correctly predicted 100% of the time, but the presence of assemblage I was only predicted 40% of the time, which was the lowest percentage for the 4 assemblages. Overall

Table 2 Quantitative samples classified by sediment type and organic matter level, where low = <1%, medium = 1–7.5% and high = >7.5%

	Sediment type	Organic matter	Number of sites
1	Slightly gravelly muddy sand	Low	1
2	Slightly gravelly muddy sand	Medium	5
3	Slightly gravelly sand	Low	3
4	Slightly gravelly sand	Medium	6
5	Gravel	Medium	2
6	Gravelly muddy sand	Medium	5
7	Gravelly sand	Low	12
8	Gravelly sand	Medium	20
9	Mud	High	1
10	Muddy sand	Low	1
11	Muddy sand	Medium	1
12	Muddy sand	High	22
13	Muddy sandy gravel	Medium	2
14	Sand	Low	29
15	Sand	Medium	28
16	Sandy gravel	Low	4
17	Sandy gravel	Medium	31
18	Sandy mud	Medium	5
19	Sandy mud	High	5
	Total		183

Table 3 BIOMÔR sites, divided by the BGS-modified Folk classification system and subdivided by organic matter, where low = <1%, medium = 1–7.5% and high = >7.5%. (Jeffrey & Wilson, 1985) Biological assemblages are taken from Mackie et al. (1995, p. 80) which identified assemblages using total species abundance data for all sites

	Sediment type	Organic matter level	No. of sites	Biological assemblage
1	Slightly gravelly muddy sand	Low	1	B1
2		Medium	1	B3
3	Slightly gravelly sand	Low	1	B4
4	Gravelly muddy sand	Medium	3	C1
5			1	B1
6	Gravelly sand	Low	3	C1
7		Medium	1	C1
8			1	B3
9	Muddy sand	Medium	2	A1
10			2	A2
11			5	B1
12	Sand	Low	2	B2
13			2	B4
14			1	C1
15		Medium	2	A2
16			1	B3
17			4	B4
18	Sandy gravel	Low	2	C1
19		Medium	5	C1
20			3	C2
21	Sandy mud	Medium	1	B1
22			3	A1
23		High	1	A1

predictions for assemblages II, III, and IV were better than those for assemblage I.

Predictions for assemblages at Levels 2 and 3 were much more solid, with the absence and presence of assemblages II.1, IV.1, II.2.1 and IV.2.3 being predicted correctly 100% of the time. However, some caution is necessary due to the low number of sites in certain assemblages: while II.1 contained 19 sites, II.2.1 contained only 2 sites. Although the results were generally better at Levels 2 and 3, there were nevertheless some causes for concern, especially in the cases of assemblages II.4 and II.2.2 where the environmental variables were still poor predictors for the presence of these assemblages.

Table 5 shows the top five contributing species for each of the group II assemblages. The results show that communities cannot be defined by one characterizing species as Petersen (1913) suggested but are a combination of species more similar to the associations of Jones (1950). The main contributing species change as similarity increases at lower levels (e.g. from II to II.2 and II.2.1). Depending on the level of discrimination, a different number of communities can be said to exist in the southern Irish Sea (4 at level 1, 10 at Level 2 and 20 at Level 3). Different levels of discrimination may need to be applied to different biological assemblages depending on the number of sites, e.g. assemblage I at Level 1 and II at Level 3. This approach is similar to the EUNIS classification system where different levels from 1 to 7 are deemed appropriate for marine habitats based on the available information.

Discussion

While predictions of the presence and absence or the assemblages were not always 100% valid, the percentage of correct predictions increased at higher levels of similarity for the biological assemblages. It was clear that biological assemblages were not only linked to particle size but also to the chemical

potential to be used to distinguish both distinct (those atypical of the surrounding area) and representative (those typical of the surrounding area) habitats and biotopes for protection as Marine Protected Areas in the southern Irish Sea. It is recommended that this approach be tested in other marine areas to see whether the general principles can apply to different zoogeographic zones. It is also recommended that an attempt be made to integrate chemical variables into existing classification systems such as EUNIS.

Acknowledgements The authors would like to thank the crew of the research boats the *RV Prince Madog* and the *RV Celtic Voyager* and Mark Kavanagh for assistance in sampling and analysis, the Marine Institute and also their colleagues in BIOMÔR (National Museum of Wales), SWISS (INTERREG-II) and HABMAP (INTERREG-IIIA) for much helpful advice and discussion.

References

Bale, A. J. & A. J. Kenny, 2005. Sediment analysis and seabed characterisation. In Eleftheriou, A. & A. McIntyre (eds), Methods for the Study of Marine Benthos. Blackwell Science Ltd, Oxford: 43–81.

Boyle, J., 2004. A comparison of two methods for estimating the organic matter content of sediments. Journal of Paleolimnology 31: 125–127.

Buchanan, 1984. Sediment analysis. In Holme, N. A. & A. D. McIntyre (eds), Methods for the Study of Marine Benthos. Blackwell Scientific Publications, London: 41–65.

Clark, K. R. & R. M. Gorley, 2001. Change in Marine Communities: An Approach to Statistical Analysis and Interpretation. PRIMER-E LTD, Plymouth.

Connor, D. W., J. H. Allen, N. Golding, K. L. Howell, L. M. Lieberknecht, K. O. Northern & J. B. Reker, 2004. The Marine Habitat Classification for Britain and Ireland version 04.05. JNCC, Petersborough.

Dobson, M. R., W. E. Evans & K. H. James, 1971. The sediment on the floor of the southern Irish Sea. Marine Geology 11: 27–69.

Ellis, J., T. Ysebaert, T. Hume, A. Norkko, T. Bult, P. Herman, S. Thrush & J. Oldman, 2006. Predicting macrofaunal species distributions in estuarine gradients using logistic regression and classification systems. Marine Ecology Progress Series 316: 69–83.

European Environment Agency, 2004. EUNIS (European Nature Information System) Habitat Classification. Available: http://eunis.eea.eu.int/habitats.jsp (February 2007).

Folk, R. L., 1954. The distinction between grain size and mineral composition in sedimentary-rock nomenclature. Journal of Geology 62: 344–359.

Gray, J. S., 1974. Animal-sediment relationships. Oceanography and Marine Biology: An Annual Review 12: 223–261.

Jeffrey, D. W. & J. G. Wilson, 1985. A Manual for the Estimation of Estuarine Quality. National Board for Science and Technology, Dublin.

Joint Nature Conservation Committee. 2007. MESH (Mapping European Seabed Habitats). Available: http://www.searchmesh.net/ (February 2007).

Jones, N. S., 1950. Marine bottom communities. Biological Reviews 25: 283–313.

Kenny, A. J., I. Cato, M. Desprez, G. Fader, R. T. E. Schuttenhelm & J. Side, 2003. An overview of seabed-mapping technologies in the context of marine habitat classification. ICES Journal of Marine Science 60: 411–418.

Mackie, A. S. Y., 1990. Offshore benthic communities of the Irish Sea. In Irish Sea Study Group (ed.), The Irish Sea: An Environmental Review. Part 1. Nature conservation. Liverpool University Press, Liverpool: 169–218.

Mackie, A. S. Y., 2004. Macrofaunal assemblages and their sedimentary habitats: working toward a better understanding. In Davies, H. (ed.), Irish Sea Sediments, University of Liverpool, 21st October 2004. Irish Sea Forum Seminar 32: 30–38.

Mackie, A. S. Y., P. G. Oliver & E. I. S. Rees, 1995. Benthic biodiversity in the southern Irish Sea. Studies in Marine Biodiversity and Systematics from the National Museum of Wales. BIOMÔR Reports 1. 1–263.

Mackie, A. S. Y., C. Parmiter, & L. K. Y. Tong, 1997. Distribution and diversity of Polychaeta in the southern Irish Sea. In Reish, D. J. & P. Y. Qian (eds), Proceedings of the Fifth International Polychaete Conference, Qingdao, China 1995. Bulletin of Marine Science 60: 467–481.

Mackie, A. S. Y., J. W. C. James, E. I. S. Rees, T. Darbyshire, S. L. Philpott, K. Mortimer, G. O. Jenkins & A. Morando, 2006. The Outer Bristol Channel Marine Habitat Study. BIOMÔR Reports 4: 1–249 pp. & Appendix A1–A228.

Olenin, S. & J.-P. Ducrotoy, 2006. The concept of biotope in marine ecology and coastal management. Marine Pollution Bulletin 53: 20–29.

Peterson, C. G. J., 1913. Valuation of the sea. II. The animal communities of the sea bottom and their importance for marine zoogeography. Report of the Danish Biological Station 16: 229–311.

Rhoads, D. C., 1974. Organism-sediment relations on the muddy sea floor. Oceanography and Marine Biology: An Annual Review 12: 263–300.

Roff, J. C. & M. E. Taylor, 2000. National frameworks for marine conservation - a hierarchical geophysical approach. Aquatic Conservation: Marine and Freshwater Ecosystems 10: 209–223.

Roff, J. C. & S. M. J. Evans, 2002. Frameworks for marine conservation- non-hierarchical approaches and distinctive habitats. Aquatic Conservation: Marine and Freshwater Ecosystems 12: 635–648.

Roff, J. C., M. E. Taylor & J. Laughren, 2003. Geophysical approaches to the classification, delineation and monitoring of marine habitats and their communities. Aquatic Conservation: Marine and Freshwater Ecosystems 13: 77–90.

Snelgrove, P. V. R. & C. A. Butman, 1994. Animal-sediment relationships revisited: cause versus effect. Oceanography and Marine Biology: An Annual Review 32: 111–177.

Thorson, G., 1957. Bottom communities (sublittoral or shallow shelf). Memorandum of the Geographical Society of America 67: 461–534.

Thrush, S. F., J. E. Hewitt, A. Norkko, P. E. Nicholls, G. A. Funnell & J. I. Ellis, 2003. Habitat change in estuaries:

predicting broad-scale responses of intertidal macrofauna to sediment mud content. Marine Ecology Progress Series 263: 101–112.

Verardo, D. J., P. N. Froelich & A. McIntyre, 1990. Determination of organic carbon and nitrogen in marine sediments using the Carlo Erba NA-1500 Analyser. Deep-Sea Research 37: 157–165.

Vincent, M. A., S. M. Atkins, C. M. Lumb, N. Golding, L. M. Lieberknecht & M. Webster, 2004. Marine nature conservation and sustainable development—the Irish sea pilot. Report to Defra by the Joint Nature Conservation Committee, Peterborough.

Wilson, J. G., A. S. Y. Mackie, B. D. S. O'Connor, E. I. S. Rees & T. Darbyshire, 2001. Benthic Biodiversity in the Southern Irish Sea 2: the South-West Irish Sea Survey. Studies in Marine Biodiversity and Systematics from the National Museum of Wales. BIOMÔR Reports 2: 1–143.

Ysebaert, T., P. Meire, P. M. J. Herman & H. Verbeek, 2002. Macrobenthic species response surfaces along estuarine gradients: prediction by logistic regression. Marine Ecology Progress Series 225: 79–95.

Zacharias, M. A. & J. C. Roff, 2000. A Hierarchical Ecological Approach to Conserving Marine Biodiversity. Conservation Biology 14: 1327–1334.

Hydrobiologia (2008) 606:105–116
DOI 10.1007/s10750-008-9342-z

CHALLENGES TO MARINE ECOSYSTEMS

Alien species in the Mediterranean Sea—which, when, where, why?

B. S. Galil

© Springer Science+Business Media B.V. 2008

Abstract A critical evaluation of more than 2,200 publications, some dating back to the late 1800s, established the presence, and traced the spatio-temporal spread, of 558 alien metazoan species in the Mediterranean Sea. The majority of aliens in the eastern Mediterranean entered through the Suez Canal, whereas mariculture and shipping are powerful means of introduction in the northwestern Mediterranean and in the Adriatic Sea. Most aliens are thermophilic species. The possible causes for the epic scale of invasion in the Mediterranean Sea are discussed.

Keywords Bioinvasion · Alien species · Mediterranean · Dataset · Stressors · Global warming

Introduction

The first alien species recorded in the Mediterranean were a pair of fouling serpulid polychaetes, *Hydroides dianthus* Mörch, 1863 and *H. diramphus* (Verrill, 1873), collected in the harbors of Izmir and Naples in 1865 and 1870, respectively (Carus, 1889; Zibrowius, 1973). The next pair were Red Sea mollusks, *Pinctada radiata* (Leach, 1814) and *Cerithium scabridum* Philippi, 1848, collected in Alexandria and Port Said in 1874 and 1883, respectively (Monterosato, 1878; Keller, 1883), heralding the Erythrean invasion of the Mediterranean Sea. The influx of alien species has continued ever since. However, no Mediterranean-wide targeted effort to survey the presence, abundance, and impacts of alien species was ever undertaken. The one targeted regional survey—a joint program by the Smithsonian Institution, the Hebrew University of Jerusalem, and the Sea Fisheries Research Station, Haifa—investigated the spread of the Erythrean biota along the coasts of Israel, Cyprus, Rhodes in the late 1960s and early 1970s (Steinitz, 1970; Por, 1978). Even when faced with the preponderance of Erythrean biota in the southeastern Mediterranean, marine bioinvasions were perceived until the last decades of the 20th C as singular and largely benign, and the risks of aliens were dismissed "Other than… *Asterina gibbosa* there is no known case in which a Lessepsian migrant species has completely replaced a local one" (Por, 1978, p. 149) or entirely denied "The Lessepsian migrants may be considered, in a figurative sense, "welcome guests" in the impoverished, subtropical cul-de-sac" (Por, 1978, p. 123).

Growing awareness worldwide that bioinvasions constitute one of the most significant components of global change, with often harmful effects on

Guest editors: J. Davenport, G. Burnell, T. Cross, M. Emmerson, R. McAllen, R. Ramsay & E. Rogan
Challenges to Marine Ecosystems

Electronic supplementary material The online version of this article (doi:10.1007/s10750-008-9342-z) contains supplementary material, which is available to authorized users.

B. S. Galil (✉)
National Institute of Oceanography, Israel Oceanographic & Limnological Research, P.O. Box 8030, Haifa 31080, Israel
e-mail: galil@post.tau.ac.il

biodiversity, economy, and human health, coupled with the opportune misfortune of a handful of invasive aliens noted for their conspicuous impacts on the native Mediterranean biota, have drawn the attention of scientists, management, and media. It was widely perceived that the littoral and infralittoral biota of the sea has been undergoing a rapid and profound change. A series of Atlases (fishes, decapod and stomatopod crustaceans, mollusks) summarized the extant knowledge of the scale and impact of 'Exotic species in the Mediterranean' (www.ciesm.org/atlas/). Recently, as part of an EC-sponsored comprehensive inventory of alien species in Europe, a voluminous body of literature, including research papers, surveys, and conference abstracts, was assembled and critically examined to construct an authoritative dataset tracing the origin, date, and mode of introduction, current distribution, rate of spread, and actual and potential impacts of the 558 alien species recorded in the Mediterranean Sea.

Materials and methods

An alien species is defined as an organism occurring outside its known or consensual range, as documented in scientific publications (Occhipinti-Ambrogi & Galil, 2004). The surge in shipping and the opening of the Suez Canal predate modern studies of Mediterranean marine taxa, save mollusks and fishes, by at least half a century. Since the beginning of the 20th century extensive biological surveys have been conducted in the Mediterranean, allowing for a reasonable measure of confidence in separating the alien from the native biota in the better-known taxa.

Since the likelihood of encountering a stray incursion in the sea is diminishingly small, most recorded alien species are considered as 'established' species that have self-maintaining populations of some duration. It is recognized that some alien species may fail to maintain populations over time and thus a single record dating back several decades may be considered an ephemeral entry. The distinction between the 'established' and 'ephemeral' aliens can vary spatially and temporally, and is sometimes difficult to discern and circumscribed in large part by our ignorance of the marine environment.

This study is geographically restricted to Mediterranean Sea *s. str.*, between the Straits of Gibraltar and the Dardanelles. The list of species was culled from myriad sources. The primary sources include research papers, biota inventories and surveys, ecological studies, and fisheries management reports. Additional data were occasionally gleaned from M.Sc. and Ph.D. theses, and unpublished monitoring studies conducted over the past 20 years by the author. These records were supplemented and amended by personal communications with scientists of particular taxonomic expertise. Earlier records were reassessed and some were deleted, such as the seven Erythrean sponge species and three species of hydroids listed by Por (1978), that, as tactfully suggested by Zibrowius (1992), were in need of "critical reevaluation," or the seven species of Erythrean ascidiaceans, that are either widely distributed, or of "problematic identification" (Zibrowius, 1992, p. 96). Records of "Lusitanian" and "Mauritanian" species collected in the Mediterranean are not considered as aliens but as natural entries, vagrants, or simply rare species, especially in cases of deep-water species such as *Trachyscorpia cristulata echinata* (Koehler, 1896) and *Beryx splendens* Lowe, 1934.

The date of the establishment of the first population in a new locality is significant for the study of the patterns and processes of invasion, so for each species, the first record in each country was entered, with its place of collection. As research efforts vary greatly along the coasts of the Mediterranean, and even the better-studied locales suffer temporal and taxonomical lacunae, we accept that the date of collection (or, when missing, the date of publication) may be years behind the actual entry dates, and that identification and publication may lag behind collection: the pyramidellid gastropods *Chrysallida maiae* (Hornung & Mermod, 1924) and *Syrnola fasciata* (Jickeli, 1882) were collected off the Israeli coast in 1935 and 1949 and identified only in 1992 (van der Linden & Eikenboom 1992) and 1995 (Mienis, 1995), respectively; the tube worm *Hydroides brachyacanthus* Rijoa, 1941, was collected in Jaffa, Israel, in 1933, but identified only nearly 60 years later (Ben-Eliahu, 1991).

The dataset includes the native range of the alien, and its means of introduction, whether through the Suez Canal, vessels (hull transport of boring, fouling, crevicolous or adherent species, and ballast), mariculture (intentional or unintentional), and other commercial introductions (ornamental, bait, edible species).

Results

Which?

The 558 metazoan species identified in the present work as alien in the Mediterranean Sea are listed in Supplementary Table 1. A taxonomic classification of the list (Fig. 1) shows that the alien phyla most frequently recorded are Mollusca (34%), Arthropoda (18%), Chordata (14%), and Rhodophyta (12%). Phyla not represented in the list include the little-studied Porifera, Nemertea, Priapula, Nematoda, Entoprocta, Pogonophora, Sipuncula, Echiura, Brachiopoda, and Phoronida. The data are presumably most accurate for large and conspicuous species, which are easily distinguished from the native biota, occurring along a frequently sampled or fished coast and for which taxonomic expertise is readily available.

A geographic classification of the list (Fig. 2) shows that the native range of the alien taxa recorded in the Mediterranean is most commonly the Indo-Pacific Ocean (39%), the Indian Ocean (18%), the Red Sea (10%), and pantropical (10%). Caution should be exercised when using these data: the true origin of Mediterranean populations of a species widely distributed in the Indo-Pacific Ocean may be its population in the Red Sea, or much further afield; an Erythrean alien off Turkey may have been introduced directly from the Red Sea, or secondarily from established southern Levantine populations. With few notable exceptions, the source populations of alien species in the Mediterranean have not be ascertained by molecular means (Jousson et al., 1998; Meusnier et al., 2004; Terranova et al., 2006; Andreakis et al., 2004, 2007). However, even taking into account these caveats, it is quite clear that most of the alien species in the Mediterranean are thermophilic, originating in tropical seas. This flies against the widely held assumption that successful bioinvasions originate in similar latitudes (Carlton, 1985).

A classification according to the means of introduction shows that the majority of aliens in the Mediterranean entered through the Suez Canal, followed by vessels and mariculture. Here too, caution should be exercised when using these data: only rarely the means and route of introduction are known from direct evidence. Mostly they are deduced from the biology and ecology (if known) of the species, the habitats and locales it occupies in both the native and introduced range, and its pattern of dispersal (if known), i.e. for a fouling species frequently recorded from ports, shipping is assumed to be the most probable vector. The means of introduction differ greatly among the phyla: whereas of the 102 alien macrophytes, 37%, 22% and 11% were introduced with mariculture, vessels, or both, respectively, the majority of alien crustaceans and mollusks are Erythrean aliens (64% and 65%, respectively), and mariculture introductions are few (3% and 5%, respectively) (Fig. 3).

When?

The date of collection (or publication, see above) may lag years behind the date of introduction and is

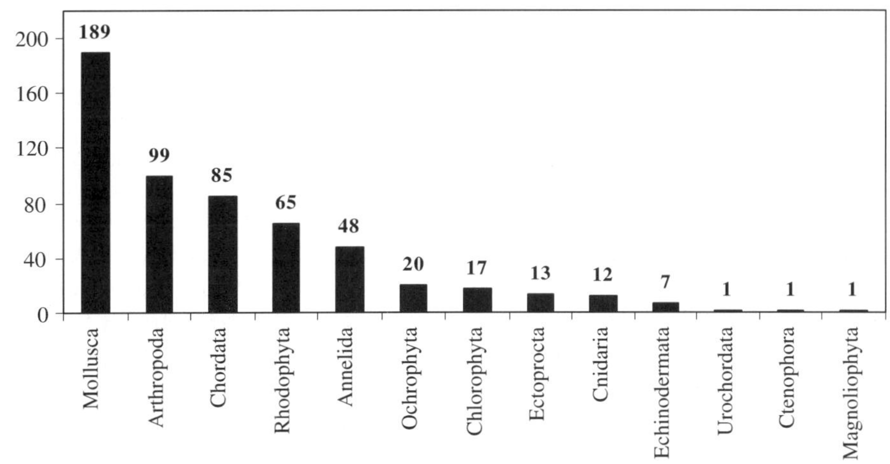

Fig. 1 Number and percentage of alien species in the Mediterranean Sea, presented by phylum

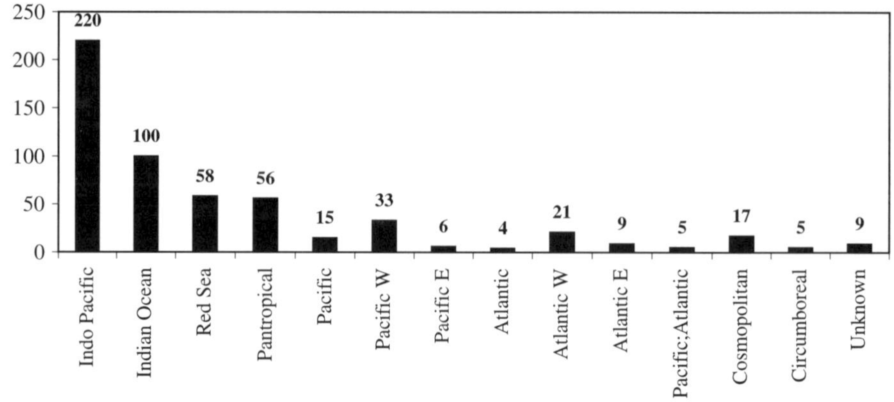

Fig. 2 Number and percentage of alien species in the Mediterranean Sea, presented by their native range. W = western, E = eastern

largely dependant on local scientific interest; i.e. *Marsupenaeus japonicus* (Bate, 1888) was collected off the Mediterranean coast of Egypt already in 1924, and off Syria and Turkey a few years later, but only in 1946 and 1975 off Israel and Lebanon, respectively. Clearly, the last two records reflect the state of local research rather than the sequence of the prawn's spread. Nevertheless, the number of alien species which have been recorded in the Mediterranean each decade over the past century is given in Figs. 4 and 5. With the exception of a gap in the 1910s, that probably indicates the devastation of the First World War, the number of introductions has been increasing steadily. A surge in the records in the 1920s and 1970s reflects the publication of the results of 'The Cambridge Expedition to the Suez Canal,' and the joint program by the Smithsonian Institution, the Hebrew University of Jerusalem, and the Sea Fisheries Research Station, Haifa, respectively. There seem to be more introductions recorded in the first years of the 21st C than in the 1980s and 1990s.

Since the 1950s major political, economic, and societal changes have occurred in the peri-Mediterranean countries. Their impact on the rate and means of introduction is apparent (Fig. 6): though the Suez Canal remained the main access route for alien species, the rate of introductions by vessels and mariculture has been higher. Of the 124 alien species known in 1950, 82% entered the Mediterranean through the Suez Canal, 10% and 4% were vessel-transported and mariculture introductions, respectively; of the 434 alien species recorded since 1950, 58% were Erythrean aliens, and 24% and 12% were vessel-transported and mariculture introductions (not counting secondary translocations, i.e. canals/vessels), respectively. The increase in vessel-transported aliens may be attributed to the increase in shipping volume throughout the region, due to the development of the Middle Eastern oil fields and later, the ascendance of the southeast Asian economies, the changing trade patterns that result in new shipping routes, improved water quality in port environments, augmented opportunities for overlapping with other introduction vectors, and rising research effort. Similarly, the increase in intentional and unintentional commercial introductions (primarily mariculture, but also ornamental, bait, and edible species) follows the rise of shellfish production (UNEP/MAP/MED POL, 2004).

Once established in the Mediterranean the temporal dynamics of the alien species are markedly varied. In some cases the interval between the initial establishment and spread had been exceedingly short: the spiny oyster, *Spondylus spinosus* Schreibers, 1793, was first collected in Haifa Bay, Israel, in 1988, and in the Gulf of Iskenderun, Turkey, in 1991; the conch, *Conomurex persicus* (Swainson, 1821), spread throughout the Levant between 1978 and 1985; the sweeper, *Pempheris vanicolensis* Cuvier, 1821, was first collected off Israel and Lebanon in 1978–79, and in 1983 off Mersin, Turkey; *Caulerpa taxifolia* (Vahl) C. Agardh, spread from a tiny patch in 1984 off Monaco, to Majorca, Elba, Sicily, and Tunisia, by 2000; *C. racemosa* var. *cylindracea* (Sonder) Verlaque, Huisman, and Boudouresque, 2003, spread in 1990s throughout much of the Mediterranean; the cornetfish, *Fistularia commersonii* Rüppell, 1835, spread, within two years, from Israel to Lampedusa Is. and Tunisia; and the grapsid crab *Percnon gibbesi* (H. Milne Edwards, 1853), first

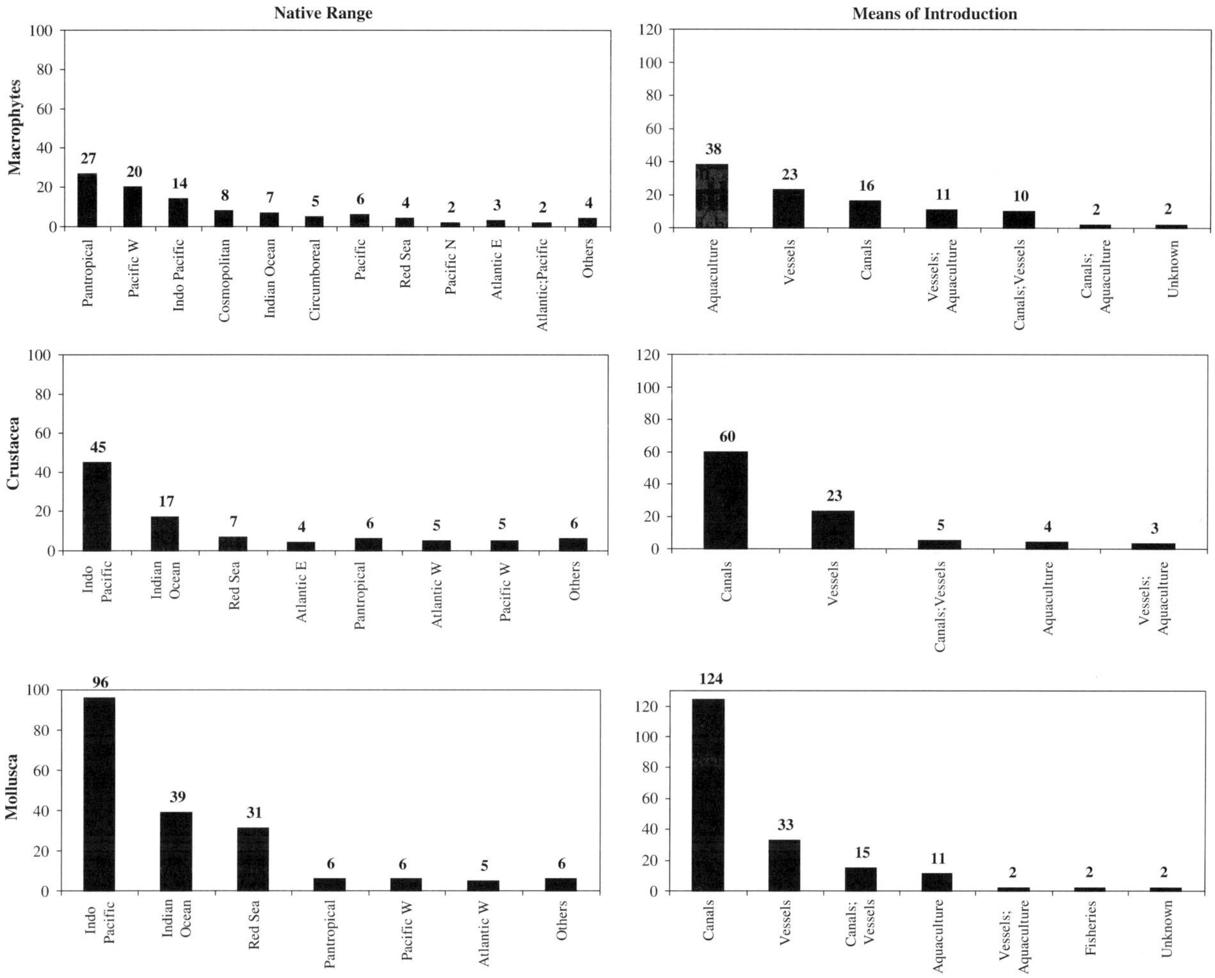

Fig. 3 Number of alien macrophytes, crustaceans, and mollusks recorded in the Mediterranean Sea, presented by their native range and means of introduction

Fig. 4 Number of alien species recorded in the Mediterranean Sea, 1900-2007

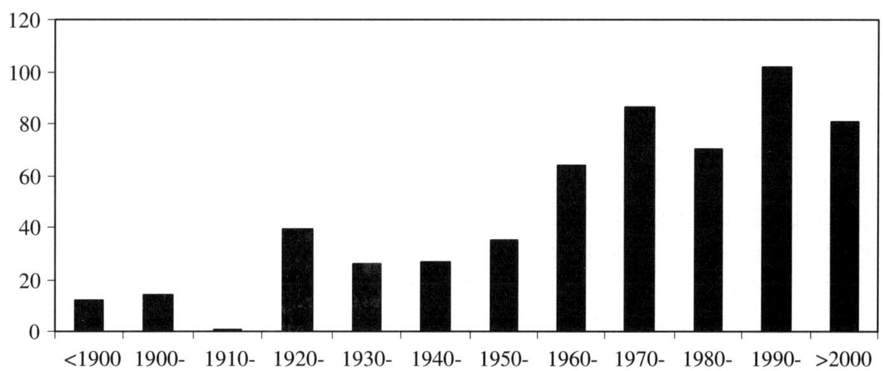

collected off the Balearic Is. in 1999, was sighted off the southern coast of Turkey in 2005. However, a time lag, sometimes extending over half a century, is a more common pattern: the jack, *Alepes djedaba* (Forsskål, 1775), and the swimming crab, *Charybdis hellerii* (A. Milne Edwards, 1867), were both collected in Haifa Bay in 1924, but nearly three decades and six decades later in Iskenderun,

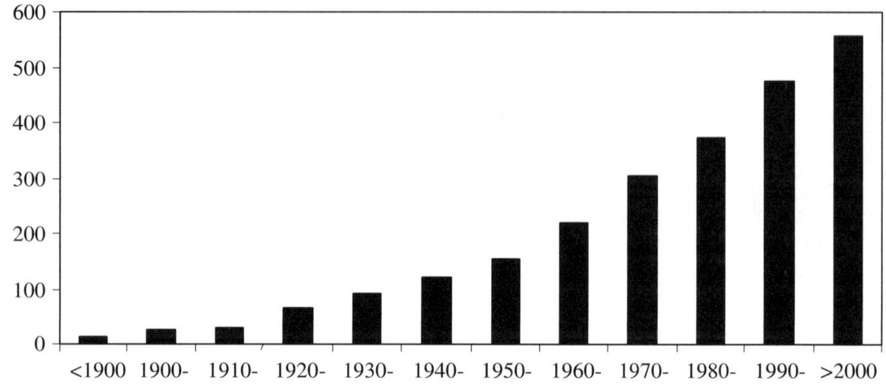

Fig. 5 Cumulative number of alien species recorded in the Mediterranean Sea, 1900–2007

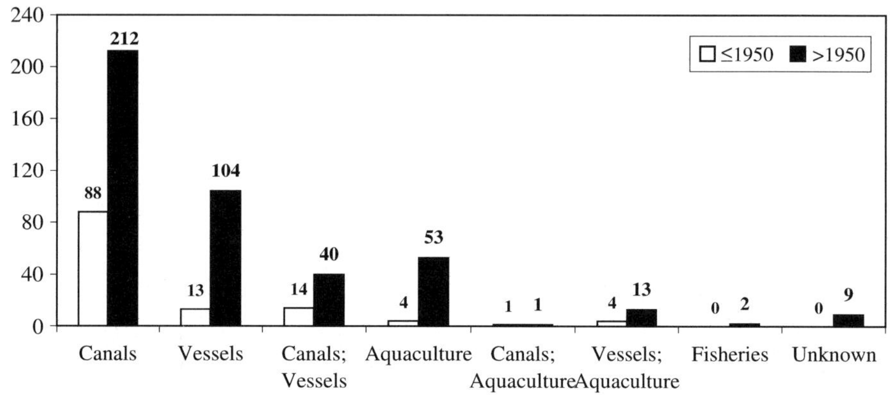

Fig. 6 Number of alien species in the Mediterranean Sea, presented by means of introduction, before and after 1950

respectively. Those are all conspicuous species, easily distinguished from the native Mediterranean taxa, and collected along intensively fished coasts, and so unlikely to be overlooked, yet the actual time lag is probably longer than the records imply, as chances of detecting the earliest members of the colonizing population are slim. The gastropods *C. scabridum* and *Thais lacera* (Born, 1778) were collected off Jaffa in 1899 and 1928, respectively, but 90 and 64 years later along the southeastern coast of Turkey; the hammer oyster, *Malvufundus regulus* (Forsskål, 1775), and the mussel, *B. pharaonis*, both sighted off Israel in 1937, were collected off Turkey in 1973 and 1975, respectively. The lantern clam, *Laternula anatine* (Linnaeus, 1758), described from Port Said a century ago, is a large (75 mm), readily distinguished clam, and it is highly unlikely one would fail to notice it, yet it was collected in Haifa Bay only in 1958 and from the Gulf of Iskenderun in 1992. The jewel box oyster, *Chama pacifica* Broderip, 1834, another massive (70 mm) and distinctive species, was collected in Alexandria in 1905, but only in 1988 in Haifa Bay, and then within a decade to Lebanon, Turkey, and Cyprus. These hiatuses cannot be considered artifacts stemming from poor collection and identification, as the mollusks off the Levantine coast had been assiduously studied throughout that period (www.ciesm.org/atlas).

Where?

The bulk of the alien species recorded in the Mediterranean are littoral and sublittoral benthic or demersal species. Since the shallow coastal zone, and especially the benthos, has been extensively studied, and is more accessible, the chances that new arrivals will be encountered and identified are higher. Also, the species most likely to be introduced by the predominant means of introduction (Suez Canal, vessels, mariculture) are shallow water species.

A comparison of alien species between the Mediterranean coasts of Spain and France, and an equivalent stretch along the Levantine coastline from Egypt to Turkey, shows marked differences in their numbers, origin, and means of introduction (Fig. 7A, B). There are almost four times as many alien species along the Levantine coast (409) as in the westernmost Mediterranean (110). The majority of aliens in the easternmost Mediterranean entered through the Suez Canal (81%, 13% vessel-transported, 3% mariculture), whereas mariculture (42%), vessels (34%), or both (9%) are the main means of introduction in the western Mediterranean (see also Verlaque, 2005). Consequently, the native ranges of the western Mediterranean aliens are spread over 'the seven seas,' while the Levantine aliens mostly originate in the tropical Indo-Pacific or parts thereof (Indo-Pacific 46%, Indian Ocean 23%, Red Sea 13%). Indeed, the Erythrean biota are mostly confined to the eastern Mediterranean, and it had been believed it will be contained east of Sicily (Por, 1990). However, of late, some Erythrean mollusks were recorded from the western Mediterranean—*B. pharaonis* and *P. radiata* in Corsica (Boudouresque, 1999), and *F. fragilis* in Valencia, Spain (Gofas & Zenetos, 2003). On the other hand, efforts to promote shellfish farming in the Levant mean the importation of commercially important alien species that had been previously established in the western and central Mediterranean: lately, *C. gigas* had been deliberately introduced as a possible mariculture crop to Syria and Turkey (Çevik et al., 2001; Saker, 2002).

Perusal of the dates of collection and spatial spread discloses a distinct pattern: many of the farthest spread species are early introductions. The six species of Erythrean decapod crustaceans recorded in Tunis or Sicily (*Metapenaeus monoceros* (Fabricius, 1798), *Trachysalambria palaestinensis* (Steinitz, 1932), *Alpheus inopinatus* Holthuis & Gottlieb, 1958, *Portunus pelagicus* (Linnaeus, 1758), *Eucrate crenata* de Haan, 1835, *Pilumnopeus vauquelini* (Audouin, 1826) have first-established populations in the Mediterranean an average of 85 years ago, and the nine fish species recorded in Tunis, Malta, or Sicily (*Parexocoetus mento* (Valenciennes, 1847), *Hemiramphus far* (Forsskål, 1775), *Fistularia commersonii* Rüppell, 1835, *Leiognathus klunzingeri* (Steindachner, 1898), *Pempheris vanicolensis* Cuvier, 1831, *Sphyraena chrysotaenia* Klunzinger, 1884, *Siganus rivulatus* Forsskål, 1775, *S. luridus* (Rüppell, 1829), *Stephanolepis diaspros* Fraser-Brunner, 1940) 62 years ago, as compared with an average

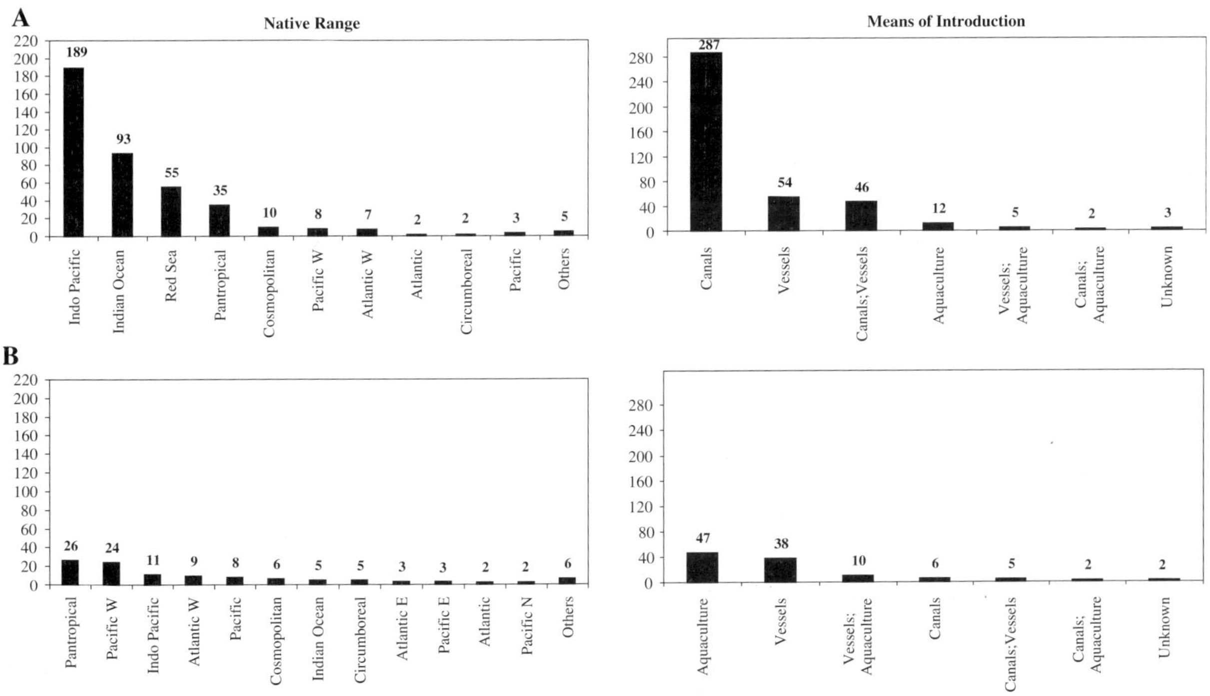

Fig. 7 Number and percentage of alien species in the easternmost (Egypt to Turkey) (**A**) and westernmost (Spain and France) (**B**) Mediterranean Sea, presented by their native range and means of introduction

Mediterranean residence of 43 and 33 years for alien decapods and fish, respectively, that are recorded only within the Levantine Sea. Similar figures are given by Gofas & Zenetos (2003) for mollusks. Eleven of the 13 vessel-transported alien species that had been recorded in the Mediterranean before 1950 are widely spread, several (*Asparagopsis armata* Harvey, 1855, *Lophocladia lallemandii* (Montagne) F. Schmitz, 1893, *Codium fragile* subsp. *tomentosoides* (Goor) P.C. Silva, 1955, *Callinectes sapidus* Rathbun, 1896) range from Spain to the Levant.

Discussion

Pathways

It is has been assumed that Erythrean aliens progress through the Suez Canal and along the coasts of the Levant as a result of "natural" dispersal, by autochthonous active or passive larval or adult movements, unaided further either directly or indirectly by human activity. Indeed, a temporal succession of directional ("stepping stones") records from the Red Sea, the Suez Canal, and along the coasts of the Levant confirms a species status as a naturally dispersing Erythrean alien. However, vessel-transported dispersal is feasible as well: "It is, of course, well known that ships have in more than one instance dispersed marine organisms from one part of the world to another. This must apply equally to transport through the Suez Canal. … Possibly tugs and barges permanently employed in the Canal may take a larger share than other vessels in this transport from one end of the Canal to the other. There are coal barges, for instance, which remain for some months at one end of the Canal and then are towed through to stay for some months more at the other end. The time spent at either end would permit on the one hand of the settling of larvae on the bottom of the barges, and on the other hand of the liberation of eggs or larvae from mature individuals" (Fox, 1926, p. 20). Indeed, the Suez Canal is one of the world's greatest short cuts and the densest shipping lane: about 6% of total world seaborne cargo passes through it (The Economist, 23 July, 2005). Vessels are also considered an important vector in the secondary spread of alien species: the Erythrean mytilid, *Brachidontes pharaonis* (Fischer, 1870) common in the Levantine Basin, where it settles in dense clusters on midlittoral and infralittoral rocks, piers, and debris (Barash & Danin, 1992), has spread as far west as Sicily (Di Geronimo, 1971), possibly in ship fouling. Similarly, the gastropod *Cerithium scabridum*, reported from Naples, Sicily, and Tunisia, was possibly transported there by shipping. The pearl oyster, *Pinctada radiata*, was one of the first Erythrean mollusks recorded in the Mediterranean (see above). It is abundant in the Levantine Basin and has spread as far west as the Tyrrhenian Sea, Sicily, Malta, Pantellaria Is., and France. Its rapid dispersal is attributed to shipborne individuals (Zibrowius, 1992) or marine turtles—*P. radiata* was recorded as an epibiont on a loggerhead turtle off Lampedusa Island (Oliverio et al., 1992). The bivalve *Fulvia fragilis* (Sowerby, 1834) has been recorded from the Suez Canal, the Levantine Basin, and Tunisia. However, its recent finding in the Italian ports of Livorno and Naples suggests it may be ship-transported as well (Crocetta, 2005). Even where records are consistent with long-shore autochthonous dispersal, there might be a degree of uncertainty where fouling organisms (such as serpulid polychaetes or mussels) are concerned, as they are more susceptible to shipping-mediated transfer. In some cases we suspect simultaneous mechanisms of transport: the widely invasive algae *Sargassum muticum* (Yendo) Fensholt, 1955, *Caulerpa taxifolia* (Vahl) C. Agardh, 1817, and *Caulerpa racemosa* var. *cylindracea* (Sonder) Verlaque, Huisman & Boudouresque, 2003, spread across the Mediterranean by ships, fishing boats, and recreational craft (Knoepffler-Péguy et al., 1985; Meinesz, 1992; Verlaque et al., 2003). Slower-moving and frequently moored vessels, such as drilling platforms employed in offshore exploration of oil and gas in the Mediterranean, pose a high risk of alien species transmission. The oil platform 'Discovery II' arriving in Genoa, Italy, in 1977 from the Indian Ocean carried "una ricca fauna tropicale vivente tra cui Teleosti Blennidi e Scorpenidi e Decapodi" (Relini Orsi & Mori, 1979). The oil-drilling platform 'Southern Cross' originating in Australia was brought to Haifa Bay, Israel, in 2003 for maintenance work including in-water scraping of its extensive fouling. The local divers employed described unfamiliar fish and crustaceans among the dense fauna, and from the shells that had been collected by the divers twelve species of mollusks were identified as new records for the Mediterranean (Mienis, 2004).

In the last twenty years higher demands for exotic fish and shellfish coupled with the crisis in wild fisheries created a surge in marine aquaculture (mariculture) farming along the shores of the Mediterranean. The Pacific oyster, *Crassostrea gigas* (Thunberg, 1793), and the Manila clam, *Ruditapes philippinarum* (Adams & Reeve, 1850), were intentionally introduced to the Mediterranean in 1950 and 1980, respectively (Bodoy et al., 1981; Madhioub & Zaouali, 1988), and their transplantation has resulted in numerous unintentional introductions of alien camp-followers (Ribera & Boudouresque, 1995): 49 of the 62 alien macrophytes recorded from the Mediterranean coast of France were introduced with oyster farming.

Why?

The present biota of the Mediterranean Sea is the issue of the region's geological and climatic history, and anthropogenic activities. The Late Pliocene cooling affected the sea's warm stenothermal biota, prompting the substitution of temperate for tropical species. During the Quaternary glacial periods temperate-cold biota dominated the Mediterranean, with mostly ephemeral entry of warmer 'Senegalian' elements during the brief interglacials. Thus, the Holocene Mediterranean had been largely inhabited by a northeastern Atlantic boreal biota. Long before the onset of the anthropogenic "global warming," the post-glacial rise in temperatures in the southern and eastern basins of the Mediterranean Sea rendered them unsuitable for many of the native boreal taxa, and the species present there were presumably at the limit of their ecological tolerance. At the same time, the entry of the warm-water West African biota was hindered by the Mauritanian cold-water barrier and the narrow Straits of Gibraltar. Williamson (1996) postulated that a successful invader occupies a vacant niche, be it a trophical niche or a functional niche—the extraordinary physiography and climatic history of the Mediterranean left it with vast vacant "climatic" niches. Modern shipping, the Suez Canal opening in 1869, and mariculture breached the biotic isolation of the Mediterranean and provided the major pathways for alien marine organisms to enter the sea and occupy those niches. Undeniably, most alien species established in the Mediterranean had not originated in similar latitudes like marine bioinvasions elsewhere, but are thermophilic, mostly tropical, species.

The last decades of the 20th C saw pronounced thermal fluctuations and "a significant increase in the average temperature of the waters" in the Mediterranean (Francour et al., 1994, p. 523) and a growing concern over the "tropicalization" of its fauna (Bianchi & Morri, 2003). The biological indicators comprised temporal changes in seasonal migration patterns (i.e. the amberjack, *Seriola dumerili* (Risso, 1810), and the bluefin tuna, *Thunnus thynnus* (Linnaeus, 1758)), a simultaneous increase in the numbers of native warm-water "vagrants" extending their range into the northern Tyrrhenian and Adriatic, some apparently forming self-sustaining populations (i.e. the ornate wrasse, *Thalassoma pavo* Linnaeus, 1758, and the parrotfish *Sparisoma cretense* (Linnaeus, 1758)), and the mass invasion of thermophilic alien biota (Francour et al., 1994; Dulcic & Lipej, 1997; Galil, 2000; Bombace, 2001; Bello et al., 2004). Already in the 1950s it was suggested that the establishment of marine alien species was related to a rise in sea surface temperature: the sudden escalation in the populations of the Red Sea lizard fish *Saurida undosquamis* (Richardson, 1848), the Red Sea goldband goatfish, *Upeneus moluccensis* (Bleeker, 1855), and other Erythrean alien species had been attributed to a rise of 1–1.5°C in the Levantine surface seawater temperature during the winter of 1955 (Ben Yami, 1955; Chervinsky, 1959; Ben Yami & Glaser, 1974). Ben Tuvia (1966, p. 254) contended that the thermophilic aliens require "temperatures high enough for the reproductive processes and development of eggs, and minimum winter temperatures above their lethal limits" to establish populations in the Mediterranean. As a matter of fact, the appearance of six Erythrean fish species in the Adriatic was concurrent with a rise in the sea surface temperatures (Dulčić & Grbec, 2000; Dulčić & Lipej, 2002). Similarly, the timing of the initiation of a significant increase in the number of Erythrean aliens along the southwestern Anatolian and the southern Aegean coasts was positively correlated with a more extensive inflow of the warm-water Asia Minor Current (Galil & Kevrekidis, 2002; Kumulu et al., 2002; Bilecenoglu et al., 2002; Corsini et al., 2002; Yokes & Galil, 2004, 2006; Yokes & Rudman, 2004; Katagan et al., 2004; Pancucci-Papadopoulou et al., 2005). Continuation of the warming trend would likely have a significant influence on the establishment and distribution of thermophilic species. Rising seawater temperature may change the pool of species which could establish themselves in the Mediterranean, enable the warm stenothermal species (native and alien) to expand beyond their present distributions, and may impact on a suite of population characteristics (reproduction, survival) that determines interspecific interactions, and, therefore, the dominance and prevalence patterns of both native and alien species, and provide the thermophilic aliens with a distinct advantage over the native Mediterranean biota.

However, concurrent alien invasion and rise in the sea surface temperature do not in themselves imply causation: the complexity of the ecosystem and the difficulty in assessing and understanding the biological response to climate forcing caution against simplistic unidimensional interpretations. The epic scale of invasion in the Mediterranean Sea may be explained as the consequence of a synergetic complex of stressors—pollution, eutrophication, destruction and fragmentation of habitats, fisheries overexploitation, fast-expanding mariculture, enlargement of the Suez Canal, increase in maritime traffic, and climate change—a catastrophic anthropogenic ecosystem shift—superimposed on a unique physiographical and climatic history.

Acknowledgments I am grateful to my colleagues, C. Bogi, M. Çinar, J. Corbera, R. Huys, H. Mienis, F. Mineur, A. Occhipinti-Ambrogi, P. Schembri, M. Verlaque, and B. Yokes, for generously sharing with me information on alien species. M. Cooper, L. Shoval, and S. Usvyatsov provided much-needed help throughout the project. Special thanks to the librarians of the American Museum of Natural History, New York; and Israel Oceanographic & Limnological Research, Haifa. This review was supported by the European Commission's 6th Framework Programme DAISIE (contract SSPI-CT-2003-511202).

References

Andreakis, N., G. Procaccini & W. H. C. F. Kooistra, 2004. *Aspargopsis taxiformis* and *Aspargopsis armata* (Bonnemaisoniales, Rhodophyta): genetic and morphological definition of Mediterranean populations. European Journal of Phycology 39: 273–283.

Andreakis, N., G. Procaccini, C. Maggs & W. H. C. F. Kooistra, 2007. Phylogeography of the invasive seaweed *Aspargopsis* (Bonnemaisoniales, Rhodophyta) reveals cryptic distribution. Molecular Ecology 16(11): 2285–2299.

Barash, A. & Z. Danin, 1992. Fauna Palestina: Mollusca I. Annotated list of Mediterranean mollusks of Israel and Sinai. The Israel Academy of Sciences and Humanities, Jerusalem.

Bello, G., N. Casavola & E. Rizzi, 2004. Aliens and visitors in the southern Adriatic Sea: effects of Tropicalization. Rapports et procès-verbaux des réunions Commission Internationale pour l'expolration scientifique de la Mer Méditerranée 37: 491.

Ben Eliahu, N. M., 1991. Red Sea Serpulids (Polychaeta) in the Eastern Mediterranean. Ophelia supplement 5: 515–528.

Ben Tuvia, A., 1966. Red Sea fishes recently found in the Mediterranean. Copeia 2: 254–275.

Ben Yami, M., 1955. 1954–1955: over-fishing or bad season? Fishermen's Bulletin, Haifa, 6: 10–14 [Hebrew].

Ben Yami, M. & T. Glaser, 1974. The invasion of *Saurida undosquamis* (Richardson) into the Levant Basin—an example of biological effect of interoceanic canals. Fishery Bulletin (Wash. D.C.) 72: 359–373.

Bianchi, C. N. & C. Morri, 2003. Global sea warming and "tropicalization" of the Mediterranean Sea: biogeographic and ecological aspects. Biogeographia 24: 319–327.

Bilecenoğlu, M., E. Taşkavak & K. B. Kunt, 2002. Range extension of three lessepsian migrant fish (*Fistularia commersoni, Sphyraena flavicauda, Lagocephalus suezensis*) in the Mediterranean Sea. Journal of the Marine Biological Association of the United Kingdom 82: 525–526.

Bodoy, A., T. Maitre-Allain & A. Riva, 1981. Croissance comparee de la palourde européenne *Ruditapes decussatus* et de la palourde japonaise *Ruditapes philippinarum* dans un ecosysteme artificiel méditerranéen. Vie Marine 2: 39–51.

Bombace, G., 2001. Influence of climatic changes on stocks, fish species and marine ecosystems in the Mediterranean Sea. Archivio Oceanografia e Limnologia 22: 67–72.

Carlton, J. T., 1985. Transoceanic and interoceanic dispersal of coastal marine organisms: the biology of ballast water. Oceanography and Marine Biology: an Annual Review 23: 313–371.

Carus, J. V., 1889–1893. Prodromus Faunae mediterraneae sive descriptio animalium maris Mediterranei incolarum, 2: Brachiostomata. Mollusca. Tunicata. Vertebrata. E. Schweitzerbart, Stuyygart [pp. 1–272 issued 1889].

Çevik, C., B. Ozturk & G. Buzzuro, 2001. The presence of *Crassostrea virginica* (Gmelin, 1791) and *Saccosterea commercialis* (Iredale & Roughley, 1933) in the Eastern Mediterranean Sea. La Conchiglia 32(298): 25–28.

Chervinsky, J., 1959. A systematic and biological comparison between the lizardfish (*Saurida grandisquamis*) from the Mediterranean and the Red Sea. Fishermen's Bulletin, Haifa, 19: 10–14 [English Abstract].

Corsini, M., G. Kondilatos & P. S. Economidis, 2002. Lessepsian migrant *Fistularia commersonii* from the Rhodes marine area. Journal of Fish Biology 61: 1061–1062.

Crocetta, F., 2005. Prime segnalazioni di *Fulvia fragilis* (Forsskål in Niehbur, 1775) (Mollusca: Bivalvia: Cardiidae) per i mari italiani. Bollettino Malacologico 41(5–8): 23–24.

Di Geronimo, I., 1971. Prima segnalazione sulle coste italiane di *Brachiodontes variabiles* Krauss. Bollettino delle

Sedute dell Accademia Gioenia di Scienze Naturali in Catania 10: 847–852.
Dulčić, J. & B. Grbec, 2000. Climatic change and Adriatic ichthyofauna. Fisheries Oceanography 9: 187–191.
Dulčić, J. & L. Lipej, 1997. New records of marine fishes from the Slovenian coastal waters. Falco 12: 35–40.
Dulčić, J. & L. Lipej, 2002. Rare and little-known fishes in the Eastern Adriatic during last two decades (1980–2001). Periodicum Biologorum 104: 185–194.
Fox, H. M., 1926. General part. Zoological results of the Cambridge expedition to the Suez Canal, 1924. 1. Transactions of the Zoological Society of London 22: 1–64.
Francour, P., C. F. Boudouresque, J. G. Harmelin, M. L. Harmelin-vivien & J.P. Quignard, 1994. Are the Mediterranean waters becoming warmer? Information from biological indicators. Marine Pollution Bulletin 28(9): 523–526.
Galil, B. S., 2000. A sea under siege - alien species in the Mediterranean. Biological Invasion 2: 77–186.
Galil, B. S. & K. Kevrekidis, 2002. Exotic decapods and a stomatopod off Rhodes Island (Greece) and the Eastern Mediterranean Transient. Crustaceana 75: 925–930.
Gofas, S. & A. Zenetos, 2003. Exotic molluscs in the Mediterranean basin: current status and perspectives. Oceanography and Marine Biology: an Annual Review 41: 237–277.
Jousson, O., J. Pawlowski, L. Zaninetti, A. Meinesz & C. F. Boudouresque, 1998. Molecular evidence for the aquarium origin of the green alga *Caulerpa taxifolia* introduced to the Mediterranean Sea. Marine Ecology Progress Series 172: 275–280.
Katagan, T., A. Kocatas, M. Zengin & A. S. Ates, 2004. An Indo-pacific Stomatopod from the sea of Marmara: *Erugosquilla massavensis* (Kossmann, 1880). Crustaceana 77: 381–383.
Keller, C., 1883. Die Fauna im Suez-Kanal und die Diffusion der Mediterranean und Erythräischen Tierwelt. Neue Denkschriften der allgemeinen Schweizerischen Gesellschaft für die gesamten naturwissenschaften, ser. 3, 28: 1–39.
Knoepffler-Peguy, M., T. Belsher, C.-F. Boudouresque & M. Lauret, 1985. *Sargassum muticum* begins to invade the Mediterranean. Aquatic Botany 23: 291–295.
Kumulu, M., O. T. Eroldogan, M. Aktas & M. Gocer, 2002. A new shrimp record for the Turkish Seas: *Melicertus hathor* (Burkenroad, 1959) (Penaeidae: Crustacea). Israel Journal of Zoology 48: 246–247.
Madhioub, M. N. & J. Zaouali, 1988. Captage de l'huitre *Crassostrea gigas* dans le lac Ichkeul. Bulletin de l'Institut National Scientifique et Technique d'Oceanographie et de Peche de Salambo 15: 47–60.
Meinesz, A., 1992. Modes de dissémination de l'algue *Caulerpa taxifolia* introduite en Méditerranée. Rapports et procès-verbaux des réunions Commission Internationale pour l'exploration scientifique de la Mer Méditerranée 33: 44.
Meusnier, I., M. Valero, J. L. Olsen & W. T. Stam, 2004. Analysis of rDNA ITS1 indels in *Caulerpa taxifolia* (Chlorophyta) supports a derived, incipient species status for the invasive strain. European Journal of Phycology 39: 83–92.
Mienis, H. K., 1995. Further records of the Erythrean gastropod *Syrnola fasciata* from the Mediterranean coast of Israel. Levantina 82: 7–8.
Mienis, H. K., 2004. New data concerning the presence of Lessepsian and other Indo-Pacific migrants among the mollusks in the Mediterranean Sea with emphasize on the situation in Israel. In Öztűrk, B. & A. Salman (eds), Proceedings 1st National Malacology Congress, 1–3 September 2004, Izmir. Turkish Journal of Aquatic Life 2(2): 117–131.
Monterosato di, T. A., 1878. Enumerazione e sinonima delle conchiglie mediterranee. Giornale di scienze naturali ed economiche di Palermo 13: 61–115.
Occhipinti-Ambrogi, A. & B. S. Galil, 2004. A uniform terminology on bioinvasions: a chimera or an operative tool? Marine Pollution Bulletin 49: 688–694.
Oliverio, M., G. Gerosa & M. Cocco, (1992) First record of *Pinctada radiata* (Bivalvia, Pteriidae) epibiont on the loggerhead sea turtle *Caretta caretta* (Chelonia, Cheloniidae). Bollettino Malacologico 28(5–12): 149–152.
Pancucci-Papadopoulou, M. A., K. Kevrekidis, M. Corsini-Foka & N. Simboura, 2005. Changes in species: invasion of exotic species. In Papathanassidou, E. & A. Zenetos (eds), State of the Hellenic Marine Environment. HCMR Publications, Athens, pp 336–342.
Por, F. D., 1978. Lessepsian Migration—The Influx of Red Sea Biota into the Mediterranean by Way of the Suez Canal. Ecological Studies, Vol. 23. Springer-Verlag, Berlin-Heidelberg-New York.
Por, F. D., 1990. Lessepsian migration. An appraisal and new data. Bulletin de l'Institut oceanographique. Monaco 7: 1–10.
Relini Orsi, L. & M. Morri, 1979. Due reperti mediterranei di *Thalamita gloriensis* Crosnier, 1962 (Crustacea: Decapoda: Portunidae). Oebalia 5: 7–13.
Ribera, M. A. & C. F. Boudouresque, 1995. Introduced marine plants with special reference to macroalgae: mechanisms and impacts. Progress in Phycological Research 11: 187–268.
Saker, F., 2002. A contribution to the study of the species composition of the benthos in the waters of Lattakia. Journal of the Union of Arab Biologists, Cairo 18(A): 287–310. [in Arabic].
Steinitz, H., 1970. A critical List of Immigrants via the Suez Canal. Biota of the Red Sea and Eastern Mediterranean, pp. 59–63. [mimeo.].
Terranova, M. S., S. Lo Brutto, M. Arculeo & J. B. Mitton, 2006. Population structure of *Brachidontes pharaonis* (P. Fischer, 1870) (Bivalvia, Mytilidae) in the Mediterranean Sea, and evolution of a novel mtDNA polymorphism. Marine Biology 150: 89–101.
UNEP/MAP/MED POL, 2004. Mariculture in the Mediterranean. MAP Technical Reports Series No. 140. 80 p.
van der Linden, J. & J. C. A. Eikenboom, 1992. On the taxonomy of the recent species of the genus Chrysallida Carpenter from Europe, the Canary Islands and the Azores (Gastropoda: Pyramidellidae). Basteria 56: 3–63.
Verlaque, M., 2005. Algal Introductions to European Shores. WP4.2 Propagule Pressure: Shellfish Industry. Final report, 5th PCRD European Program. 134 pp.

Verlaque, M., C. Durand, J. M. Huisman, C. F. Boudouresque & Y. Le Parco, 2003. On the identity and origin of the Mediterranean invasive *Caulerpla racemosa* (Caulerpales, Chlorophyta). European Journal of Phycology 38: 325–339.

Williamson, M., 1996. Biological Invasions. Chapman & Hall, London.

Yokes, B. & B. S. Galil, 2004. New record of alien decapods from the southwestern coast of Turkey. Rapports et procès-verbaux des réunions Commission Internationale pour l'expolration scientifique de la Mer Méditerranée 37: 556.

Yokes, B. & B. S. Galil, 2006. New records of alien decapods (Crustacea) from the Mediterranean coast of Turkey, with a description of a new palaemonid species. Zoosystema 28(3): 747–755.

Yokes, B. & W. B. Rudman, 2004. Lessepsian opisthobranchs from southwestern coast of Turkey; five new records for Mediterranean. Rapports et procès-verbaux des réunions Commission Internationale pour l'expolration scientifique de la Mer Méditerranée 37: 557.

Zibrowius, H., 1973. Serpulidae (Annelida Polychaeta) des côtes ouest de l'Afrique et des Archipels Voisins. Annales Musee Royal de l'Afrique Centrale 207: 1–93.

Zibrowius, H., 1992. Ongoing modifications of the Mediterranean marine fauna and flora by the establishment of exotic species. Mesogee 51: 83–107.

Hydrobiologia (2008) 606:117–127
DOI 10.1007/s10750-008-9339-7

CHALLENGES TO MARINE ECOSYSTEMS

The effect of high flow events on mussels (*Mytilus edulis*) in the Conwy estuary, North Wales, UK

Lucie R. Oliver · Raymond Seed · Brian Reynolds

© Springer Science+Business Media B.V. 2008

Abstract One of the predicted consequences of climate change is an increase in the occurrence of extreme rainfall and a subsequent increase in frequency of high flow events in rivers. High flow events have the potential to impact estuarine communities like mussel assemblages due to decreased salinity and/or the transport of sediments, organic matter and nutrients from the terrestrial environment to the estuary. The impact of two high flow events was investigated using mussels located within the Conwy estuary, North Wales, using the 'Beyond BACI' approach. Three study sites were chosen, the potentially impacted site (Conwy) and two control sites located outside the estuary. Sampling took place over 18 months with samples being collected before and after each event. On each sampling occasion, the following data were collected: the total haemocyte count (THC) and condition index (CI) of the mussels and the diversity ($Hlog_e$) of their associated macrofauna. A significant effect of the first event (22nd October 2004) was found on the CI of the Conwy mussels, whereas a significant effect of the second event (10th October 2005) was found on mussel THC. No effect of either event was found on the diversity of the associated fauna. The results of this study suggest that any increase in the number or intensity of heavy precipitation or high flow events have potential implications for the health and resilience of estuarine mussel populations.

Keywords Estuary · Mussels · Climate change · High flow · Immunology · Diversity

Guest editors: J. Davenport, G. Burnell, T. Cross, M. Emmerson, R. McAllen, R. Ramsay & E. Rogan
Challenges to Marine Ecosystems

L. R. Oliver (✉) · R. Seed
School of Ocean Sciences, University of Wales, Bangor, Menai Bridge, Anglesey LL59 5AB, UK
e-mail: l.oliver@ccw.gov.uk

B. Reynolds
Centre for Ecology and Hydrology, Bangor, Orton Building, Deiniol Road, Bangor, Gwynedd LL57 2UP, UK

Introduction

One of the predicted consequences of global climate change is an increase in the occurrence of extreme rainfall leading to a subsequent increase in the frequency of high flow events in rivers and estuaries (Booij, 2005). High flow events led to increased sediment loading and reduced salinity in the receiving estuaries (Kimmerer, 2002), both of which have the potential to impact estuarine organisms. The response of an estuary to these events can reveal a great deal about its resilience and the role of disturbance in shaping its structure and function (Davis et al., 2004). In addition, an understanding of the effects of natural events on estuarine organisms is a basic tool in discriminating between natural changes and human disturbances (Chainho et al., 2006).

The blue mussel (*Mytilus edulis* Linnaeus) is a commercially important species; mussel production in the UK alone is worth ~£5 million per annum and mussel farms are often located either inside or close to the mouths of estuaries (Laing & Spencer, 1997). Therefore, if high flow events have a negative impact on mussels, this could have important economic implications for mussel farmers. Mussels also have several important ecosystem functioning roles including nutrient cycling (Prins & Smaal, 1994), providing a complex habitat for diverse epi- and in-faunal assemblages (Seed, 1996) and, in common with other invertebrate species, are a food source for a range of animals including shorebirds, crabs and starfish (see Saier, 2001) as well as commercially important fisheries species e.g. plaice (McNeish et al., 1997). Consequently, if high flow events are found to negatively impact mussels, this could also have implications for the wider ecosystem.

Both reductions in salinity (Davenport, 1979) and increased flow velocity (Newell et al., 2001) are known to provoke an isolation response in mussels, a three-step process with the exhalant siphon being closed first followed by the inhalant siphon and then the shell valves being closed until acclimation occurs or conditions return to normal (Berger & Kharazova, 1997). It was predicted that the reduction in feeding that would occur during such a period of isolation, coupled with the energetic cost of adjusting to the external change in salinity (Sadok et al., 1997), would cause mussel condition to decrease following the high flow events.

Reduced salinity has also been shown to influence several metabolic functions and physiological conditions in mussels; these include heart rate (Bakhmet et al., 2005; Braby & Somero, 2006); protein and nucleic acid content of tissues (de Bravo, 2003); energy acquisition and growth rate (Gardner & Thompson, 2001); byssal thread production (Pelc & Alexander, 1999) and acetylcholinesterase activity (Pfeifer et al., 2005). Bivalve defensive cells (haemocytes) are also influenced by acute changes in salinity with increases being shown to inhibit haemocyte circulation and locomotion (Fisher, 1988) and decreases resulting in an increase in haemocyte mortality (Gagnaire et al., 2006). Thus, high flow events might also be predicted to result in a decline in mussel immunocompetence as shown by a decrease in total haemocyte count (THC). This is important for two reasons; first any reduction of immune function may leave mussels at risk from disease and parasitism (Martello et al., 2000) and second there is continued interest in the use of mussel immunocompetence as a biomarker for a range of potential pollutants and such applications require a better understanding of the influence of natural environmental stressors on these defence mechanisms (Pfeifer et al., 2005).

In rivers, high flow events are considered a major form of disturbance and are often instrumental in structuring benthic invertebrate communities (Hancock, 2006). However, less is understood about their effects in estuaries (Matthews & Constable, 2004). The use of benthic invertebrate assemblages to monitor the ecological quality of estuaries is a major tool in the implementation of the Water Framework Directive (2000/60/CE) but it is vital to identify the impact of extreme natural events so that they can be discriminated from human impacts (Chainho et al., 2006). It was predicted that the high flow events would cause changes in the diversity of the associated epi- and in-fauna of the estuarine mussel beds due to mortality of more sensitive species, colonisation by opportunistic species and/or physical transportation of species from further up the estuary (e.g. Currie & Small, 2005; Ritter et al., 2005; Chainho et al., 2006; Silva et al., 2006).

Methods

Study sites

The estuary chosen for this study is that of the Afon Conwy, in North Wales, UK (see Fig. 1), which drains into Liverpool Bay and is 55 km in length. The catchment covers an area of ~400 km^2 and is wet, mountainous and prone to flooding. The upper reaches of the catchment include extensive sheep farms, moorland and forestry; whereas the lower Conwy valley is a fertile, alluvial flood plain with mixed beef, dairy and sheep farms. The estuary supports a number of naturally settled mussel beds, which are subject to artesanal harvesting. The two control sites: Rhos on Sea (Rhos) and Llanfairfechan

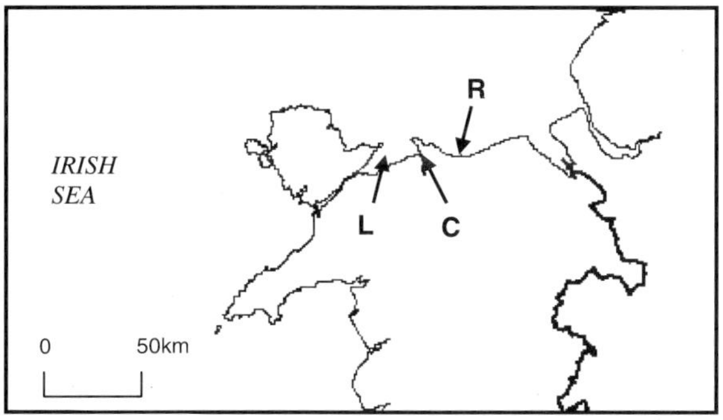

Fig. 1 Map showing the locations of the three study sites: L, Llanfairfechan; C, Conwy; R, Rhos on Sea (map reproduced with kind permission of the Ordnance Survey)

(Llanfair) are marine mussel beds situated ~25 km either side of the Conwy estuary.

High flow events

The first high flow event occurred on the 22nd–23rd of October 2004 when 83 mm of rainfall were recorded in the Conwy Valley over 24 h and the flow rate of the river increased from a daily mean of 72 $m^3 s^{-1}$ on 21st October to 306 $m^3 s^{-1}$ on 22nd October. The second high flow event occurred 1 year later on the 10th–11th of October 2005 when 71 mm of rain fell in 24 h and the mean daily flow rate of the river increased from 7 $m^3 s^{-1}$ on 10th October to 156 $m^3 s^{-1}$ on 11th October (data reproduced with kind permission of the Environment Agency).

Experimental design

Since the potentially impacted site (the estuary) was unreplicated, the experimental design used was the asymmetric 'Beyond Before After Control Impact (BACI)' design proposed by Underwood (1992). This involves two ANOVA's, one including data from all three sites and the second using only the controls. The variability contributed by the 'impacted' site is calculated by subtracting the sums of squares of the second analysis from those of the first (Roberts et al., 1998).

Samples were collected from each site on 12 occasions, three 'before' and three 'after' each of the two high flow events (see Figs. 2 and 3 for dates). Since the timing of the two events could not be predicted the 'before' samples were collected at the time of year estimated to have the lowest rainfall and flow rate i.e. summer. The following data were collected on each occasion. Five replicate 10 × 10 cm quadrats were placed at metres apart and the mussels and associated fauna scraped off the rock using a wallpaper scraper. Samples were placed in plastic bags and transported back to the laboratory where the mussels were removed and the remaining animals sorted and identified. The Shannon diversity of the mussel assemblages ($n = 5$) was then calculated. The field total haemocyte count (THC) of 15 randomly selected mussels (shell length 3–5 cm) was also measured using a method adapted from Auffret & Oubella (1995). The 15 mussels were divided into three replicate groups of five animals and 0.05 ml of haemolymph collected from the posterior adductor muscle using a 2.5-ml syringe, with a 25-gauge needle, to withdraw the fluid from a notch made in the shell using a scalpel. The haemolymph was aspirated into a 1.5-ml Eppendorf tube containing 0.75 ml of Alsever's formalin (pH 7.5) to fix the haemocytes and prevent clumping. Haemolymph from five mussels was added to each Eppendorf tube, giving a final volume of 1 ml and a dilution factor of 3:1. The fixed haemolymph samples were then packed on ice and transported back to the laboratory where THC's were carried out within 1 h of returning using a haemocytometer. The Condition Index (CI) of the same 15 mussels ($n = 15$) was also measured in the laboratory using the formula: dry flesh weight/shell length^3 (Seed & Suchanek, 1992).

Results

The results of the asymmetric analysis of variance of the univariate data collected are shown in Tables 1–3. The results of the analysis of the mussel total haemocyte counts (THC) revealed no effect of the first high flow event but a significant effect (B × I versus residual, $F = 10.85$**) of the second event (Table 1). Figure 2 shows that this effect took the form of a significant decrease in the THC of the Conwy mussels after the second event, which gradually increased over the next 5 weeks until it reached a similar level to the average of the 'before' samples. Interestingly, although this was not reflected in the analysis because of the variability in THC at the different sites, it is clear that the same pattern was

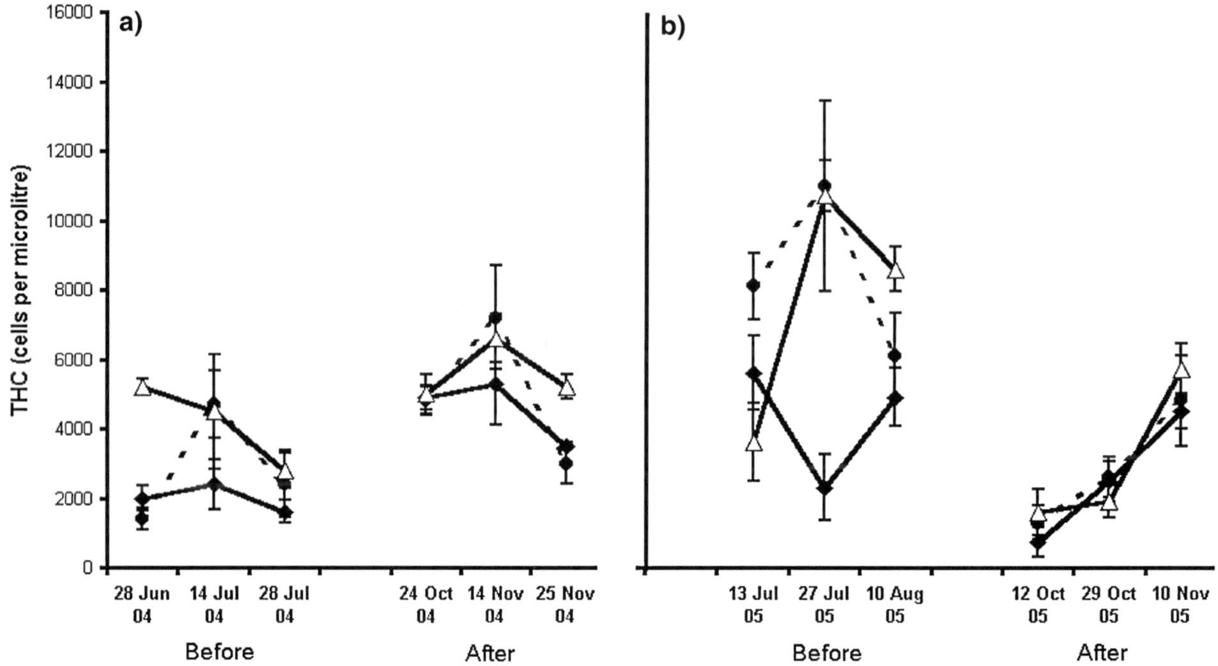

Fig. 2 Mean total haemocyte counts, expressed as cells per μl, (±SE) of mussels collected from the Conwy estuary and two control locations (Llanfairfechan and Rhos on Sea) before and after the two high flow events ($n = 3$). (**a**) Event 1, (**b**) Event 2; diamonds, Conwy; circles, Llanfairfechan; triangles, Rhos on Sea

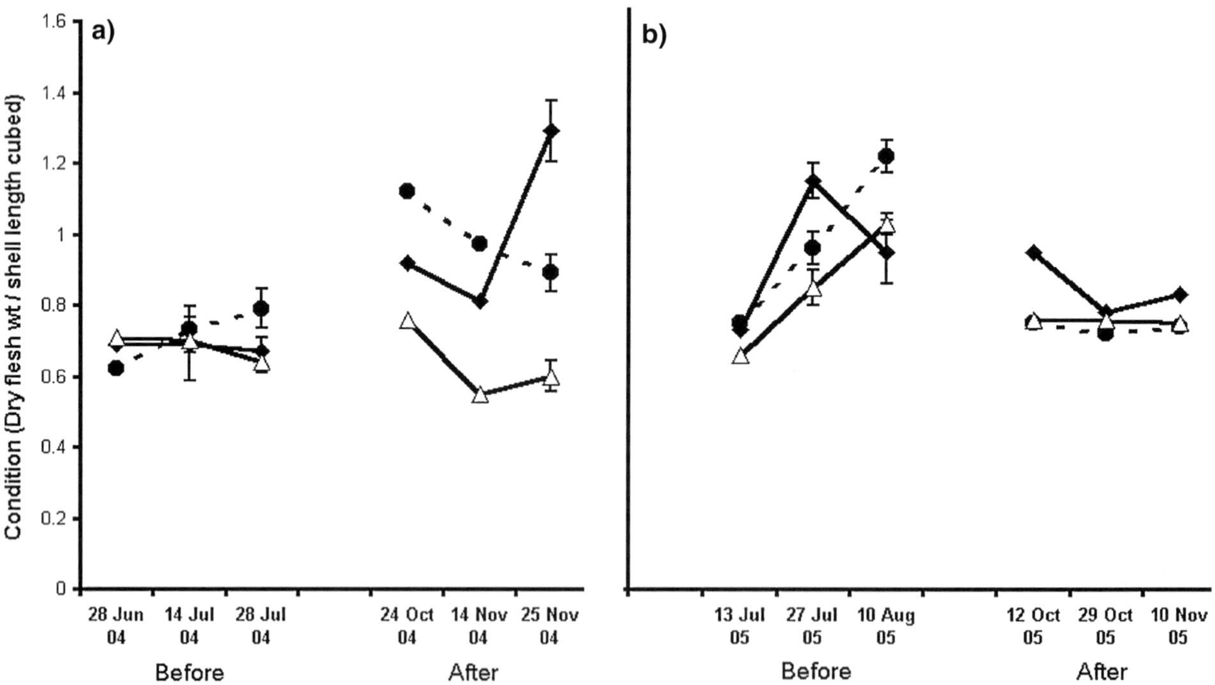

Fig. 3 Mean condition index (±SE) of mussels collected from the Conwy estuary and two control locations (Llanfairfechan and Rhos on Sea) before and after the two high flow events ($n = 15$). (**a**) Event 1, (**b**) Event 2; diamonds, Conwy; circles, Llanfairfechan; triangles, Rhos on Sea

Table 1 Analysis of the total haemocyte count (THC) of mussels from the Conwy estuary (I) and two control locations (C—Llanfairfechan and Rhos on Sea) at six times of sampling ($n = 3$)

Source of variation	Event 1				Event 2			
	DF	*F*-ratio vs.	MS	*F*	DF	*F*-ratio vs.	MS	*F*
Before vs. after = B	1		5.6590E + 13	5.03	1		2.09E + 14	7.76
Times (B) = T(B)	4		1.0284E + 13	4.25*	4		2.30E + 13	1.59
Among locations = L	2		1.5912E + 13	4.69	2		2.63E + 13	1.44
Impact vs. control = I	1		1.54E + 13		1		5.17E + 13	
Among controls = C	1		1.6457E + 13		1		8.53E + 11	
B × among locations	2		3.3950E + 12	1.4	2		1.83E + 13	1.27
B × I	1	Residual	1.70E + 12	1	1	Residual	3.46E + 13	10.85**
B × C	1	Residual	5.0925E + 12	3	1	Residual	1.99E + 12	0.62
Locations × T(B)	8		2.4185E + 12	1.42	8		1.44E + 13	4.52**
Locations × T(before)	4		2.54E + 12		4		2.80E + 13	
I × T(before)	2		9.00E + 11		2		3.72E + 13	
C × T(before)	2		4.20E + 12		2		1.88E + 13	
Locations × T(after)	4		2.28E + 12		4		7.50E + 11	
I × T(after)	2	Residual	1.43E + 12	0.84	2	Residual	5.50E + 11	0.17
C × T(after)	2	Residual	3.12E + 12	1.84	2	Residual	9.48E + 11	0.3
Residual	36		1.70E + 12		36		3.19E + 12	
Total	53				53			
Two tailed *F* tests					T(after) × I/T(before) × I = 0.01			

* $P < 0.05$, ** $P < 0.01$, repartitioned sources of variation are offset

shown by the control samples as well. This suggests that this decrease in THC was related to a rapid reduction in salinity after the high flow event which the control sites may well have experienced despite being marine rather than estuarine, due to inundation by rain during the event. This identifies one of the problems inherent in the design of this study and therefore it may have been more desirable to have control sites in completely different areas, although this could have resulted in differences due to other environmental factors. The results of the analysis of the CI data (Table 2) show that there was no effect of event 2 but that there was a significant effect of event 1 (T(after) × I versus residual, $F = 15.88$***). Figure 3 indicates that there was a sharp increase in condition of the Conwy mussels 1 month after this high flow event.

There was no significant impact of either event on the univariate diversity of the mussel assemblages (Table 3). This may have been due in part to the small number of taxa found during this study, with the average total number of taxa found inside the estuary ranging from 6.4 to 8.4 and the highest number found in any one sample being 9.

Overall, these results suggest that high flow events have the potential to affect the condition and total haemocyte count of mussels but that these effects were not consistent and may not be readily predicted. The results of the THC analyses highlight the fact that the heavy rainfall leading to the two high flow events could either have contributed to the observed effects, or may in fact explain them completely. Further experiments are needed in order to unconfound these factors.

Discussion

Much of the high temporal and spatial variability in physical, chemical and biological conditions in estuaries occurs through seasonal and interannual variability in freshwater flow (Kimmerer, 2002). Differences in the response of the mussels and their associated fauna to the two high flow events in this

Table 2 Analysis of the condition index (CI) of mussels from the Conwy estuary (I) and two control locations (C—Llanfairfechan and Rhos on Sea) at six times of sampling ($n = 15$)

Source of variation	Event 1				Event 2			
	DF	*F*-ratio vs.	MS	*F*	DF	*F*-ratio vs.	MS	*F*
Before vs. after = B	1		2.372	2.83	1		1.390	1.7
Times (B) = T(B)	4		0.189	0.69	4		0.800	5.69**
Among locations = L	2		1.045	1.13	2		0.700	0.477
Impact vs. control = I	1		0.460		1		0.220	
Among controls = C	1		1.630		1		0.120	
B × among locations	2		0.923	3.35	2		0.150	0.78
B × I	1	Residual	0.640		1	Residual	0.050	0.56
B × C	1	Residual	1.210		1	Residual	0.260	2.89
Locations × T(B)	8		0.276	4.93**	8		0.140	0.148
Locations × T(before)	4		0.059		4		0.227	
I × T(before)	2		0.015		2		0.438	
C × T(before)	2		0.103		2		0.017	
Locations × T(after)	4		0.493		4		0.056	
I × T(after)	2	Residual	0.953	15.88***	2	Residual	0.109	1.21
C × T(after)	2	Residual	0.034	0.57	2	Residual	0.002	0.02
Residual	252		0.060		252		0.090	
Total	269				269			
Two tailed *F* tests	T(after) × I/T(before) × I = 63.5				T(after) × I/T(before) × I = 0.25			

** $P < 0.01$, *** $P < 0.001$, repartitioned sources of variation are offset

study may be related to the hydrological conditions immediately preceding the events. The mean rainfall and flow rate recorded in the 7 days prior to event one were 18 mm and 46 $m^3 s^{-1}$, respectively, whereas the 7 day means prior to event two were 1.4 mm and 9 $m^3 s^{-1}$. Thus, although the magnitude of event one (maximum flow = 306 $m^3 s^{-1}$) was almost double that of event two (maximum flow = 156 $m^3 s^{-1}$) the estuary had received much higher volumes of freshwater for the week before the first high flow event than before the second event. This could have resulted in a certain amount of acclimation by the mussels and their associated fauna, an ability that is found in many euryhaline species including *M. edulis* and is brought about by efficient cellular and whole-organism compensatory mechanisms allowing the animals to adjust to a persistent change in the environmental salinity (see Sukhotin et al., 2003 and references therein). Conversely, a rapid decrease in salinity with no period of acclimation can provoke an almost simultaneous response in mussels (Davenport, 1982) and this type of change in salinity rather than the sinusoidal fluctuations that are normally experienced by intertidal organisms may have a far more detrimental effect.

After event two there was a significant effect on the THC of the Conwy mussels with the THC of mussels decreasing significantly after the event and then increasing over the next 5 weeks, a pattern that was also observed at the two control locations. This suggests that the cause of this effect was related to a decrease in salinity, which the control mussels may also have experienced during the precipitation that led to the high flow event. Fisher et al. (1996) suggest that the influence of ambient temperature and salinity on bivalve haemocytes may account for the high variability in the defence responses of oysters collected at different times of year, although salinity fluctuations may track other hydrographic changes that affect oyster biology such as turbidity, oxygen, nutrient availability and contaminant loadings. It is not clear what the exact mechanism of such a response to a drop in salinity would be. Like many other intertidal molluscs, *M. edulis* uses a complex of behavioural, physiological and biochemical mechanisms to withstand prolonged periods of unfavourable

Table 3 Analysis of the Shannon diversity ($Hlog_e$) of mussel beds in the Conwy estuary (I) and at two control locations (C—Llanfairfechan and Rhos on Sea) at six times of sampling ($n = 5$)

Source of variation	Event 1				Event 2			
	DF	*F*-ratio vs.	MS	*F*	DF	*F*-ratio vs.	MS	*F*
Before vs. after = B	1		0.037	0.22	1		0.310	2.55
Times (B) = T(B)	4		0.132	2.12	4		0.227	0.515
Among locations = L	2		0.865	8.77	2		1.326	8.76
Impact vs. control = I	1		1.720		1		2.650	
Among controls = C	1		0.010		1		0.021	
B × among locations	2		0.099	1.59	2		0.151	0.59
B × I	1		0.000		1	Residual	0.214	0.84
B × C	1		0.197		1	Residual	0.089	0.35
Locations × T(B)	8		0.062	2.34**	8		0.256	5.76**
Locations × T(before)	4		0.036		4		0.143	
I × T(before)	2		0.043		2		0.034	
C × T(before)	2		0.029		2		0.253	
Locations × T(after)	4		0.088		4		0.369	
I × T(after)	2	C × T(after)	0.027	0.118	2	Residual	0.127	0.5
C × T(after)	2	Residual	0.150	5.56**	2	Residual	0.561	2.19
Residual	72		0.027		72		0.256	
Total	89				89			
Two tailed *F* tests	T(after) × I/T(before) × I = 0.63				T(after) × I/T(before) × I = 3.74			

** $P < 0.01$, repartitioned sources of variation are offset

environmental conditions such as exposure to air or extreme salinity changes (Sukhotin et al., 2003). One possibility is that the reduction in feeding caused by the mussels' isolation response (Davenport, 1979) coupled with the energetic cost of adjusting to the external change in salinity (Sadok et al., 1997), causes an increase in autophagy which can allow cells to be self-sustaining during times when nutrients are scarce by breaking down and recycling longer-lived proteins and thereby releasing energy (Moore et al., 2006). An increase in autophagy has been shown by various molluscs in response to stress caused by changes in salinity, hypoxia or nutrient decrease and would result in lower cell concentrations if it reached the point where cell death occurred. Gagnaire et al. (2006) found that an acute decrease in salinity resulted in a 30–50% increase in cell mortality and a 70% decrease in phagocytosis activity in oysters within 24 h. Another possibility is that the decrease in THC is caused by a decrease in mussel heart rate resulting in fewer haemocytes being pumped to the peripheral parts of the body such as the adductor muscle. Bakhmet et al. (2005) noted that, when exposed to moderate hyposalinity (15 g l^{-1}), *M. edulis* showed a significant decrease in heart rate that only returned to normal after 5 days. This decrease in the number of circulating haemocytes may cause problems if the bacterial load of an estuary is increased after a high flow event (Kimmerer, 2002) because any reduction of immune function will leave mussels at risk from disease (Martello et al., 2000).

Contrary to the prediction that mussel condition would decrease after the high flow events, due to the mussels' isolating response resulting in a temporary reduction or cessation in feeding, there was no effect of event two and a significant increase in the condition of the Conwy mussels after event one. This could also have been due to the more prolonged nature of this particular event, which may have resulted in more nutrients and organic matter being washed into the river from its catchment. Increased freshwater input is shown to increase the secondary production of estuaries through stimulating primary production and directly providing food for filter feeders in the form of particulate organic matter

(Montagna & Kalke, 1992; Currie & Small, 2005), although mussel feeding can decline drastically if seston concentrations are too high (Winter, 1978). An increase in primary production due to high flow is unlikely in this case, given the timing of both events, but the increase in condition after event one could reflect changes in food quality. For example, Mudge & Norris (1997) found that the freshwater reaches of the river Conwy contained higher concentrations of long-chain fatty alcohols from terrestrial sources that, if washed into the estuary during a high flow event, could provide more high-energy food for mussels. The use of condition indices in monitoring studies has been recommended because such measures integrate stress responses on somatic growth, however, condition is not always indicative of stress and can be affected by seasonal changes and associated nutritional and reproductive states at spatially different sites (Nicholson, 1999).

Estuarine organisms can be influenced by high flow events in several ways including increased abundance and biomass due to increased inputs of nutrients and exogenous carbon. Alternatively, physical washout, salinity stress and hypoxia caused by algal blooms can reduce benthic populations (Montagna & Kalke, 1992). We know of no previous research that is directly comparable with the present study in terms of the effects of high flow events on the diversity of mussel beds, although there are a few studies relating the effects of high flows on benthic infaunal communities. For example, the studies of Montagna & Kalke (1992) and Chainho et al. (2006) have identified the importance of freshwater inputs and changes in salinity in driving seasonal changes in estuarine benthic communities by imposing strong decreases in the densities of a several species during periods of high flow.

It is worth noting, however, that the diversity of the mussel beds in this study was much lower than have previously been reported for marine mussel beds. For example, Lintas & Seed (1994) recorded a total of 59 taxa at a rocky, wave exposed shore in North Wales almost three times the total number of taxa reported for the Conwy mussel beds (22). Low diversity is a feature of estuarine communities (see Rutger & Wing, 2006) and may have contributed to the lack of any significant effect of the high flow events on mussel bed diversity. However, mussel patches are known to provide a certain degree of protection from environmental extremes due to the complex structure they provide (Seed & Suchanek, 1992) and this arguably could have contributed to the lack of difference in diversity before and after the events. It should also be emphasised that this study considered only intertidal mussel beds and the associated taxa, like the mussels themselves, will be more adapted to experiencing extremes of salinity and temperature (Bakhmet et al., 2005). The salinity of an intertidal site in the Conwy estuary, for instance, which was monitored over two 24 h periods fluctuated in a roughly sinusoidal manner and often fell to near zero salinity (Davenport, 1982).

Previous studies have also highlighted the difficulties associated with differentiating the effects of high flow events from the precipitation that precedes them. A long-term study of the Tagus estuary, Portugal found that rainfall was an essential factor in structuring intertidal soft-sediment assemblages. There was a direct effect of rainfall (and therefore reduced salinity) on the metabolism of the benthic infauna, but also an indirect effect of changes in sediment composition (less total organic matter, more coarse particles) caused by the high riverine flows (Silva et al., 2006). Observed effects of precipitation events on marine communities include alterations in the density, abundance and dominance of benthic polychaetes, (see Chollett & Bone, 2007) and a decrease in mussel predation by starfish and crabs in shallow subtidal assemblages due to lowered salinity creating an effective refuge from predation (Witman & Grange, 1998).

The results of this study suggest that high flow events have the potential to affect the condition and total haemocyte count of mussels and the composition of their associated macrofaunal communities. The lack of any field measurements such as salinity, flow rate, temperature or turbidity, and potential confounding by rainfall effects, means that it is impossible to draw any conclusions regarding the exact cause of these effects. Nevertheless, it appears that any future increase in the number or intensity of heavy precipitation or high flow events, e.g. due to climate change, could have implications for the health and resilience of estuarine mussel populations. The results also highlight the need to consider this potential source of variability when attempting to detect anthropogenic disturbance using either the immune responses of indicator species like *M. edulis*.

Although the effects of the two events were not consistent, our results highlight an area that requires further research particularly in those estuaries that are subject to more intensive agricultural and/or industrial influences than the Afon Conwy as the responses observed in this study may be further magnified in estuaries that are already under stress from other disturbances. High flow events are fundamental to the transfer of sediments from terrestrial or riverine sources to estuaries and coastal ecosystems and these can often carry large loads of particulate associated pollutants (Zonta et al., 2005) such as PCBs, PAHs and organochlorine pesticides (Foster et al., 2000). The impact of high flow events occurring at different times of year should also be assessed as these may have different effects to winter events, e.g. by disrupting reproduction and/or recruitment of estuarine species. Summer events are likely to be particularly detrimental to mussels and other commercially important bivalves, which may already be stressed by high temperatures or reproductive activities (Cajaraville et al., 1996; Bocchetti & Regoli, 2006) and any resulting inhibition of immunocompetence could be more problematic in summer when bacterial numbers are higher (Delaporte et al., 2006).

A number of modifications would improve the power of this experimental design to detect changes caused by high flow events. For instance, rather than using marine control sites located near to the potentially impacted estuarine site, it would be preferable to use different estuaries, not influenced by the same precipitation events, as controls. This could be achieved by identifying several estuaries as potential control locations and selecting the most appropriate of these after the occurrence of a high flow event. In order to avoid collecting 'before' data from all of these potential control sites, some of which may not be used, it is possible to sample only after a high flow event. This approach has been developed for use in situations where an unplanned impact or disturbance (e.g. an oil spill) occurs and there is no before data to refer to (i.e. Quieroz et al., 2006), and requires more controls than were used in the present study to avoid the risk of spatial confounding (Glasby, 1997). A further requirement of this 'after only' design is greater temporal replication than was used in the present study, preferably at a range of nested scales (Morrisey et al., 1992). This may also help to determine if the effects of high flow events on the different parameters measured in this study are operating at a range of temporal scales, e.g. hours or days for the THC of the mussels as opposed to days or weeks in the case of mussel condition.

Acknowledgements The authors would like to thank the Natural Environment Research Council and the Centre for Hydrology and Ecology for funding this project and Shelagh Malham for her assistance with the immunological aspects of the study. R. Seed and B. Reynolds were co-funded by the Biotechnology and Biological Sciences Research Council (grant number BB/C507129/1). We would also like to thank two anonymous referees for their useful comments on an earlier version of the manuscript.

References

Auffret, M. & R. Oubella, 1995. Cytological and cytometric analysis of bivalve mollusc haemocytes. In Stolen, J. S., T. C. Fletcher, S. A. Smith, J. T. Zelikoff, S. L. Kaattari, R. S. Anderson, K. Soderhall & B. A. Weeks-Perkins (eds), Techniques in Fish Immunology—4. Immunology and Pathology of Aquatic Invertebrates. SOS Publications, Fair Haven, NJ, USA: 55–64.

Bakhmet, I. N., V. J. Berger & V. V. Khalaman, 2005. The effect of salinity change on the heart rate of *Mytilus edulis* specimens from different ecological zones. Journal of Experimental Marine Biology and Ecology 318: 121–126.

Berger, V. J. & A. D. Kharazova, 1997. Mechanisms of salinity adaptations in marine molluscs. Hydrobiologia 355: 115–126.

Bocchetti, R. & F. Regoli, 2006. Seasonal variability of oxidative biomarkers, lysosomal parameters, metallothioneins and peroxisomal enzymes in the Mediterranean mussel *Mytilus galloprovincialis* from Adriatic Sea. Chemosphere 65: 913–921.

Booij, M. J., 2005. Impact of climate change on river flooding assessed with different spatial model resolutions. Journal of Hydrology 303: 176–198.

Braby, C. E. & G. N. Somero, 2006. Following the heart: temperature and salinity effects on heart rate in native and invasive species of blue mussels (genus Mytilus). Journal of Experimental Biology 209: 2554–2566.

Cajaraville, M. P., I. Olabarrieta & I. Marigomez, 1996. In vitro activities in mussel haemocytes as biomarkers of environmental quality: a case study in the Abra Estuary (Biscay Bay). Ecotoxicology and Environmental Safety 35: 253–260.

Chainho, P., J. L. Costa, M. L. Chaves, M. F. Lane, D. M. Dauer & M. J. Costa, 2006. Seasonal and spatial patterns of distribution of subtidal benthic invertebrate communities in the Mondego River, Portugal—a poikilohaline estuary. Hydrobiologia 555: 59–74.

Chollett, I. & D. Bone, 2007. Effects of heavy rainfall on polychaetes: differential spatial patterns generated by a large-scale disturbance. Journal of Experimental Marine Biology and Ecology 340: 113–125.

Currie, D. R. & K. J. Small, 2005. Macrobenthic community responses to long-term environmental change in an east Australian sub-tropical estuary. Estuarine, Coastal and Shelf Science 63: 315–331.

Davenport, J., 1979. The isolation response of mussels *Mytilus edulis* L. exposed to falling seawater concentrations. Journal of the Marine Biological Association of the United Kingdom 59: 123–132.

Davenport, J., 1982. Environmental simulation experiments on marine and estuarine animals. Advances in Marine Biology 19: 133–256.

Davis, S. E., J. E. Cable, D. L. Childers, C. Coronado-Molina, J. W. Day, C. D. Hittle, C. J. Madden, E. Reyes, D. Rudnick & F. Sklar, 2004. Importance of storm events in controlling ecosystem structure and function in a Florida Gulf Coast estuary. Journal of Coastal Research 20: 1198–1208.

de Bravo, M. I. S., 2003. Influence of salinity on the physiological conditions in mussels, *Perna perna* and *Perna viridis* (Bivalvia: Mytilidae). Revista de Biologia Tropical 51: 153–158.

Delaporte, M., P. Soudant, C. Lambert, J. Moal, S. Pouvreau & J. Samain, 2006. Impact of food availability on energy storage and defense related haemocyte parameters of the Pacific oyster *Crassostrea gigas* during an experimental reproductive cycle. Aquaculture 254: 571–582.

Fisher, W. S., 1988. Environmental influence on bivalve haemocyte function. American Fisheries Society Special Publications 18: 225–237.

Fisher, W. S., L. Oliver & P. Edwards, 1996. Hematological and serologic variability of Eastern oysters from Apalachicola Bay, Florida. Journal of Shellfish Research 15: 555–564.

Foster, G. D., E. C. Roberts Jr, B. Gruessnerb & V. J. Velinskyc, 2000. Hydrogeochemistry and transport of organic contaminants in an urban watershed of Chesapeake Bay (USA). Applied Geochemistry 15: 901–915.

Gagnaire, B., H. Frouin, K. Moreau, H. Thomas-Guyon & T. Renault, 2006. Effects of temperature and salinity on haemocyte activities of the Pacific oyster, *Crassostrea gigas* (Thunberg). Fish and Shellfish Immunology 20: 536–547.

Gardner, J. P. A. & R. J. Thompson, 2001. The effects of coastal and estuarine conditions on the physiology and survivorship of the mussels *Mytilus edulis*, *M-trossulus* and their hybrids. Journal of Experimental Marine Biology and Ecology 265: 119–140.

Glasby, T. M., 1997. Analysing data from post-impact studies using asymmetrical analyses of variance: a case study of epibiota on marinas. Australian Journal of Ecology 22: 448–459.

Hancock, P. J., 2006. The response of hyporheic invertebrate communities to a large flood in the Hunter River, New South Wales. Hydrobiologia 568: 255–262.

Kimmerer, W. J., 2002. Effects of freshwater flow on abundance of estuarine organisms: physical effects or trophic linkages? Marine Ecology Progress Series 243: 39–55.

Laing, I. & B. E. Spencer, 1997. Bivalve Cultivation: Criteria for Selecting a Site. Centre for Environmental, Fisheries and Aquaculture Science, Lowestoft, UK.

Lintas, C. & R. Seed, 1994. Spatial variation in the fauna associated with *Mytilus edulis* on a wave-exposed rocky shore. Journal of Molluscan Research 60: 165–174.

Martello, L. B., C. S. Friedman & R. S. Tjeerdema, 2000. Combined effects of pentachlorophenol and salinity stress on phagocytic and chemotactic function in two species of abalone. Aquatic Toxicology 49: 213–225.

Matthews, T. G. & A. J. Constable, 2004. Effect of flooding on estuarine bivalve populations near the mouth of the Hopkins River, Victoria, Australia. Journal of the Marine Biological Association of the United Kingdom 84: 633–639.

McNeish, A. S., M. S. Johnson & R. T. Leah, 1997. Methylated hexachlorocyclohexane in blue mussels (*Mytilus edulis*) from the Mersey and Dee estuaries. Environmental Pollution 95: 303–309.

Montagna, P. A. & R. D. Kalke, 1992. The effect of freshwater inflow on meiofaunal and macrofaunal populations in the Guadalupe and Nueces estuaries, Texas. Estuaries 15: 307–326.

Moore, M. N., J. I. Allen & P. J. Somerfield, 2006. Autophagy: role in surviving environmental stress. Marine Environmental Research 62, S420–S425.

Morrisey, D. J., A. J. Underwood, L. Howitt & J. S. Stark, 1992. Temporal variation in soft-sediment benthos. Journal of Experimental Marine Biology and Ecology 164: 233–245.

Mudge, S. M. & C. E. Norris, 1997. Lipid biomarkers in the Conwy Estuary (North Wales, U.K.): a comparison between fatty alcohols and sterols. Marine Chemistry 57: 61–84.

Newell, C. R., D. J. Wildish & B. A. MacDonald, 2001. The effects of velocity and seston concentration on the exhalant siphon area, valve gape and filtration rate of the mussel *Mytilus edulis*. Journal of Experimental Marine Biology and Ecology 262: 91–111.

Nicholson, S., 1999. Cytological and physiological biomarker responses from green mussels, *Perna viridis* (L.) transplanted to contaminated sites in Hong Kong coastal waters. Marine Pollution Bulletin 39: 261–268.

Pelc, M. & R. R. Alexander, 1999. Salinity and sediment-mediated byssal thread production by *Mytilus edulis* Linnaeus and Geukensia demissa Dillwyn from New Jersey salt marshes. Journal of Shellfish Research 18: 375–384.

Pfeifer, S., D. Schiedek & J. W. Dippner, 2005. Effect of temperature and salinity on acetylcholinesterase activity, a common pollution biomarker, in *Mytilus* sp. from the south-western Baltic Sea. Journal of Experimental Marine Biology and Ecology 320: 93–103.

Prins, T. C. & A. C. Smaal, 1994. The role of the blue mussel, *Mytilus edulis*, in the cycling of nutrients in the Ooesterschelde Estuary (the Netherlands). Hydrobiologia 283: 413–429.

Queiroz, N., F. P. Lima, P. A. Ribeiro, S. G. Pereira & A. M. Santos, 2006. Using asymmetrical designs for environmental impact assessment of unplanned disturbances. Hydrobiologia 555: 223–227.

Ritter, C., P. A. Montagna & S. Applebaum, 2005. Short-term succession dynamics of macrobenthos in a salinity-

stressed estuary. Journal of Experimental Marine Biology and Ecology 323: 57–69.
Roberts, D. E., A. Smith, P. Ajani & A. R. Davis, 1998. Rapid changes in encrusting marine assemblages exposed to anthropogenic point-source pollution: a 'Beyond BACI' approach. Marine Ecology Progress Series 163: 213–224.
Rutger, S. M. & S. R. Wing, 2006. Effects of freshwater input on shallow-water infaunal communities in Doubtful Sound, New Zealand. Marine Ecology Progress Series 314: 35–47.
Sadok, S., R. F. Uglow & S. J. Haswell, 1997. Haemolymph and mantle fluid ammonia and ninhydrin positive substances variations in salinity-challenged mussels (*Mytilus edulis* L.). Journal of Experimental Marine Biology and Ecology 212: 195–212.
Saier, B., 2001. Direct and indirect effects of seastars *Asterias rubens* on mussel beds (*Mytilus edulis*) in the Wadden Sea. Journal of Sea Research 46: 29–42.
Seed, R., 1996. Patterns of biodiversity in the macro-invertebrate fauna associated with mussel patches on rocky shores. Journal of the Marine Biological Association of the United Kingdom 76: 203–210.
Seed, R. & T. H. Suchanek, 1992. Population and community ecology of Mytilus. In Gosling, E. (ed.), The Mussel Mytilus: Ecology, Physiology, Genetics and Culture. Elsevier, Amsterdam: 589.
Silva, G., J. L. Costa, P. Raposo de Almeida & M. J. Costa, 2006. Structure and dynamics of a benthic invertebrate community in an intertidal area of the Tagus estuary, western Portugal: a six year data series. Hydrobiologia 555: 115–128.
Sukhotin, J., D. L. Lajus & P. A. Lesin, 2003. Influence of age and size on pumping activity and stress resistance in the marine bivalve *Mytilus edulis* L. Journal of Experimental Marine Biology and Ecology 284: 129–144.
Underwood, A. J., 1992. Beyond BACI: the detection of environmental impacts on populations in the real, but variable, world. Journal of Experimental Marine Biology and Ecology 161: 145–178.
Winter, J. E., 1978. A review on the knowledge of suspension-feeding in lamellibranchiate bivalves, with special reference to artificial aquaculture systems. Aquaculture 13: 1–33.
Witman, J. D. & K. R. Grange, 1998. Links between rain, salinity and predation in a rocky, subtidal community. Ecology 79: 2429–2447.
Zonta, R., F. Collavini, L. Zaggia & A. Zuliana, 2005. The effect of floods on the transport of suspended sediments and contaminants: a case study from the estuary of the Dese River (Venice Lagoon, Italy). Environment International 31: 948–958.

Hydrobiologia (2008) 606:129–137
DOI 10.1007/s10750-008-9340-1

Mobility of metals in salt marsh sediments colonised by *Spartina maritima* (Tagus estuary, Portugal)

Rosa Reboreda · Isabel Caçador · Sílvia Pedro · Pedro Raposo Almeida

© Springer Science+Business Media B.V. 2008

Abstract Chemical associations of Zn, Pb, Cu, Co and Cd were determined using a sequential extraction procedure in sediments colonised by *S. maritima* in three salt marshes within the Tagus estuary: Rosário, Corroios and Pancas. Concentrations of these metals were also analysed in above- and belowground parts of *Spartina maritima*, as well as in sediments colonised by the plant. The highest metal concentrations in sediments were found in the marshes near the industrial and urban areas, whereas metal concentrations in plants were not significantly different among sites. This was thought to be a consequence of differences observed in metal bioavailability: Metals in Pancas, the least polluted location, were largely associated to easily accessible fractions for plant uptake, probably as a result of low organic matter content and high sandy fraction in sediments. *S. maritima* was able to induce the concentration of metals between its roots in the three salt marshes. The results obtained in this study indicate that *S. maritima* could be useful to induce phytostabilisation of metals in sediments, although the effectiveness to modify chemical associations is highly dependent on existing sediment parameters, and thus different results could be obtained depending on site characteristics.

Keywords Bioavailability · Metals · Phytoimmobilisation · Salt marsh · *Spartina maritima*

Guest editors: J. Davenport, G. Burnell, T. Cross, M. Emmerson, R. McAllen, R. Ramsay & E. Rogan
Challenges to Marine Ecosystems

R. Reboreda · I. Caçador (✉) · S. Pedro · P. R. Almeida
Faculty of Sciences, Institute of Oceanography, University of Lisbon, Campo Grande, 1749-016 Lisbon, Portugal
e-mail: micacador@fc.ul.pt

Introduction

Salt marshes are complex coastal environments usually located within estuarine systems. Estuaries receive important inputs of pollutants as they are often situated in the vicinities of highly populated and industrialised areas. The release of heavy metals into the aquatic environment may cause detrimental effects to the receiving environment. Most metals entering into the aquatic system become associated with particulates and may accumulate in sediments.

Vascular plants in salt marshes are determinant to the dynamics of the estuarine ecosystem and strongly influence the processes of accumulation of heavy metals there (Alberts et al., 1990; Caçador et al., 1996a; Weis et al., 2002; Windham et al., 2003). Plant roots can alter physical and chemical properties of the sediment, influencing geochemical fractionation of metals and thus availability to the plant (Caçador et al., 1996b, 1999; Doyle & Otte, 1997; Lacerda et al., 1997; Otero & Macías, 2002; Mucha et al., 2005). Factors affecting bioavailability of metals to the marsh plants include pH, redox potential, salinity, particle size and organic matter content (Alloway, 1990; Williams et al., 1994a). The ability of some halophytes to alter these properties has drawn the interest for their potential use in phytostabilisation of metals in salt marsh sediments, as they may be able to reduce translocation to aboveground parts through immobilisation of pollutant metals in less bioavailable forms in the sediment (Weis & Weis, 2004). However, the characteristics of the sediment, where the plant is established, are expected to also have an important effect on metal fractionation, even though these are susceptible to modifications by the plant activity.

Studies in the Tagus estuary salt marshes have found that sediments colonised by *S. maritima* concentrate more metals than sediments colonised by other species like *H. portulacoides*, whereas its tissues show an opposite trend (Caçador et al., 2000; Reboreda & Caçador, 2007). Furthermore, speciation of Pb and Zn in sediments colonised by these two species has been found to be different, and such differences might be a result of both the type of colonising species and the location (own unpublished results). Studies made in several estuaries have shown that speciation of metals in salt marsh sediments depends greatly on local factors, such as sediment redox potential or grain size, in addition to metal nature or presence of roots (Allen et al., 1990; Caçador et al., 1996b; Mortimer & Rae, 2000; Otero et al., 2000; Sáenz et al., 2003; Almeida et al., 2004).

In the present work we analysed chemical associations (bioavailability) of Zn, Pb, Cu, Co and Cd using a sequential extraction procedure in sediments colonised by *S. maritima* in three salt marshes with different locations within the Tagus estuary. The aim was to evaluate the influence of site location within the estuary on metal concentration and bioavailability in sediments colonised by a single halophyte species, *Spartina maritima*, to better understand the limitations that local factors impose on the capacity of halophytes to modify metal speciation in salt marshes.

Materials and methods

Site description

The Tagus estuary is situated on the Atlantic coast of Portugal, covering an area of about 320 km^2 at extreme high tide. The southern and eastern parts of the estuary contain extensive intertidal mudflat areas colonised by halophyte plant communities dominated by *Spartina maritima* (Curtis) Fernald, *Halimione portulacoides* (L.) Aellen, *Sarcocornia fruticosa* (L.) A. J. Scott and *Sarcocornia perennis* (Mill.) A. J. Scott. Homogeneous stands of *S. maritima* colonise the muddy low marsh, whereas *H. portulacoides* occupies the banks of creeks. *S. fruticosa* and *S. perennis* colonise the high marsh. The study was carried out in three salt marshes: Pancas, located within a nature Reserve; Rosário and Corroios, both situated nearby anthropogenic sources of pollution (Fig. 1). These sites experience two tidal flushing a day.

Sampling

Five replicates of *Spartina maritima* were collected in early April 2006 at each of the three salt marshes in the Tagus estuary (Pancas, Rosário and Corroios). The samples were collected at low tide at areas with pure stands with a lump of soil, using a spade (ca. 20 × 20 × 15 cm^3). Five samples of the non-vegetated sediment in each salt marsh were also collected.

Redox potential (Eh) and pH of sediments between the roots and without vegetation were measured immediately on arrival to the laboratory, using a portable HI 9025 pH/Eh meter with an HI 1230B electrode. Values were accepted after a 15-min period of stabilisation. Organic matter was obtained by loss on ignition (LOI) in a muffle furnace (550°C, 2 h). Grain size distribution was determined by sieve analysis.

The aboveground material was clipped from the sediment lump, washed with distilled water to remove sediment particles and dried to constant weight at 80°C. Plant material was subsequently homogenised. The remaining sediment lump was divided into two parts. From one of them belowground material was sorted out by washing under a flux of water. Roots were separated, rinsed in demineralised water, dried to constant weight at 80°C and homogenised. The second piece of the sediment was air dried and the layer between 10 and 15 cm depth retained. This sub-sample was cleaned off roots with tweezers, homogenised by grinding with mortar and pestle and sieved through a 1,000-μm nylon sieve to avoid any contamination of the sample.

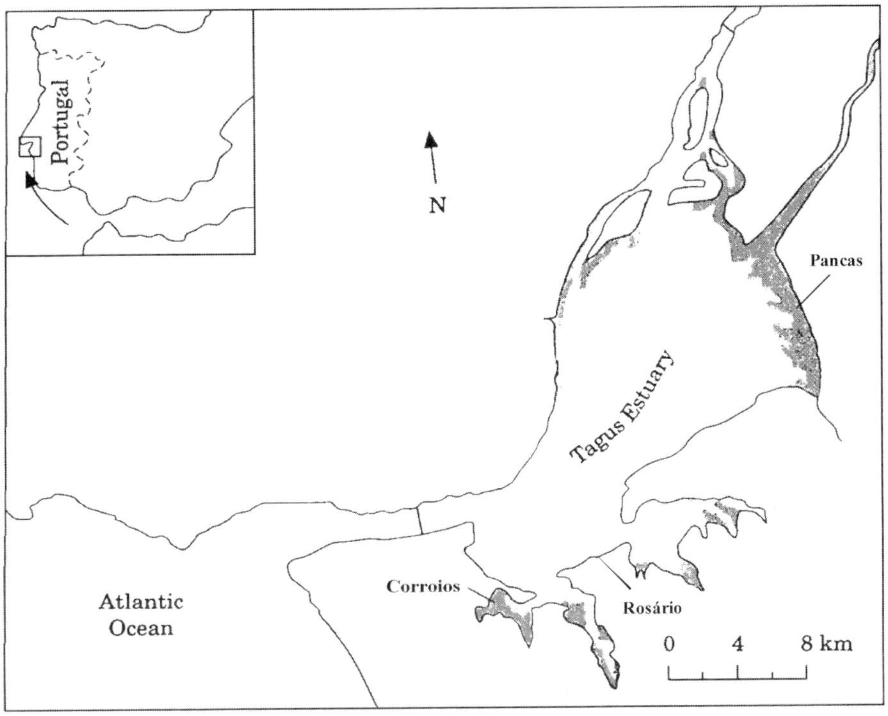

Fig. 1 Location of Corroios, Rosário and Pancas in the Tagus estuary (Portugal)

Sediment samples without vegetation followed the same procedure.

Analysis of metals

To avoid external metal contamination, all sampling and lab ware materials were soaked in 5% (v/v) HCl solution for at least 24 h and rinsed several times with bideionised water. Reagents used were pro-analysis grade or equivalent.

A composite sample of the sediment was obtained from each salt marsh by combining the five replicates of sediments colonised by *S. maritima*. Composite samples were analysed by sequential extraction to estimate the metal availability of Cd, Co, Cu, Pb and Zn in these two types of the sediment. Dry homogenised sediment (1 g) was sequentially treated with the following solutions: (1) 40 ml of 0.11 M CH_3COOH (exchangeable fraction), (2) 40 ml of 0.5 M $NH_2OH{\cdot}HCl$ (Fe and Mn oxy-hydroxides bound fraction), (3) 10 ml of H_2O_2 30% and subsequently 10 ml NH_4CH_3COO 1 M (fraction bound to organic matter and sulphide), following the sequential extraction procedure established by the Measuring and Testing Program of the European Commission (Rauret et al., 1999). The remaining sediment was digested with 10 ml of HNO_3/HCl (3:1 v/v) at 110°C for 3 h to calculate the residual fraction. The sum of the metal extracted in steps 1, 2, 3 and the residue was compared with that obtained by extraction with 10 ml of HNO_3/HCl (3:1 v/v) of a separate 1 g sample of the sediment to check for losses during the sequential extraction.

For the analysis of Cd, Co, Cu, Pb and Zn in aboveground and belowground plant material, dry homogenised plant material (100 mg) was digested by adding 2 ml $HNO_3/HClO_4$ (7:1, v/v) in a reactor and heated at 110°C for 6 h.

Concentrations of metals in the solutions were determined by graphite furnace atomic absorption spectrometry. The accuracy of this analytical method was assessed by the analysis of international certificate standards for the sediment (CRM 145, CRM 146) and the vegetal material (BCR 62). Trace metal concentrations obtained were not statistically different from their certified ones (*t*-Student; $\alpha = 0.05$).

Statistics

Statistical differences in metal concentration in plant parts were determined using two-way analysis of variance (ANOVA) (significance 0.05) considering

Fig. 2 Concentrations (mean ± SD) of Zn, Pb, Cu, Co and Cd ▶ (µg g^{-1} dry weight) in belowground (roots + rhizomes) and aboveground parts of *Spartina maritima* in Corroios, Rosário and Pancas. Significant differences ($P < 0.05$) between plant parts are indicated with *, and letters indicate significant differences between salt marshes (two-way ANOVA)

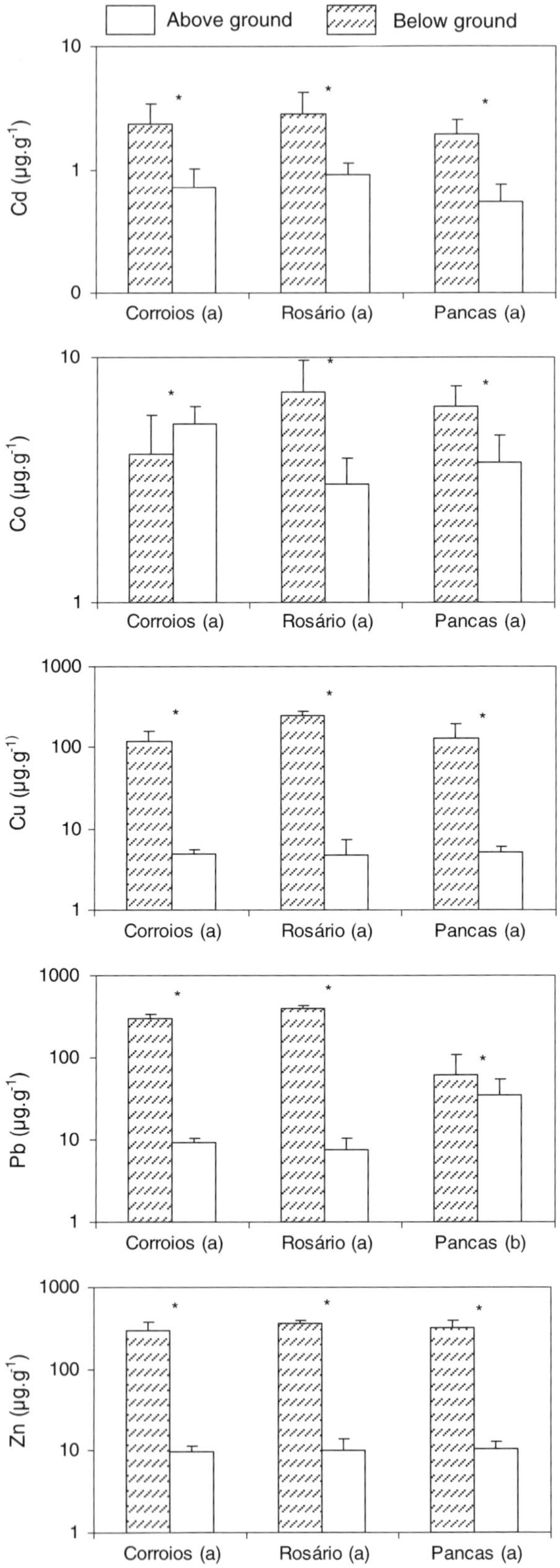

plant part (above or below ground) and site as factors of variation. The same analysis was performed to compare metal concentration in vegetated and non-vegetated sediments among salt marshes. Subsequent Tukey HSD test was used when significant differences between groups were found.

Results

Aboveground and belowground parts of *S. maritima*

Figure 2 shows concentrations of Cd, Co, Cu, Pb and Zn in aboveground and belowground parts of *S. maritima* in the three salt marshes studied. The highest metal concentration in plants was found in the root system, all metals tested showing significant differences ($P < 0.05$) between aboveground and belowground parts. For Zn, Pb and Cu the difference was up to one or two orders of magnitude higher. Only Co presented slightly higher concentration in aboveground parts than in roots at Corroios. Differences among salt marshes were not statistically significant, except for Pb ($P < 0.05$). The concentration of this metal was significantly smaller in belowground parts of *S. maritima* growing in Pancas than in Corroios and Rosário salt marshes, whereas the opposite occurred in above ground parts, i.e., higher concentrations were observed in Pancas than in Corroios and Rosário ($P < 0.05$; Tukey HSD). Cd displayed the lowest concentration values, while Pb and Zn were the most concentrated metals.

Sediment

Redox potential (Eh) was not significantly different at Corroios and Pancas ($P < 0.05$, Tukey HSD), both being significantly different from those at Rosário ($P > 0.05$, Tukey HSD) (Table 1). Organic matter content (LOI) was similar at Corroios and Rosário ($P > 0.05$), but both showed significantly higher

Fig. 3 Concentrations (mean ± SD) of Zn, Pb, Cu, Co and Cd ▶ (μg g^{-1} dry weight) in sediments colonised by *Spartina maritima* and unvegetated sediments in Corroios, Rosário and Pancas. Significant differences ($P < 0.05$) between colonised and unvegetated sediments are indicated with *, and letters indicate significant differences between sites (two-way ANOVA)

values than those at Pancas ($P < 0.05$, Tukey HSD). Values of pH were similar at the three salt marshes ($P > 0.05$) (Table 1). Pancas had significantly higher percentage of sandy sediment than Rosário and Corroios, which did not show significant differences in grain size distribution between them ($P < 0.05$, Tukey HSD) (Table 1).

All metals studied showed higher concentration in sediments between the roots of *S. maritima* than in those without vegetation ($P < 0.05$) (Fig. 3). Concentrations of Cd, Co, Cu, Pb and Zn differed significantly among Corroios, Pancas and Rosário, both in vegetated and non-vegetated sediments (two-way ANOVA; $P < 0.05$). Subsequent post-hoc Tukey HSD test analysis revealed that Zn and Co concentrations were different among all three salt marshes, whereas Cd, Cu and Pb were similar in Rosário and Corroios, both different from Pancas, which showed the lowest concentrations. Concentrations of the studied metals followed the order Zn > Co > Pb > Cu > Cd for every site and type of sediments (with and without vegetation).

Availability of the studied metals in sediments between the roots of *S. maritima* at Rosário, Corroios and Pancas is presented in Fig. 4. The exchangeable fraction was the most available for plant uptake and the residual fraction was the most strongly bound to sediments, i.e., less available. The chemical

Table 1 Characteristics of sediments colonised by *Spartina maritima* in Corroios, Pancas and Rosário salt marshes

	Corroios	Rosário	Pancas
Eh (mV)	14.2 ± 2.2	26.7 ± 8.3	12.6 ± 4.5
pH (18°C)	7.1 ± 0.1	6.9 ± 0.2	7.0 ± 1.0
L.O.I. (%)	40.3 ± 3.0	45.6 ± 8.6	5.3 ± 3.0
Grain size distribution (%)			
Sand (63–2,000 μm)	83.1 ± 3.6	83.1 ± 1.1	94.9 ± 2.3
Silt and clay (<63 μm)	17.8 ± 2.9	16.4 ± 1.8	5.2 ± 3.4

Values from the vegetated sediments correspond to mean ± SD ($n = 3$)

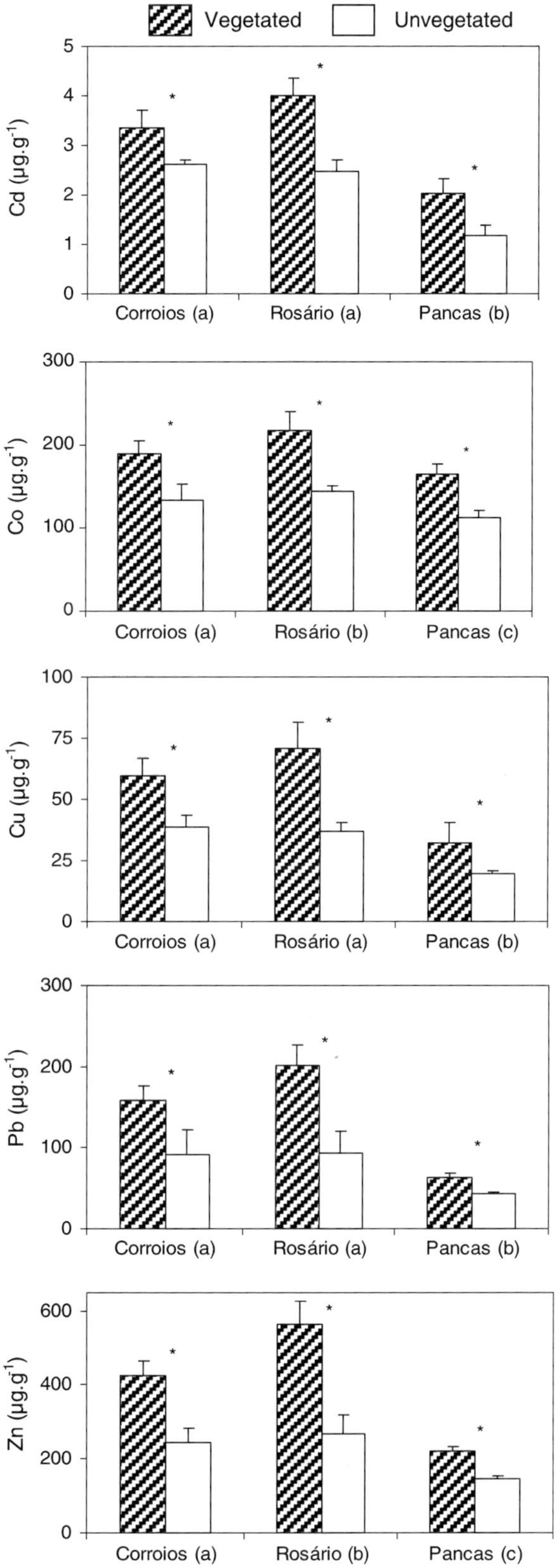

Fig. 4 Chemical associations of Zn, Pb, Cu, Co and Cd in sediments colonised by *Spartina maritima* in Corroios, Rosário and Pancas

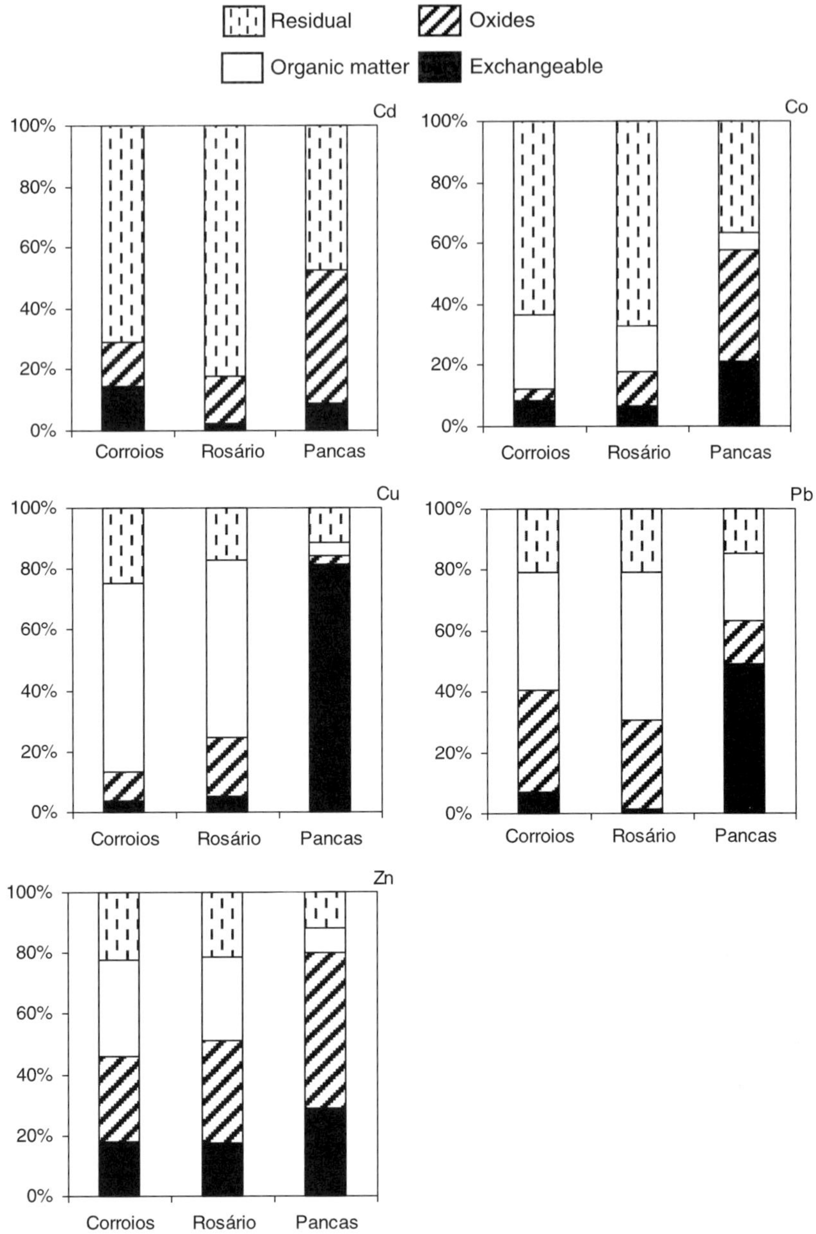

associations of Zn, Pb, Cu, Co and Cd varied with metal nature and site. Cd and Co showed the highest associations to residual fraction, so they were presumably less bioavailable than the other metals. Zn was mainly bound to exchangeable (easily accessible for plant uptake) and Fe/Mn oxide fraction. The association to organic matter/sulphide fraction was high for Cu and Pb, except in Pancas, where the presence of these metals in the exchangeable fraction was dominant. For all five metals a trend was observed for similar fractionation in Rosário and Corroios, whereas speciation in Pancas was noticeably different. In general, the percentage of the metal bound to the most available fraction (exchangeable) and the Fe/Mn fraction (potentially released by reduction) was higher in Pancas than in Rosário or Corroios, while association to fractions strongly bound to sediments (residual and organic matter/sulphide) was smaller in Pancas than in the other two salt marshes.

Table 2 Enrichment factors ([metal in roots]/[metal in sediments]) of Zn, Pb, Cu, Co and Cd in belowground parts of *Spartina maritima* in Corroios, Pancas and Rosário

	Corroios	Rosário	Pancas
Enrichment factor			
Cd	0.69 ± 0.29	0.68 ± 0.31	0.98 ± 0.35
Co	0.02 ± 0.01	0.03 ± 0.01	0.03 ± 0.01
Cu	1.99 ± 0.78	3.54 ± 0.84	4.22 ± 2.50
Pb	1.92 ± 0.47	1.97 ± 0.21	0.99 ± 0.72
Zn	0.69 ± 0.11	0.64 ± 0.09	1.47 ± 0.32

Enrichment factors were calculated to estimate the capability of *S. maritima* roots to accumulate these metals (Table 2). It was concluded that belowground parts were able to accumulate Cu at the three sites, Pb at Rosário and Corroios, and Zn at Pancas, whereas Co and Cd were not accumulated.

Discussion

The higher concentration of metals in belowground parts of the plant is in agreement with the literature about halophytes (Reboredo, 1993; Williams et al. 1994b; Caçador & Vale, 1997; Padinha et al., 2000; Reboreda & Caçador, 2007). Usually small quantities are translocated to stems and leaves, except in the case of more mobile elements as Mn, Cd and Zn (Reboredo, 1992; Williams et al., 1994b). In this study Cd and Co, the least concentrated metals in plant tissues, showed higher translocation to aboveground parts than the other metals for all three salt marshes. Although Zn is considered to be a mobile element, it concentrated mainly in the root system of *S. maritima* (Fig. 2). Metal analysis in *S. maritima* plants from Ria Formosa, in southern Portugal, also showed high translocation of Cd to aboveground parts and the majority of Zn was found in belowground parts (Padinha et al., 2000).

The lower concentrations of Cd, Co, Cu, and Zn found in sediments of Pancas were not reflected in lower concentrations in plants, as differences in concentrations of these metals among plants growing in different sites were not significant. This was presumably a consequence of the observed metal speciation in sediments of the three salt marshes. Metals in Pancas, still showing lower concentrations in sediments, were more associated to the bioavailable fraction (exchangeable) than in Rosário and Corroios. Thus, they were more available for plant uptake, resulting in plant concentrations similar to salt marshes with higher metal content in sediments. This was particularly clear for Cu. However, Pb showed a different behaviour. Belowground concentrations of this metal in plants were significantly lower in Pancas than in Corroios or Rosário, and the opposite occurred for aboveground parts. Rozema et al. (1986) found that plants translocate higher concentrations of metals to aboveground parts when growing in sediments where concentrations of these metals are low, and tend to retain them in the roots when concentrations increase, as a defence system for the plant. This might be the case for Pb in Pancas. Padinha et al. (2000) also found that metal concentrations within *S. maritima* did not show the spatial pattern showed by metal concentrations in sediments from different sites.

Cd and Co presented noticeably lower concentration in plants than the other metals, especially in belowground parts, and this seemed to be related to the strong association of these metals to the residual fraction, i.e., they were less available for plant uptake. On the other hand, Zn associated to the sum of exchangeable (easily accessible) and Fe/Mn oxides (potentially released by reduction) fractions was high, and thus it was the most concentrated metal in plants.

The described differences in metal availability between salt marshes seemed to be related to sediment properties in the three locations. In particular, grain size distribution and organic matter content appeared to play an important role. Both were similar between Rosário and Corroios, but different from Pancas (Table 1); metal fractionation followed the same trend (Fig. 4). Pancas contained low levels of organic matter and high percentage of sandy sediments. The capacity of this type of sediments to retain metal ions is poor (Williams et al., 1994a). Corroios and Rosário contained organic-matter-rich sediments with high silt and clay fractions. This type of sediment is able to bind metal ions by means of physical/chemical processes as adsorption, chelation and ion exchange (Williams et al., 1994a).

Rosário and Corroios are closer to anthropogenic sources of pollution than Pancas, which is included within the Tagus estuary natural reserve, and thus receive higher inputs of metals. This fact most likely explains the higher Cd, Co, Cu, Pb and Zn concentrations presented in these salt marshes. The presence of *S. maritima* plants also influenced the concentrations of metals in sediments, which were higher between the roots than in non-vegetated sediments (Fig. 3). These findings are in agreement with Caçador et al. (1996a, b) for the Tagus estuary. Higher sedimentation rates in areas colonised by halophytes compared to unvegetated areas are likely to be related to the observed pattern in metal concentration (1 and 0.6 cm year^{-1}, respectively, at Corroios, according to Caçador et al. (2004)). In addition, activity of roots liberates oxygen and organic matter to sediments, and thus it is able to modify characteristics as Eh, pH and microbial activity which will eventually modify metal retention capacity (Williams et al., 1994a; Doyle & Otte, 1997; Kamnev & van der Lelie, 2000; Almeida et al., 2004). *Spartina maritima* has a well-developed arenchyma and therefore presents the ability to oxidise the rhizosphere, being able to colonise anoxic environments. This activity can either mobilise or immobilise metals, depending on the existing physical/chemical properties in the salt marsh sediment (Jacob & Otte, 2003).

Conclusions

From the observed results it was concluded that *S. maritima* was able to induce higher concentrations of metals in sediments between the roots than unvegetated sediments even in low metal-polluted salt marshes, and hence it could be useful for phytostabilisation purposes (Weis & Weis, 2004). Previous results from this group estimated that belowground retention (roots + rhizosediments) of Cu, Cd and Pb in sediments colonised by *S. maritima* was 113.1, 6.8 and 681.9 Tm, respectively, for the whole Tagus estuary (Reboreda & Caçador, 2007). On the other hand, it was evident that the fractionation of metals in sediments between the roots of *S. maritima* was highly dependent on site location, driven by differences in sediment parameters. Enrichment factors also showed that the ability of belowground parts to accumulate certain metals depended on the salt marsh considered. Therefore, the effectiveness of the halophyte species to immobilise metals in salt marsh sediments is limited by local factors as grain size distribution and organic matter content, and this should be taken into account when planning phytoremediation actions.

Acknowledgement Thanks to 'Fundação para a Ciência e a Tecnologia' for funding this research.

References

Alberts, J. J., M. T. Price & M. Kania, 1990. Metal concentrations in tissues of *Spartina alterniflora* (Loisel.) and sediments of Georgia salt marshes. Estuarine Coastal and Shelf Science 30: 47–58.

Allen, J. R. L., J. E. Rae & P. E. Zanin, 1990. Metal speciation (Cu, Zn, Pb) and organic matter in an oxic salt marsh, Severn Estuary, southwest Britain. Marine Pollution Bulletin 21: 574–580.

Alloway, B. J., 1990. Soil processes and the behaviour of metals. In Alloway, B. J. (ed.), Heavy Metals in the Soils. John Willey and Sons, Glasgow, 7–27.

Almeida, C. M., A. P. Mucha & M. T. Vasconcelos, 2004. Influence of the sea rush *Juncus maritimus* on metal concentration and speciation in estuarine sediment colonized by the plant. Environmental Science and Technology 38: 3112–3118.

Caçador, I., A. L. Costa & C. Vale, 2004. Carbon storage in Tagus salt marsh sediments. Water, Air and Soil Pollution: Focus 4: 701–714.

Caçador, I. & C. Vale, 1997. The role of halophytes in the Tagus Estuary environmental quality. In Lieth, H., A. Hamdy & H.-W. Koyro (eds), Water Management, Salinity and Pollution Control Towards Sustainable Irrigation in the Mediterranean Region. Valenzano-Bari, Italy: 85–99.

Caçador, I., C. Vale & F. Catarino, 1996a. Accumulation of Zn, Pb, Cu and Ni in sediments between roots of the Tagus estuary salt marshes, Portugal. Estuarine Coastal and Shelf Science 42: 393–403.

Caçador, I., C. Vale & F. Catarino, 1996b. The influence of plants on concentration and fractionation of Zn, Pb, and Cu in salt marsh sediments (Tagus Estuary, Portugal). Journal of Aquatic Ecosystem Health 5: 193–198.

Caçador, I., C. Vale & F. Catarino, 1999. Relationships between metal concentrations and organic matter content in the Tagus estuary salt marsh sediments. In Lieth, H., M. Moschenko, M. Lohmann, H.-W. Koyro & A. Hamdy (eds), Progress in Biometeorology (Volume 13): Halophyte Uses in Different Climates I. Backhuys Publishers, Leiden, The Netherlands: 105–112.

Caçador, I., C. Vale & F. Catarino, 2000. Seasonal variation of Zn, Pb, Cu and Cd concentrations in the roots-sediment system of *Spartina maritima* and *Halimione portulacoides* from Tagus estuary salt marshes. Marine Environmental Research 49: 279–290.

Doyle, M. O. & M. L. Otte, 1997. Organism-induced accumulation of Fe, Zn and AS in wetland soils. Environmental Pollution 96: 1–11.

Jacob, D. L. & M. L. Otte, 2003. Conflicting processes in the wetland plant rhizosphere: Metal retention or mobilization? Water, Air and Soil Pollution 3: 91–104.

Kamnev, A. A. & D. van der Lelie, 2000. Chemical and biological parameters as tools to evaluate and improve heavy metal phytoremediation. Bioscience Reports 20: 239–258.

Lacerda, L. D., J. L. Freixo & S. M. Coelho, 1997. The effect of *Spartina alterniflora* Loisel on trace metals accumulation in inter-tidal sediments. Mangroves and Salt Marshes 1: 201–209.

Mortimer, R. J. G. & J. E. Rae, 2000. Metal speciation (Cu, Zn, Pb, Cd) and organic matter in oxic to suboxic salt marsh sediments, Severn Estuary, Southwest Britain. Marine Pollution Bulletin 40: 377–386.

Mucha, A. P., C. M. Almeida, A. A. Bordalo & M. T. Vasconcelos, 2005. Exudation of organic acids by a marsh plant and implications on trace metal availability in the rhizosphere of estuarine sediments. Estuarine, Coastal and Shelf Science 65: 191–198.

Otero, X. L., M. A. Huerta-Diaz & F. Macías, 2000. Heavy metal geochemistry of salt marsh soils from Ría of Ortigueira (mafic and ultramafic areas, NW Iberian Peninsula). Environmental Pollution 110: 285–296.

Otero, X. L. & F. Macías, 2002. Spatial and seasonal variation in heavy metals in interstitial water of salt marsh soils. Environmental Pollution 120: 183–190.

Padinha, C., R. Santos & M. T. Brown, 2000. Evaluating environmental contamination in Ria Formosa (Portugal) using stress indexes of *Spartina maritima*. Marine Environmental Research 49: 67–78.

Rauret, G., J. F. López-Sánchez, A. Sahuquillo, R. Rubio, C. Davidson, A. Ure & Ph. Quevauviller, 1999. Improvement of the BCR three-step sequential extraction procedure prior to the certification of new sediment and soil reference materials. Journal of Environmental Monitoring 1: 57–61.

Reboredo, F., 1992. How differences in the field influence Cu, Fe and Zn uptake by *Halimione portulacoides* and *Spartina maritima*. The Science of the Total Environment 133: 11–132.

Reboredo, F., 1993. How differences in the field influence Cu, Fe and Zn uptake by *Halimione portulacoides* and *Spartina maritima*. The Science of The Total Environment 133: 111–132.

Reboreda, R. & I. Caçador, 2007. Halophyte vegetation influences in salt marsh retention capacity for heavy metals. Environmental Pollution 146: 147–154.

Rozema J., M. L. Otte, R. A. Broekman & J. M. Wezenbeek, 1986. The uptake and translocation of heavy metals by salt marsh plants from contaminated estuarine salt marsh sediment: Possibilities for bioindication. In Proceedings of the International Conference on Environmental pollution, Amsterdam. CEP Consultants, Edinburgh (UK), 123–125.

Sáenz, V., J. Blasco & A. Gómez-Parra, 2003. Speciation of heavy metals in recent sediments of three coastal ecosystems in the Gulf of Cádiz, Southwest Iberian Peninsula. Environmental Toxicology and Chemistry 22: 2833–2839.

Weis, J. S. & P. Weis, 2004. Metal uptake, transport and release by wetland plants: Implications for phytoremediation and restoration. Environment International 30: 685–700.

Weis, P., L. Windham, D. J. Burke & J. S. Weis, 2002. Release into the environment of metals by two vascular salt marsh plants. Marine Environmental Research 54: 325–329.

Williams, T. P., J. M. Bubb & J. N. Lester, 1994a. Metal accumulation within salt marsh environments: A review. Marine Pollution Bulletin 28: 277–290.

Williams, T. P., J. M. Bubb & J. N. Lester, 1994b. The occurrence and distribution of trace metals in halophytes. Chemosphere 28: 1189–1199.

Windham, L., J. S. Weis & P. Weis, 2003. Uptake and distribution of metals in two dominant salt marsh macrophytes, *Spartina alterniflora* (cordgrass) and *Phragmites australis* (common reed). Estuarine, Coastal and Shelf Science 56: 63–72.

Hydrobiologia (2008) 606:139–151
DOI 10.1007/s10750-008-9338-8

Ecological hindcasting of biogeographic responses to climate change in the European intertidal zone

David S. Wethey · Sarah A. Woodin

© Springer Science+Business Media B.V. 2008

Abstract Intertidal organisms are often assumed to live close to their thermal limits, and have emerged as potential early indicators of the effects of climate change. We compared our survey of the 2004–2006 geographic distribution of the barnacle *Semibalanus balanoides* to its distribution in 1872, 1955, 1963, 1971, and 1985, from surveys by Fischer, Crisp, Fischer-Piette, Barnes, Powell, and Southward. The southern geographic limit has retreated 300 km in France since 1872, at a rate of 15 to 50 km per decade. We compared our 2006 survey of the geographic distribution of the polychaete *Diopatra neapolitana* to its distribution in 1893–1923, from surveys by Saint-Joseph and Fauvel, and its distribution in 1969–1976 from surveys by Glémarec. The northern geographic limit of this species has advanced 300 km in France since 1893 at similar rates to *Semibalanus*. We used NOAA weather reanalysis data and our mechanistic simulation model of intertidal animal body temperatures to hindcast the thermal environmental change near historical geographic limits in Europe for the past 55 years. Results indicate that changes in the southern limit of *S. balanoides* are due to intolerance of winter body temperatures above 10°C, leading to reproductive failure. Results for *Diopatra* are ambiguous: based on the northern extension of its range, either cold winters or cool summers limit its range, while gaps in its distribution are consistent with limitation by cooler summer conditions. The parallel shifts of *D. neapolitana* on sedimentary shores and *Semibalanus* on rocky shores suggest that similar climatic factors control the geographic limits of both species. The intertidal zone is a model system for examining the effects of climate change on biogeographic change both because of the rapidity of its response, and because the rich historical record allows direct tests of hypotheses.

Keywords Biogeography · Climate change · Barnacle · Polychaete · *Semibalanus balanoides* · *Diopatra neapolitana*

Guest editors: J. Davenport, G. Burnell, T. Cross, M. Emmerson, R. McAllen, R. Ramsay & E. Rogan
Challenges to Marine Ecosystems

D. S. Wethey (✉) · S. A. Woodin
Department of Biological Sciences, University of South Carolina, 715 Sumter St., Columbia, SC 29208, USA
e-mail: wethey@biol.sc.edu

Introduction

The European littoral zone presents an excellent opportunity to study the effects of climate change on the geographic distribution of organisms, because of the rich historical record of species distributions in Iberia, France, and the British Isles during the late nineteenth and twentieth centuries (e.g., Lafont,

1871; Fischer, 1872; Saint-Joseph, 1898; Fauvel, 1923; Fischer-Piette, 1932; Evans, 1954; Fischer-Piette, 1955; Crisp & Southward, 1958; Southward & Crisp, 1954; Crisp & Fischer-Piette, 1959; Lewis, 1964; Barnes & Barnes, 1966; Glémarec, 1979; Southward et al., 1995, 2005; Simkanin et al., 2005; Mieszkowska et al., 2006). In addition, the oceanographic context of Europe provides a complex mosaic of climatic zones from Portugal to the English Channel. Here, we examine the climate change in relation to the present and past distributions of the intertidal barnacle *Semibalanus balanoides* (L.) and the infaunal polychaete *Diopatra neapolitana* Delle Chiaje 1841. *Semibalanus* is a northern species with its southern limits in France and Spain, and *Diopatra* is a Mediterranean species with its northern limit in France.

Numerous hypotheses have been proposed to explain the geographic limits of species. Hutchins (1947) organized several of these hypotheses into the general categories of factors that limit adult distribution and factors that limit recruitment. He proposed, for example, that equatorward limits could be set by high summer temperatures causing adult mortality, or by winter temperatures being so warm as to limit reproduction. For *Semibalanus*, the summer thermal extreme hypothesis has been elaborated by such authors as Southward (1958) and Foster (1969) who identified the heat coma (37°C) and thermal death (42°C) limits for the species. The warm winter hypothesis has support in that temperatures above 10°C have been shown in the laboratory to inhibit reproduction (Barnes, 1958; Crisp & Patel, 1969; Barnes & Stone, 1972). More recently Davenport et al. (2005) proposed that day length might set the southern geographic limit of *Semibalanus*.

Here we test the summer heat death, day length, and winter cold limitation of reproduction hypotheses by examining the changes in the southern geographic limit of the species since 1872 in relation to climate change and to the geographic and temporal changes in day length.

For *Diopatra*, the northern limit is likely to be set either by cold winter temperatures or by summer temperatures too cold for reproductive success (Hutchins, 1947; Glémarec, 1979). In contrast to *Semibalanus*, little is known of its physiological limitations. However, extensive data exist for its congener *Diopatra cuprea*, which is common on the shores of the western Atlantic and has a very similar life habit. Both are large tube-building worms that construct a 30–50 cm tube in sediments with a tube-cap decorated with debris that extends above the sediment surface. Both species sit near the mouth of the tube and extend outward to feed. Nothing is known about *D. cuprea* reproductive failure relative to temperature. However, its northern limit is immediately north of Cape Cod (Mangum et al., 1968), reflecting the species' behavioral limitations in cold temperature conditions: *D. cuprea* does not feed below 5°C (Mangum & Cox, 1971) and does not build new tubes or repair tubes below 1.8°C (Myers, 1972). Low-temperature inhibition of tube building will be limiting in sedimentary habitats with cold winters where storm-driven erosion and deposition are common, necessitating constant tube maintenance. Given the similarity between the two species of *Diopatra*, we expected *D. neapolitana* to have similar behavioral limitations during cold periods. An alternative hypothesis is that geographic distribution is limited by cool summer conditions acting on adults or juveniles at the northern geographic limit (e.g., Hutchins, 1947). Here we examine the distribution of *D. neapolitana* to ask whether both the location of the northern limit as well as gaps within its distribution are consistent with the winter cold limitation hypothesis or the cool summer hypothesis, or both.

Biophysical context

Intertidal organisms are exposed to terrestrial conditions at low tide and to oceanic conditions at high tide, therefore, they experience enormous thermal extremes over the tidal cycle. At low tide their body temperatures can rise 10°C or more above sea water temperature (e.g., Southward, 1958; Vermeij, 1971; Etter, 1988; Helmuth, 1999; Wethey, 2002). At high tide they equilibrate rapidly to seawater temperature. Therefore, one should not assume that either air temperature or sea surface temperature is a good proxy for body temperature. We use a validated biophysical model of the transient dynamics of rocky-shore intertidal animal body temperature (Gilman et al., 2006) to hindcast the conditions experienced by rocky-shore intertidal animals over the past 50 years. Meteorological variables (air temperature, wind speed, humidity, cloud cover, solar radiation, and sea surface temperature) were obtained from the

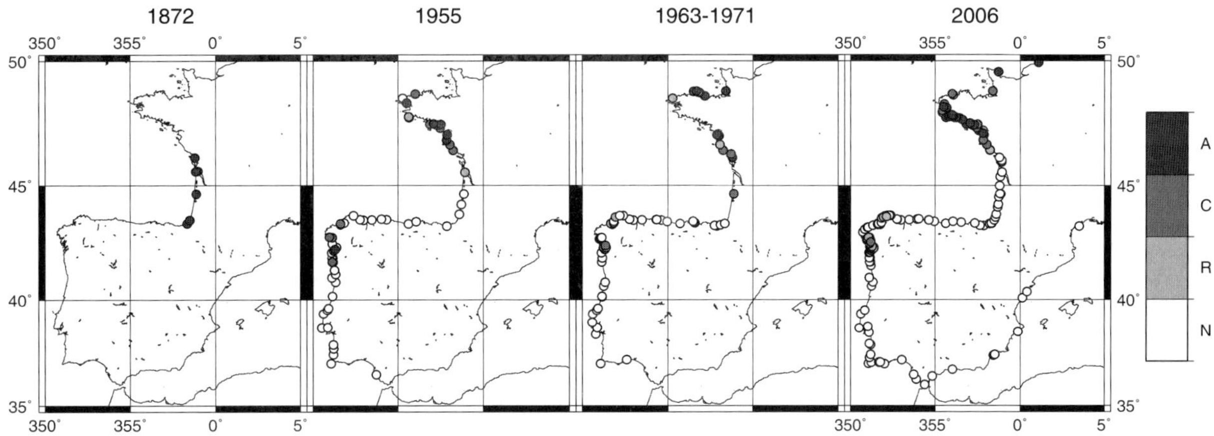

Fig. 1 Distribution of *Semibalanus balanoides* in 1872 (Fischer, 1872), 1955 (Crisp & Fischer-Piette, 1959), 1963 (Barnes & Barnes, 1966; Barnes & Powell, 1966), 1971 (Barnes et al., 1972) and 2006 (this study). Density Scale: A = Abundant, C = Common, R = Rare, N = Absent (ACFOR density scale: Crisp & Southward 1958)

NOAA global reanalysis project which estimated these variables on a worldwide 2° grid every 6 h back to 1948 (Kalnay et al., 1996). We used the reanalysis variables as inputs to our biophysical model, and calculated the average daily maximum and minimum temperature experienced by rocky-shore intertidal animals each month at sites from southern Portugal to northern France, encompassing the geographic limits of our species of interest.

Oceanographic and biogeographic context

Numerous authors beginning with Savageau (1897) have remarked that the shores and biota of northwest Spain have more in common with the shores and biota of northwest France than they do with the shores in between. This occurs because of strong upwelling in Galicia (NW Spain) and Brittany (NW France) which generates a cool oceanic and terrestrial climate. The shores in between the zones of upwelling on the northern Spanish coast and southern French Biscay coast are more like Morocco than other parts of Europe (Fischer-Piette, 1955). Floristic and faunistic surveys in the mid nineteenth to mid twentieth centuries confirmed the temperate characteristics of Galicia which contrast strongly with the rest of Iberia (e.g., Savageau, 1897; Fischer-Piette, 1955). For example, the northern algae *Fucus serratus* and *Ascophyllum nodosum* disappear in southern France and northern Spain, but reappear in Galicia and disappear again in Portugal (Fischer-Piette, 1955; Crisp & Fischer-Piette, 1959; Margalet et al., 1992, 1993).

Biogeographic surveys

The geographic distribution of *S. balanoides* (Fig. 1) was reported by Fischer (1872), Fischer-Piette (1955), Fischer-Piette & Prenant (1956, 1957), Crisp & Fischer-Piette (1959), Barnes & Barnes (1966), Barnes & Powell (1966), and Barnes et al., (1972). The geographic distribution of *D. neapolitana* (Fig. 4) was reported by Lafont (1871), Saint-Joseph (1898), Fauvel (1923), Faure (1969), Lagardère (1972a, b), Glémarec (1979) and Montaudouin et al. (2000). We resurveyed the Iberian and French coasts from Tarifa to Brittany in 2005 and 2006 at a spatial scale of 50–100 km (Figs. 1, 4). We used the geographic limits from these surveys to test the geographic hypotheses outlined in the Introduction.

Climate reconstructions

In order to hindcast the body temperature fluctuations experienced by rocky-shore intertidal animals since 1950, we used our validated biophysical model of intertidal body temperature (Helmuth, 1999; Wethey, 2002; Gilman et al., 2006). The model predictions are within 3°C of data logger averages, and 95% of predictions are within the range of datalogger observations over the US west coast from southern California to the Canadian border, for the period 2000–2004 (Gilman et al., 2006). These results indicate that the model is as good at predicting animal body temperature on rocky shores as the data loggers are at measuring it. At low tide, the model calculates the surface temperature of the organism and bare substratum based on a balance between short-wave solar radiation, convection, conduction to the substratum, evaporation, and long-wave radiation. Body temperature in the core of the animal and in the substratum is calculated by a finite-difference approximation to the thermal-conduction process. At high tide, the organismal body temperature is set to sea surface temperature. The model calculates all conditions every 30 s, so the daily cycles of heating and cooling are represented (Wethey, 2002). Historical tide levels were calculated using X-Tide (Flater, 2006). Meteorological variables (solar radiation, wind speed, air temperature, relative humidity, cloud cover, and sea surface temperature) were obtained from the global reanalysis carried out by the US National Centers for Environmental Prediction (Kalnay, 1996). The reanalysis data are available on a 2° grid every 6 h from 1948 to the present. Sea surface temperatures were obtained from the Hadley Centre GISST which has monthly worldwide values on a 1° grid from 1870 (Rayner et al., 2003). Modern sea surface temperatures were obtained from the NASA Jet Propulsion Laboratory Physical Oceanography Data Archive Center (podaac.jpl.nasa.gov). Monthly averages of 4 km grid-scale observations were obtained from the MODIS-Aqua sensor (2002–2006).

We calculate Average Daily Maximum and Average Daily Minimum body temperature for each month in our model runs (Gilman et al., 2006). These statistics give an indication of the monthly high and low temperatures experienced by rocky-shore intertidal animals. The daily maximum value is dependent largely on warming by short-wave solar radiation. The daily minimum temperature depends partly on sea surface temperature, and partly on night time conditions when animals lose heat by long wave radiation and evaporative cooling. For the equatorward limit of *Semibalanus*, the daily minimum value in the coldest month is a measure of the winter conditions associated with reproduction, and the daily maximum value in the hottest month is a measure of the summer conditions associated with adult survival.

Tests of geographic hypotheses

We examined hindcasts of body temperature maxima and minima at locations near the geographic limits of the species over the period 1950–2005. Short-term (annual scale) and long-term (decadal scale) changes in geographic limits were compared to changes in body temperature maxima and minima in order to test the hypothesis that high or low temperature could explain the species' geographic shifts. If climate change was not associated with change in the species' geographic distribution, then climate mechanisms would be falsified. If latitudinal distribution did not change over the past century then the day-length hypothesis would be corroborated.

Results

Geographic distribution and hindcasts of climate

Semibalanus balanoides

Semibalanus has a disjunct distribution, with one southern limit in France, and another in Galicia (Fig. 1). On the French coast in 1872, the southern limit was at least as far south as St. Jean de Luz, on the Spanish Border at 43.39° N (Fischer, 1872), in 1955 it was at the Pointe de Grave, 45.57° N (Crisp & Fischer-Piette, 1959), from 1963 to 1971 it expanded slightly southward from Arcachon, 44.66° N to Petit Nice, 44.57° N (Barnes & Powell, 1966; Barnes et al., 1972), and in 2006 it was at Les Sables d'Olonne, 46.49° N (Fig. 1).

On the Iberian coast in 1955 its southern limit was Teis near Vigo, Spain, 42.26° N, and its north-eastern limit was A Coruña, 43.37° N (Fischer-Piette & Prenant, 1956). In 1956 its southern limit was Viana do Castelo, Portugal, 41.70° N but the individuals there all died by 1957 (Fischer-Piette & Prenant, 1957). In 1963 its southern limit was Baiona, Spain, 42.12° N and its north-eastern limit was Serrantes, 43.38° N (Barnes & Barnes, 1966). In the mid to late 1960s its density in the Muros–Noia estuary was large enough that Barnes & Barnes (1976) used the Portosín (42.76° N) population for experiments on reproductive timing. In 1985, it was seen in Cambados, Spain, 42.52° N (A. J. Southward, personal communication). By 2006, its distribution in Galicia

Fig. 2 Hindcasts of body temperatures of mid intertidal *Semibalanus* 1950–2004. *Top line*: average daily maximum temperature in the hottest month, *bottom line*: average daily minimum temperature in the coldest month. *Top graph*: Les Sables d'Olonne, France; *middle graph*: Vigo, Spain; *bottom graph*: Viana do Castelo, Portugal. The daily minimum body temperature in Viana do Castelo was 10°C only in 1956

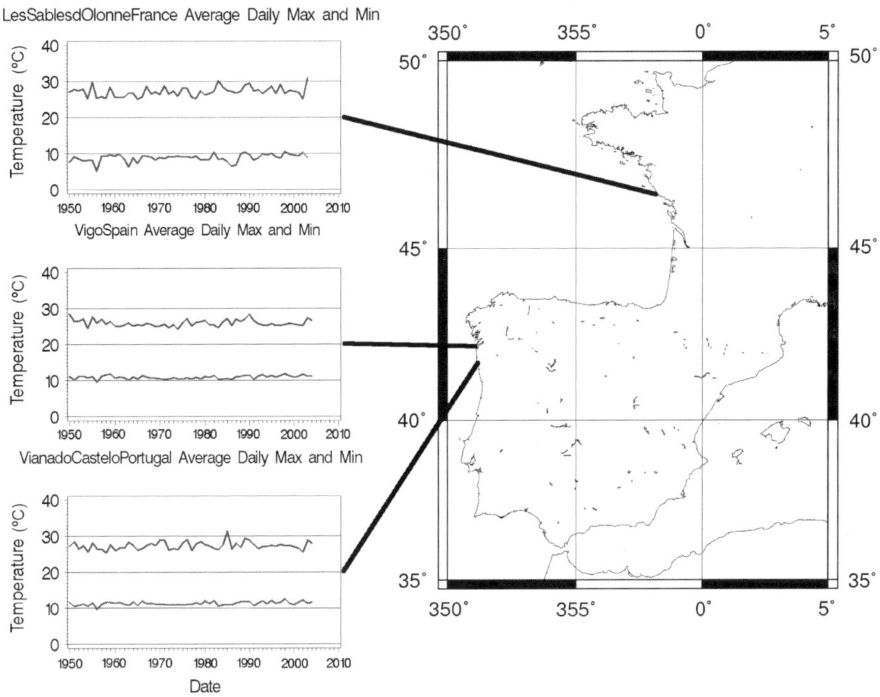

consisted of an area where very rare, 10- to 15 year-old single adults were found, between Vigo, 42.20° N, and Ortiguera, 43.68° N. The only area where there was evidence of reproduction was in the Ría de Arousa between O Grove, 42.50° N, and Vilagarcía de Arousa, 42.58° N. In this region there were young adults within copulation distance of one another, and there was a broad age-structure including young of the year and older individuals. This represents a compression of the reproductive range of the species by 250 km within the past 10–15 years.

These records indicate that *Semibalanus* on the French coast shifted more than 300 km north between 1872 and 2006, half of which occurred between 1972 and 2006. The shift during the past 30 years was at a rate of 50 km per decade. On the Iberian coast, *Semibalanus* has contracted from reproductive populations throughout Galicia in the 1950s and 1960s to a single reproductive population in the Ría de Arousa in 2006. The survivors of the 1980s cohorts in Galicia have become so rare that no pairs of individuals within copulation distance remain, except in the Ría de Arousa. This is a contraction of 50 km from the south to 250 km from the north since the 1960s–1990s.

Model hindcasts based on Hadley SST and NOAA Reanalysis variables (1° and 2° grid scale) indicate that the one-year incursion of *Semibalanus* into Portugal in 1956 (Fischer-Piette & Prenant, 1956, 1957) coincided with a one-year cooling event when temperatures should have been low enough for reproduction in Viana do Castelo, i.e., below the 10°C winter minimum (Fig. 2).

Hindcasts indicate that Galicia in NW Spain has been a marginal habitat for *Semibalanus* for at least the past 50 years. The average daily minimum body temperature in the intertidal has remained slightly above 10°C for most of the period, indicating that reproduction would be unlikely based on large-scale oceanographic conditions (Fig. 2). In addition, there has been an increase in average daily minimum body temperature since the mid 1990s at all three locales modeled (Fig. 2). The grid scale of these values is problematic for an area with intense upwelling along a deeply indented coastline. Finer scale (4 km grid scale) analysis of sea surface temperature conditions can be carried out using NASA MODIS data which indicate that local upwelling occurs very close to the coast, at a much lower spatial scale than the sea surface temperature values available in the reanalysis data and the Hadley Centre GISST data. Individual Rías can be several degrees colder than the offshore water during upwelling events (Fig. 3).

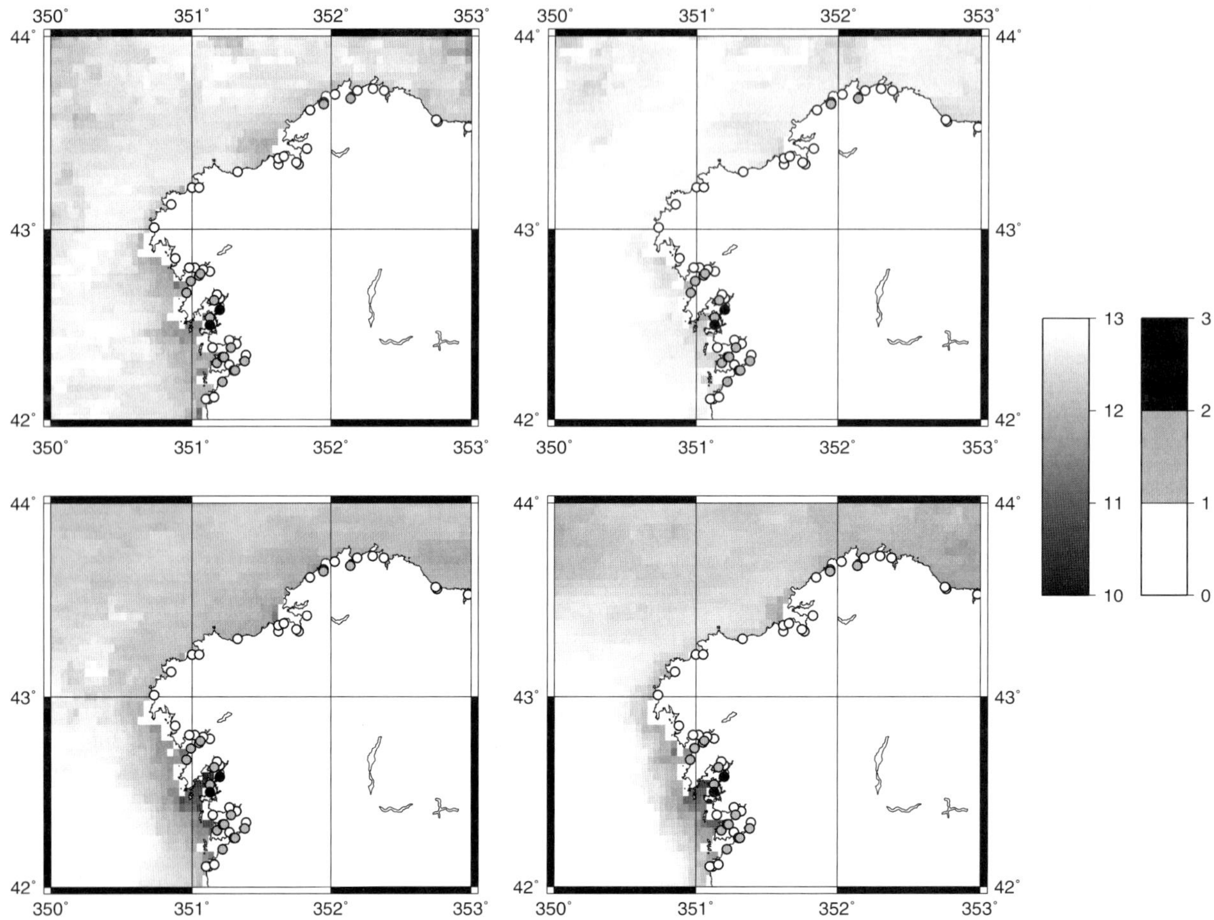

Fig. 3 Ocean temperatures in Galicia 2003–2006 from MODIS Aqua, and 2006 distribution of *Semibalanus*. Temperature scale on right. White areas in ocean next to coast indicate missing data. *Top left*: February 2003, *top right*: February 2004, *bottom left*: February 2005, *bottom right*: February 2006. *Semibalanus* abundance scale: 2–3: population with adults and juveniles; 1–2: population of old non-reproductive adults only; 0–1: absent

The 2006 age structure and restriction of *Semibalanus* reproduction to the Ría de Arousa in Galicia, NW Spain can be understood in relation to the fine scale temperature distribution in the region. The Ría de Arousa is the only estuary in Galicia where it has been cold enough for *Semibalanus* to reproduce during the years 2002–2006 (Fig. 3, dark gray color). This observation is consistent with physical oceanographic studies of the Rías Baixas estuaries in Galicia, which found that the Ría de Arousa is 1–2°C colder than the neighboring Ría de Vigo and Ría de Pontevedra (Alvarez et al., 2005).

The shift of the southern limit of *Semibalanus* on the French coast is consistent with changes in local marine conditions. The ocean climate of the Bay of Biscay has warmed during the past century. *Semibalanus* was common in Arcachon in 1972 (southern France, 44.66° N, Barnes et al., 1972), but strong warming occurred in the region between 1972 and 1993. Mean winter sea surface temperature (SST) increased from 10.6°C in 1971 to 11.8°C in 1993, and mean summer SST increased from 18.6°C to 19.9°C during the same period (Koutsikopoulos et al., 1998). By 2006, the *Semibalanus* southern limit had moved 150 km north (Fig. 1), where conditions have remained cooler (Fig. 2).

Diopatra neapolitana

Diopatra neapolitana is a Mediterranean species that occurs throughout Iberia (Fauvel, 1923). It was recorded from Arcachon, 44.66° N, in southern

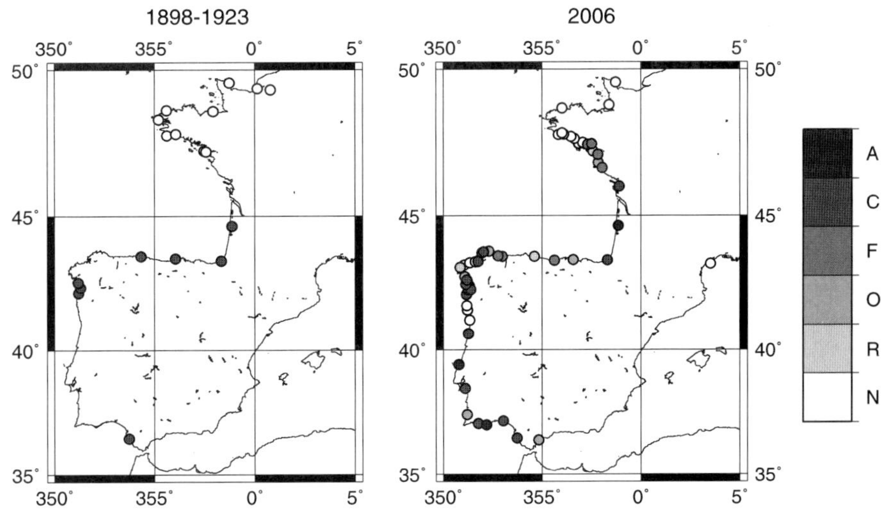

Fig. 4 Distribution of *Diopatra neapolitana* in 1898–1923 (Saint-Joseph, 1898; Fauvel, 1923) and 2006 (this study). Density scale: A = Abundant (>24 m^{-2}), C = Common (10–24 m^{-2}), F = Frequent (1–10 m^{-2}), O = Occasional (0.1–1 m^{-2}), R = Rare (<0.1 m^{-2}), N = Absent. Only presence or absence data were available from earlier work.

France, in 1869–70 (Lafont, 1871). Until 1923, the northern limit was at Arcachon; it was absent from Brittany (Saint-Joseph, 1898; Fauvel, 1923, Fig. 4). By 1966, the limit had shifted 100 km north to Châtelaillon-Plage, 46.06° N (Faure, 1969). It was also recorded from the southern tip of the Île d'Oléron, 45.85° N, (Lagardère, 1972a, b) in the late 1960s and at the mouth of the Gironde estuary, 45.57° N, (Glémarec, 1979) in 1969. By 1976 it had moved 60–130 km north to the region of St. Gilles Croix de Vie, 46.70° N (Glémarec, 1979).

Our surveys in 2006 show that the northern limit of *Diopatra* has shifted 300 km north since 1923, to Pointe de Penvins (47.50° N) in Brittany (Fig. 4). We observed the species at locations where Saint-Joseph (1898) was unable to find it although he searched for it, and we have confirmed Glémarec's (1979) observations of the species in the region between the Gironde and the Loire rivers. In addition, we identified a gap in the *Diopatra* distribution on the Portuguese coast, between Porto (41.14° N) and Viana do Castelo (41.70° N). In this gap, we searched Porto (41.14° N), Vila do Conde (41.34° N), Ofir (41.52° N), Esposende (41.54° N), Darque (41.69° N), and Viana do Castelo (41.70° N) without success. We searched semi-exposed and sheltered shores where we would expect to see *Diopatra*, based on its distribution elsewhere in Europe. This gap is in the same region of coastal upwelling where there have historically been gaps in the distribution of tropical species like *Patella rustica* (Fischer-Piette, 1955; Boaventura et al., 2002; Lima et al., 2006). The distribution within the remainder of Galicia in Spain continues as a mosaic, perhaps reflecting the hydrographic conditions. There is another mosaic distribution near the geographic limit on the French coast. In the mosaic regions, localities with abundant *Diopatra* are interspersed with localities, where it is absent (Fig. 4). For example in southern Brittany, *Diopatra* is rare but present at La Baule, 47.27° N, absent at Piriac, 47.38° N, but frequent at the Dolmen des Crapauds, 47.52° N. In southern Galicia *Diopatra* is common at Baiona, 42.12° N (Anadon & Anadon, 1973; Moreira et al., 2006; Xunta de Galicia, 2006), absent at the Playa de Foz in Vigo, 42.20° N, and is common again across the Ría from Vigo in Meira, 42.29° N and is common further up the Ría in Redondela (Xunta de Galicia, 2006). In northern Galicia, it is present in Noia, 42.79° N, absent in Baldaio, 43.30° N and Santa Cristina, 43.34° N, but common at Sada, 43.35° N. We searched extensively in appropriate habitats and did not find *Diopatra* in the locations where we recorded it as absent.

Discussion

Tests of biogeographic hypotheses for *Semibalanus balanoides* by modeling and observation

Summer heat death hypothesis

The summer heat death hypothesis (Hutchins, 1947) postulates that adult survival is restricted at the

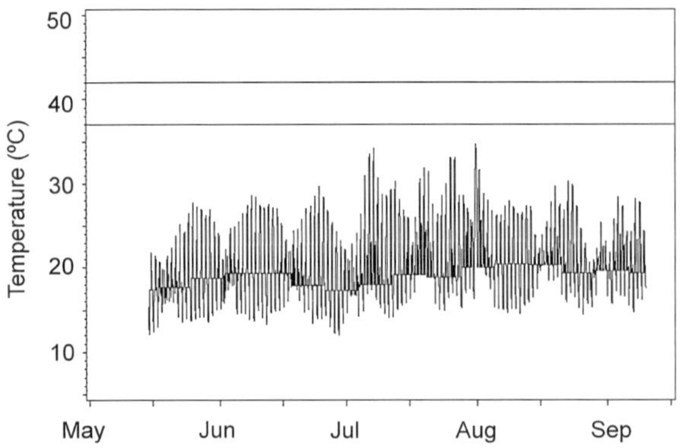

Fig. 5 Modeled summer body temperatures of *Semibalanus* on north-facing mid-shore surfaces near Lisbon, south of the southern limit of the species on the Portuguese coast. Horizontal lines are 37°C (coma) and 42°C (lethal) temperatures

southern geographic limit by high temperatures. Our hindcasts from 1950 to 2005 indicate that it is not hot enough in southern Portugal to kill adult *Semibalanus*. In shaded locations such as mid shore level north-facing slopes, temperatures are likely to be below the coma threshold of 37°C (Fig. 5). This temperature does not cause adult death (Foster, 1969) or reduction in juvenile survival (Wethey, 2002). For this reason, shaded surfaces would serve as thermal refuges for the species. We therefore reject the summer heat death hypothesis.

Day-length hypothesis

The day-length hypothesis (Davenport et al., 2005) postulates that reproduction is dependent upon shortening day length, and that there is geographic variation in response. Extrapolation of the latitude-day of year regressions indicated that the southern limit of *Semibalanus* should be either in southern France or in Galicia (Davenport, 2005). This hypothesis does not explain the geographic changes in the distribution of the species over the past 100 years, unless one hypothesizes that day length has changed over that period. It also does not explain the 1957 incursion of *Semibalanus* into northern Portugal. We, therefore, reject the day-length hypothesis.

Winter cold limitation of reproduction hypothesis

The winter cold limitation of reproduction hypothesis (Barnes, 1963; Barnes & Stone, 1972; Crisp & Patel, 1969) postulates that reproduction fails if temperatures remain above 10°C during winter months. In laboratory culture experiments, animals held at temperatures above 10°C failed to reproduce. In addition, on the Normandy coast of France, the recruitment of *Semibalanus* over the period 1983–2004 was negatively correlated with temperature in the coldest winter month, with minimum recruitment occurring in 1990 when average air temperature was 8.7°C, and maximum recruitment occurring in 1986 when average air temperature was 2.2°C (Drévès, 2001; Drévès et al., 2005). Our climate hindcasts are consistent with this hypothesis. The 1956 incursion of *Semibalanus* into Portugal was coincident with a cooling event with minimum winter body temperatures around 10°C (Fig. 2). Populations at the 2006 southern limit in France do not appear to have been at risk of reproductive failure over the past 50 years (Fig. 2). In contrast, the populations in NW Spain appear to have been very close to the reproductive failure limit over the past 50 years (Fig. 2). In 2006, the geographic distribution of populations with multiple age classes in NW Spain, and geographic distribution of successful reproduction are coincident with locations where water temperatures are in the 10–11°C range (Fig. 3). The range shift on the French coast between 1972 and 2006 was coincident with winter warming in the Bay of Biscay (Garcia-Soto et al., 2002) and a shift in the position of the 10–11°C isotherms (Koutsikopoulos et al., 1998). We, therefore, believe that the winter cold limitation of reproduction hypothesis has been corroborated by multiple lines of evidence.

Tests of biogeographic hypotheses for *Diopatra neapolitana*

Winter cold limitation hypothesis

Our prediction of limitation due to winter cold temperatures for *D. neapolitana* was by analogy to its congener *D. cuprea*. Based on data for *D. cuprea* we expected *D. neapolitana* to be limited by winter cold temperatures that interfere with normal behaviors, such as tube repair and feeding (Mangum & Cox, 1971; Myers, 1972). Thus the movement of its distribution northward, corresponding to the displacement northward of the winter 10–11°C isotherms (Koutsikopoulos et al., 1998), is consistent with this

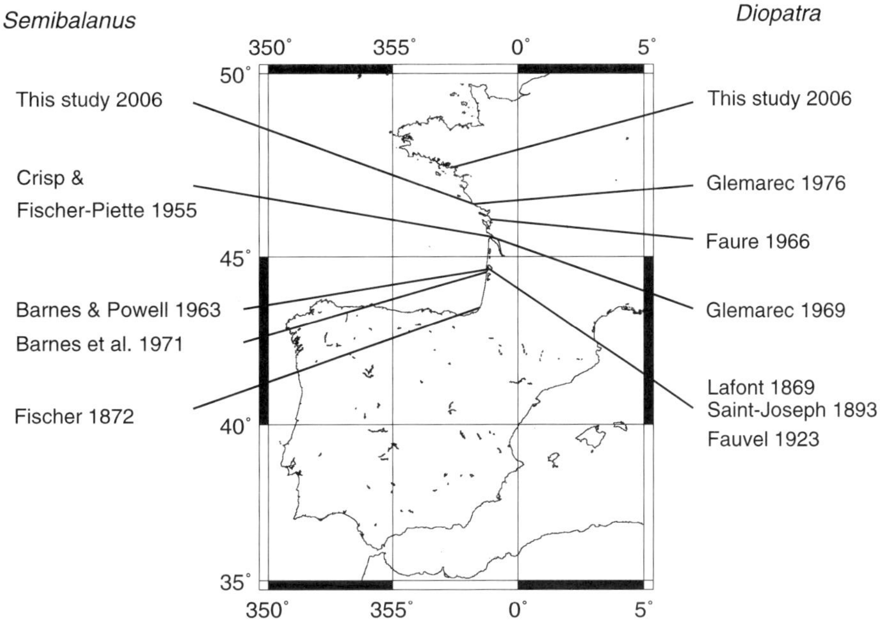

Fig. 6 Shifts in the southern limit of *Semibalanus* and the northern limit of *Diopatra* in France between 1869 and 2006. The two species overlapped by 100 km in the early twentieth century, and overlapped by the same amount both in the late 1960s and again in the early twenty-first century, but the zone of overlap has shifted 300 km north. Dates of surveys are listed, which are usually earlier than the publication dates of the papers: Fischer 1872 (Fischer, 1872), Crisp & Fischer-Piette 1955 (Crisp & Fischer-Piette, 1959), Barnes & Powell 1963 (Barnes & Powell, 1966), Barnes et al., 1971 (Barnes et al., 1972), Lafont 1869 (Lafont, 1871), Saint-Joseph 1893 (Saint-Joseph, 1898), Fauvel 1923 (Fauvel, 1923), Glémarec 1969 (Glémarec, 1979), Glémarec 1976 (Glémarec, 1979)

prediction and is consistent with the patterns of movement of both *Semibalanus* and *D. neapolitana* which have been associated in distribution pattern for over 100 years (Fig. 6).

Summer cold-limitation hypothesis

An alternative hypothesis is that *Diopatra* is restricted by cool summer conditions at its northern limit. In order to explore the possibility of a cool summer limitation, we examined MODIS sea surface temperature records for August 2005 from southern Portugal to Brittany in NW France. August is the time of warmest sea surface temperatures in this region. We used 2005 because it was the year prior to our sampling, and therefore probably had the greatest influence on the population densities that we measured. There is a summer thermal threshold in the geographic distribution of *Diopatra* near 17°C. *Diopatra* was absent in 2006 from all locations where August 2005 sea surface temperatures were below 17°C except for three sites in Galicia in NW Spain (Fig. 7). The distributional gap in Portugal occurs in locations where temperatures were between 15.9°C and 17.2°C (Fig. 7, solid triangles). This area on the central Portuguese coast has been cooler than the regions north and south at least since the 1950s (Lima et al., 2006). The northern limit of *Diopatra* in NW France also occurs where temperatures were between 16°C and 17.4°C (Fig. 7, open circles). Other locations where *Diopatra* was absent within its broadscale geographic range had temperatures between 16.8°C and 17.9°C (Fig. 7, open triangles). Clearly, we need additional data on *Diopatra* to understand these patterns, particularly asking if cool summers influence reproduction or recruitment (Figs. 4, 7). Nevertheless, cool summers appear to be stronger limiting factors than cold winters, in that they provide an explanation for both the distributional gaps and the northern limit.

Concluding remarks

The rates of climate-related geographic change that we have documented for species in the intertidal zone are

Fischer-Piette, E., 1932. Repartition des principales espèces fixées sur les rochers battus des côtes et des îles de la Manche, de Lannion à Fécamp. Annales de l'Institut Océanographique de Monaco 12: 105–213.

Fischer-Piette, E., 1955. Repartition, le long des côtes septentrionales de l'Espagne, des principales espèces peuplant les rochers intercotidaux. Annales de l'Institut Océanographique de Monaco 31: 37–124.

Fischer-Piette, E. & M. Prenant, 1956. Distribution des cirripedes intercotidaux d'Espagne septentrionale. Bulletin du Centre d'Etudes et Recherches Scientifiques Biarritz 1: 7–19.

Fischer-Piette, E. & M. Prenant, 1957. Quelques données ecologiques sur les cirripedes intercotidaux du Portugal, de l'Espagne du sud et du nord du Maroc. Bulletin du Centre d'Etudes et Recherches Scientifiques Biarritz 1: 361–368.

Flater, D., 2006. X-Tide. http://www.flaterco.com/xtide.

Foster, B.A., 1969. Tolerance of high temperatures by some intertidal barnacles. Marine Biology 4: 326–332.

Garcia-Soto, C., R. D. Pingree & L. Valdés, 2002. Navidad development in the southern Bay of Biscay: climate change and swoddy structure from remote sensing and in situ measurements. Journal of Geophysical Research 107: C8, 3118, doi:10.1029/2001JC001012.

Gilman, S. E., D. S. Wethey & B. Helmuth, 2006. Variation in the sensitivity of organismal body temperature to climate change over local and geographic scales. Proceedings of the National Academy of Sciences 103: 9560–9565.

Glémarec, M., 1979. Les fluctuations temporelles des peuplements benthiques liées aux fluctuations climatiques. Oceanologica Acta 2: 365–371.

Hansen, J., M. Sato, R. Ruedy, K. Lo, D. W. Lea & M. Medina-Elizade, 2006. Global temperature change. Proceedings of the National Academy of Sciences 103: 14288–14293.

Harrison, S. J. & A. P. Phizacklea, 1987. Vertical temperature gradients in muddy intertidal sediments in the Forth estuary, Scotland. Limnology and Oceanography 32: 954–963.

Helmuth, B. S. T., 1999. Thermal biology of rocky intertidal mussels: quantifying body temperatures using climatological data. Ecology 80: 15–34.

Helmuth, B., N. Mieszkowska, P. Moore & S. J. Hawkins, 2006. Living on the edge of two worlds: forecasting the response of rocky intertidal ecosystems to climate change. Annual Review of Ecology, Evolution, and Systematics 37: 373–404.

Hutchins, L. W., 1947. The bases for temperature zonation in geographical distribution. Ecological Monographs 17: 325–335.

Jansen, J. M., A. E. Pronker, S. W. Bonga & H. Hummel, 2007. *Macoma balthica* in Spain, a few decades back in climatic history. Journal of Experimental Marine Biology and Ecology 334: 161–169.

Kalnay, E., M. Kanamitsu, R. Kistler, W. Collins, D. Deaven, L. Gandin, M. Iredell, S. Saha, G. White, J. Woollen, Y. Zhu, M. Chelliah, W. Ebisuzaki, W. Higgins, J. Janowiak, K.C. Mo, R. Ropelewski, J. Wang, A. Leetmaa, R. Reynolds, R. Jenne & D. Joseph, 1996. The NCEP/NCAR 40–year reanalysis project. Bulletin of the American Meteorological Society 77: 437–471.

Kendall, M. A., M. T. Burrows, A. J. Southward & S. J. Hawkins, 2004. Predicting the effects of marine climate change on the invertebrate prey of the birds of rocky shores. Ibis 146: 40–47.

Koutsikopoulos, C., P. Beillois, C. Leroy & F. Taillefer, 1998. Temporal trends and spatial structures of the sea surface temperature in the Bay of Biscay. Oceanologica Acta 21: 335–344.

Lafont, A., 1871. Note pour servir a la faune de la Gironde contenant la liste des animaux marins dont la présence a été constatée a Arcachon pendant les annés 1869–1870. Actes de la Societé Linnéene de Bordeaux 28: 237–280.

Lagardère, F., 1971(1972a). Les fonds de pêche de la côte ouest de l'ile d'Oléron. Cartographie bionomique. II. Remarques systématiques, biologiques et écologiques. Tethys 3: 265–281.

Lagardère, F., 1971(1972b). Les fonds de pêche de la côte ouest de l'ile d'Oléron. Cartographie bionomique. III.—Les peuplements benthiques. Tethys 3: 507–538.

Lewis, J. R., 1964. The Ecology of Rocky Shores. English Universities Press. London.

Lima, F., N. Queiroz, P. A. Ribeiro, S. J. Hawkins & A. M. Santos, 2006. Geographic expansion of a marine gastropod, *Patella rustica* Linnaeus, 1758, and its relation with unusual climatic events. Journal of Biogeography 33: 812–822.

Mangum, C. P. & C. D. Cox, 1971. Analysis of the feeding response in the onuphid polychaete *Diopatra cuprea* (Bosc). Biological Bulletin 140: 215–229.

Mangum, C. P., S. L. Santos & W. R. Rhodes Jr., 1968. Distribution and feeding in the onuphid polychaete, *Diopatra cuprea* (Bosc). Marine Biology 2: 33–40.

Margalet, J. L. & M. J. Navarro, 1992. Mapas de distribución de algas marinas de la Península Ibérica. II. *Ascophyllum nodosum* (L.) Le Jolis, *Pelvetia canaliculata* (L.) Decne. et Thur. e *Himanthalia elongata* (L.) S. F. Gray. Botanica Complutensis 17: 117–132.

Margalet, J. L., T. Almaraz, M. J. Navarrro & I. M. Pérez-Ruzafa, 1993. Mapas de distribución de algas marinas de la Península Ibérica. III. *Fucus ceranoides* L., *F. serratus* L., *F. spiralis* L. y *F. vesiculosus* L. (Fucales, Fucophyceae). Botanica Complutensis 18: 267–290.

Mieszkowska, N., M. A. Kendall, S. J. Hawkins, R. Leaper, P. Williamson, N. J. Hardman-Mountford & A. J. Southward, 2006. Changes in the range of some common rocky shore species in Britain – a response to climate change? Hydrobiologia 555: 241–251.

Montaudouin, X. & P.-G. de Sauriau, 2000. Contribution to a synopsis of marine species richness in the Pertuis Charentais Sea with new insights in soft-bottom macrofauna of the Marennes-Oléron Bay. Cahiers de Biologie Marine 41: 181–222.

Moreira, J., P. Quintas & J. Troncoso, 2006. Spatial distribution of soft-bottom polychaete annelids in the Ensenada de Baiona (Ría de Vigo, Galicia, north-west Spain). Scientia Marina 70S3: 217–224.

Myers, A. C., 1972. Tube-worm-sediment relationships of *Diopatra cuprea* (Polychaeta, Onuphidae). Marine Biology 17: 350–356.

Parmesan, C., N. Ryrholm, C. Stefanescu, J. K. Hill, C. D. Thomas, H. Descimon, B. Huntley, L. Kaila, J. Kullberg,

T. Tammaru, W. J. Tennet, J. A. Thomas & M. Warren, 1999. Poleward shifts in geographical ranges of butterfly species associated with regional warming. Nature 399: 579–583.

Parmesan, C. & G. Yohe, 2003. A globally coherent fingerprint of climate change impacts across natural systems. Nature 421: 37–42.

Rayner, N. A., D. E. Parker, E. B. Horton, C. K. Folland, L. V. Alexander, D. P. Powell, E. C. Kent & A. Kaplan, 2003. Global analysis of sea surface temperature, sea ice, and night marine air temperatures since the late nineteenth century. Journal of Geophysical Research 108, D14, 4407.

Saint-Joseph, 1898. Les annelids polychetes des côtes de France (Manche et Océan). Annales des Sciènces Naturelles Zoologie Ser 8(5): 209–464.

Savageau, C., 1897. Note préliminaire sur les Algues marines du golfe de Gascogne. Journal de Botanique 11: 166, 175, 202, 207, 252, 263, 275, 301, 307.

Simkanin, C., A. Power, A. Myers, D. McGrath, A. Southward, N. Mieszkowska, R. Leaper & R. O'Riordan, 2005. Using historical data to detect temporal changes in the abundances of intertidal species on Irish shores. Journal of the Marine Biological Association of the United Kingdom 85: 1329–1340.

Southward, A. J., 1958. Note on the temperature tolerance of some intertidal marine animals in relation to environmental temperatures and geographical distribution. Journal of the Marine Biological Association of the United Kingdom 37: 49–66.

Southward, A. J., 1963. Distribution of some plankton animals in the English Channel and approaches. III. Theories about long term biological changes, including fish. Journal of the Marine Biological Association of the United Kingdom 43: 1–29.

Southward, A. J., 1991. 40 years of changes in species composition and population-density of barnacles on a rocky shore near Plymouth. Journal of the Marine Biological Association of the United Kingdom 71: 495–513.

Southward, A. J. & D. J. Crisp, 1954. Distribution of certain intertidal animals around the Irish Coast. Proceedings of the Royal Irish Academy 57: 1–29.

Southward, A. J., S. J. Hawkins & M. Burrows, 1995. Seventy years' observations of changes in distribution and abundance of zooplankton and intertidal organisms in the western English Channel in relation to rising sea temperature. Journal of Thermal Biology 20: 127–155.

Southward, A. J., O. Langmead, N. J. Hardman-Mountford, J. Aiken, G. T. Boalch, P. R. Dando, M. J. Genner, I. Joint, M. A. Kendall, N. C. Halliday, R. P. Harris, R. Leaper, N. Mieszkowska, R. D. Pingree, A. J. Richardon, D. W. Sims, T. Smith, A. W. Walne & S. J. Hawkins, 2005. Long-term oceanographic and ecological research in the Western English Channel. Advances in Marine Biology 47: 1–105.

Vermeij, G. J., 1971. Temperature relationships of some tropical Pacific intertidal gastropods. Marine Biology 10: 308–314.

Wethey, D. S., 2002. Biogeography, competition, and microclimate: the barnacle *Chthamalus fragilis* in New England. Integrative and Comparative Biology 42: 872–880.

Woodin, S. A., 1974. Polychaete abundance patterns in a marine soft-sediment environment: the importance of biological interactions. Ecological Monographs 44:171–187.

Xunta de Galicia, 2006. Consellería de Pesca y Asuntos Marítimos. Orden de 18 de abril de 2006 por la que se reglan los planes de explotación específicos para poliquetos y se prueban los planes para la explotación de poliquetos en el litoral de Galicia para el año 2006. Diario Oficial de Galicia 80: 6.615. http://www.xunta.es/Doc/Dog2006.nsf/FichaContenido/C452?OpenDocument.

Hydrobiologia (2008) 606:153–160
DOI 10.1007/s10750-008-9351-y

Long-term changes in the status of Sevastopol Bay and the Crimean coast: anthropogenic and climatic influences

J. G. Wilson · T. S. Osadchaya · S. V. Alyomov

© Springer Science+Business Media B.V. 2008

Abstract The Black Sea as a whole is subject to considerable anthropogenic and natural pressures, and these can be intensified in enclosed Bays such as those around the city of Sevastopol. The overall biological status of Sevastopol Bay has improved considerably in the last 30 years or so. The number of macrobenthic species in the Bay has increased from just over 20 in 1973 to around 70 in 2000, and there have been similar enrichments in abundance, in biomass and in diversity (H'). The same pattern is shown in other locations such as the Southern Bay (the most highly polluted of the embayments) from which macrofauna were completely absent in 1976 and 1982. Over the same period oil hydrocarbons in sediments of the Southern Bay have actually increased from 8760 mg kg^{-1} in 1976 to 12,020 mg kg^{-1} in 2002, while the levels in the open sea sediments have decreased slightly over the same period (450 mg kg^{-1} and 280 mg kg^{-1}, respectively). Moreover, this pattern of biological improvement is also seen outside the Bay, suggesting a common factor in the changes. Since the effect of the local factors (i.e. pollution) can be readily seen in the gradient of biological response among sites, yet they have not significantly changed over the period of study, the overall improvement must involve some over-arching driver such as climatic influences. This point is discussed further in relation to the wider changes in the Black Sea system.

Keywords Long-term monitoring · Macrofauna · Black Sea · Abundance · Biomass · Diversity

Guest editors: J. Davenport, G. Burnell, T. Cross, M. Emmerson, R. McAllen, R. Ramsay & E. Rogan
Challenges to Marine Ecosystems

J. G. Wilson (✉)
Zoology Department, TCD, Dublin 2, Ireland
e-mail: jwilson@tcd.ie

T. S. Osadchaya · S. V. Alyomov
Institute of Biology of the Southern Seas, 2 Nakhimov Avenue, Sevastopol 99011, Ukraine

Introduction

The Black Sea system has long been recognised as under significant stresses, both from the anthropogenic impacts of contaminant inputs and from its susceptibility to disruption and invasion (GESAMP, 1997). Contaminant inputs into the Black Sea come from a number of sources including the loads carried by rivers, of which the Danube is by far the largest (Milliman et al., 1995). There are also local inputs as a result of urbanisation and industry, both of which have been implicated in the anthropogenic impacts in Sevastopol Bay on the southern Crimea coast. High concentrations of organic matter in parts of Sevastopol Bay have facilitated the accumulation of hydrocarbons and also the induction of periodic anoxic events (Osadchaya et al., 2003). Monitoring of the plankton of Sevastopol Bay has suggested that these impacts have been established for several decades and that the quality status of the Bay waters is, if anything, still declining (Gordina et al., 2001).

The benthos of the northern Black Sea generally is somewhat limited with for example only 128 species listed from the whole of the Georgian coastline to a depth of 100 m (Komakhidze & Mazmanidi 1998). Within Sevastopol Bay, the list is further restricted, firstly by depth and available habitat, and secondly by the supply and the survival of the meroplanktonic larvae (Gordina et al., 2001). The latter is linked to the hydrographics of the Bay and to the pollution status, both of which vary from station to station (Gordina et al., 2001).

In addition, climate changes have impacted on the Black Sea, with Oguz (2005) suggesting that the effects are imposed through a bottom-up forcing of the system. Yunev's analyses of chlorophyll and nutrient long-term data (Yunev et al., 2002, 2005) show that not only have global climatic processes changed the pelagic system of the Black Sea, with the shift being most pronounced in the latter part of the 1980s, but that the response of coastal systems differed from that of the open waters. Since then, the general opinion (e.g. Kideys, 2002) has been that the system has been recovering, although there is still considerable debate as to whether the changes could be ascribed directly to nutrient enrichment or to hydrographic changes on the foot of winter temperature anomalies (Konovalov & Murray, 2001). However, there was no suggestion of an overall temperature trend in either direction in the system (Konovalov & Murray, 2001).

The interaction of the anthropogenic impacts of climate change and pollution on the benthos of coastal waters makes it difficult to ascribe the rather often diffuse changes unequivocally to climate. Clark & Frid's (2001) review of long-term changes in the North sea ecosystem at a number of levels of organisation, suggested a lag, or periodicity of response of the macrobenthos between 2 and 7 years, depending on the driver, and also pointed out that the drivers such as pollution or climate change may themselves be channelled through a secondary factor such as food availability. The exceptions can be ascribed to catastrophic local extinctions, consequent on conditions such as a severe winter, producing a very strong signal similar to that of a severe pollution incident (Beukema, 1988). Nevertheless, there have been some instances recorded. For example the recent northward spread of *Abra tenuis* into the North Sea (Dekker & Beukema, 1999) and into the Irish Sea (Wilson, 2001) has been ascribed to climate warming.

In this paper we set out the changes in the benthic fauna of Sevastopol Bay with the known changes in the anthropogenic pressures, to test to what extent the faunal changes can be related to local pollution or to general climatic conditions.

Materials and methods

The major sampling areas in and around Sevastopol Bay are shown in Fig. 1. Two sites were sampled in the Southern Bay (which opens out into Sevastopol Bay but is here treated separately), and five in the rest of Sevastopol Bay, while two sites were also sampled in Streletskaya Bay, Kamishovaya Bay and at an open coast site (Uchkuevka Bay). Sampling was carried out at the same time of year and at the same stations throughout the study.

The methodologies for the different elements of this study have largely been set out elsewhere (Wilson et al. (2008) and references therein). Briefly:

- Macrofauna was sampled by 0.038 m^2 grab (three replicates per station) and identified after sieving through a 1-mm mesh sieve;
- Sediments were likewise sampled by grab, and passed through a 2-mm mesh to remove shells and large stones before analysis. Hydrocarbons were extracted by using CCl_4 and determined on infrared spectrophotometer IRS-29 (Osadchaya et al. 2003). Reference sediment MESS-3 was used as the standard.

Diversity measures (Shannon–Weiner Index (H') and Evenness ($J' = H'/H_{max}$)) were calculated with the PRIMER© package along with the Shannon–Weiner Evenness proportion (= $H'_{biomass}/H'_{abundance}$, (McManus & Pauly, 1990)). PRIMER© was also used to calculate the expected number of species (ES100) using Sanders' rarefaction method calculated for a sample of 100 individuals to correct for differences in sampling effort or in community abundance. PRIMER© was also used to calculate (SIMPER routine) the contributions of the individual species

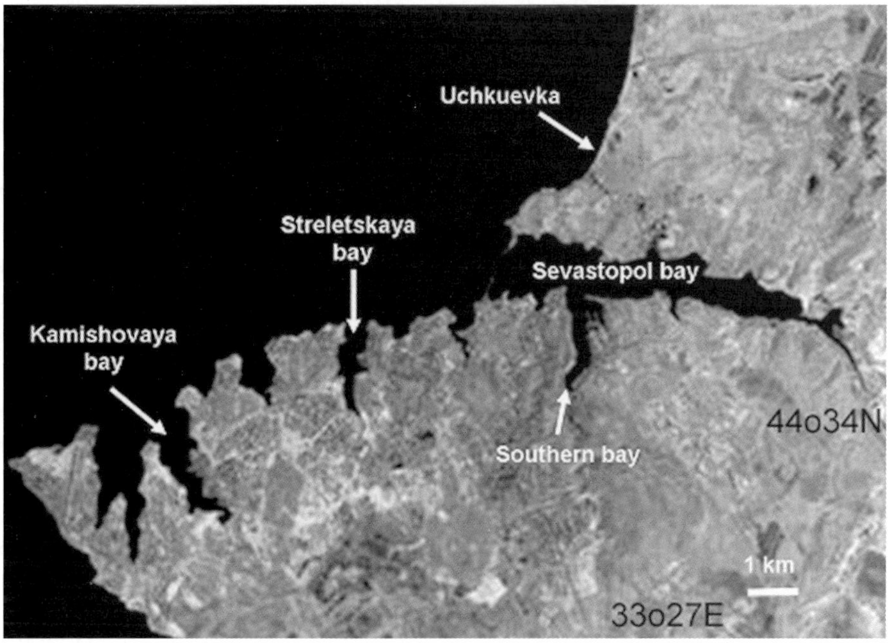

Fig. 1 Crimean coast, showing location of other Bays sampled 1973–2003

to the assemblages in each of the locations (Clarke & Warwick, 2001).

Results

Table 1 sets out the major contributors to the macrofaunal assemblages in each of the locations as determined by the SIMPER routine. Overall, the only Bay to stand out was Uchkuevka, both in terms of the species concerned (*Chamelea gallina*) and the degree of dominance (92%). In all the rest, *Cerastoderma glaucum* and *Tritia reticulata* accounted for over 60% of the similarity.

Hydrocarbon contamination is a major stressor in Southern Bay and also in Sevastopol Bay and to a lesser extent in the Bays around. Figure 2 shows that levels of oil in the sediment in general have been more elevated in the Southern Bay than in the rest of the system and that these levels declined over the period from 1982 (when the level was 14720 mg kg^{-1} (=1.47%)) to 2000, but seem to have risen back almost to their starting, 1982, levels in 2003 (1.29%). In Sevastopol Bay itself, the levels showed a slight decline, as they did in Streletskaya Bay, while those in Kamishovaya Bay were both very much lower, and rather variable (0.46–0.18%) over time (Fig. 2). Hydrocarbon concentrations at the open sea control site (Uchkuevka) were close to the limit of detection (Fig. 2). Page et al. (2003) have suggested a toxic threshold value for total PAHs of

Table 1 Contribution (%) of individual species to similarity (SIMPER) 1973–2003 within location (Bay)

	Southern	Sevastopol	Streletskaya	Kamishovaya	Uchkuevka
Cerastoderma glaucum	54.3	18.8	4.6	42.2	
Tritia reticulata	24.6	68.2	55.0	23.6	
Nereis diversicolor	12.7				
Mytilus galloprovincialis		5.0			
Abra ovata			29.5		
Bittium reticulatum			2.7		
Chamelea gallina				20.2	92.4
Politapes sp.				3.5	
Pitar rudis				3.1	

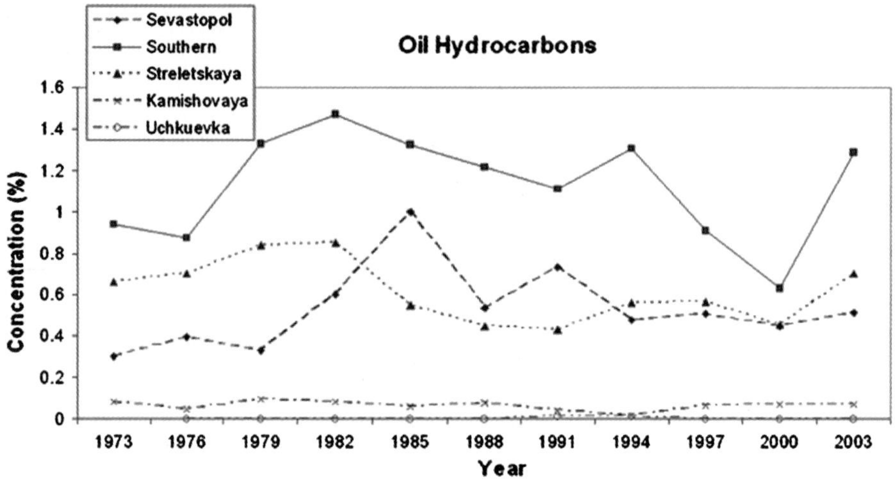

Fig. 2 Concentrations (%, g 100 g^{-1} dry sediment) of total hydrocarbons in five sample locations around the Crimean coast

2.6 μg g^{-1}, which, while recognising the influences of hydrocarbon chemical make-up and of weathering, still suggests that there should be some level of toxicity at almost all the sites in this present study.

In contrast to the rather slight changes in pollution pressures over time, there have been marked changes in the benthic macrofaunal community properties. At the most affected site (Southern Bay) the number of species steadily increased to a total of 20 species in 2003 from 1986, until when no more than one species (*T. reticulata*) had been found (Fig. 3a). The pattern of individual abundance and biomass shows a clear peak in 1994 with a decline to 2003 (Fig. 3b), and this latter pattern is mirrored, although the maxima spread from 1991 to 1997, in the diversity measures H', J and SEP (Fig. 3c).

At the next two sites in terms of anthropogenic impacts, Sevastopol and Streletskaya Bays, the patterns over time were also similar (Figs. 4 and 5, respectively). However, the improvement started sooner, with the numbers of species peaking at both locations in 2000 (72 and 51 species, respectively) while numbers of individuals also broadly showed an increase, although again with a fall in the latter part of the period (Fig. 4a, b, 5a, b). However, as with the Southern Bay site, biomass again peaked in the 1990s (Figs. 4b, 5b), even if the diversity measures (H', J and SEP) did suggest an overall improvement to 2003, an improvement which was slightly more pronounced in Sevastopol Bay (Fig. 4c) than in Streletskaya Bay (Fig. 5c).

At the two least-contaminated sites, Kamishovaya Bay (Fig. 6) and Uchkuevka Bay (Fig. 7) the patterns were different again. While there was an increase in the numbers of species, especially from the early 1980s, to maxima of 68 (1994) and 27 (1997) respectively, in 2003 (Figs. 6a, 7a), abundance at Kamishovaya (Fig. 6b) showed the same trend as at the other sites, while at Uchkuevka the peak of abundance was in 1998, since when it has been in decline (Fig. 7b). Both sites revealed a biomass peak in the middle of the monitoring period (Figs. 6b, 7b) and overall diversity (H', J and SEP) showed little trend or change in either (Figs. 6c, 7c).

The final (2003) ES100 in all the Bays was remarkably similar, yielding 10–12 species per 100 individuals (Figs. 3a–7a) and, as with the total number of species (S), showed an increasing trend across all sites from the start of the monitoring period.

Over all the locations, the H' values reflected the generally low numbers of species, rarely exceeding 3.0, but the evenness (J), while again generally low (indicating high dominance), did rise above 0.5 in Uchkuevka and also to a lesser degree at Kamishovaya and Streletskaya. Southern Bay and Sevastopol Bay were the only two Bays where the SEP exceeded unity (1.0).

Discussion

The spatial trend in all the macrobenthic community measures (S and ES100; and H', J and SEP) reported here supports the original rankings of the locations in terms of anthropogenic influence, such that Southern Bay was the most heavily impacted and Uchkuekva

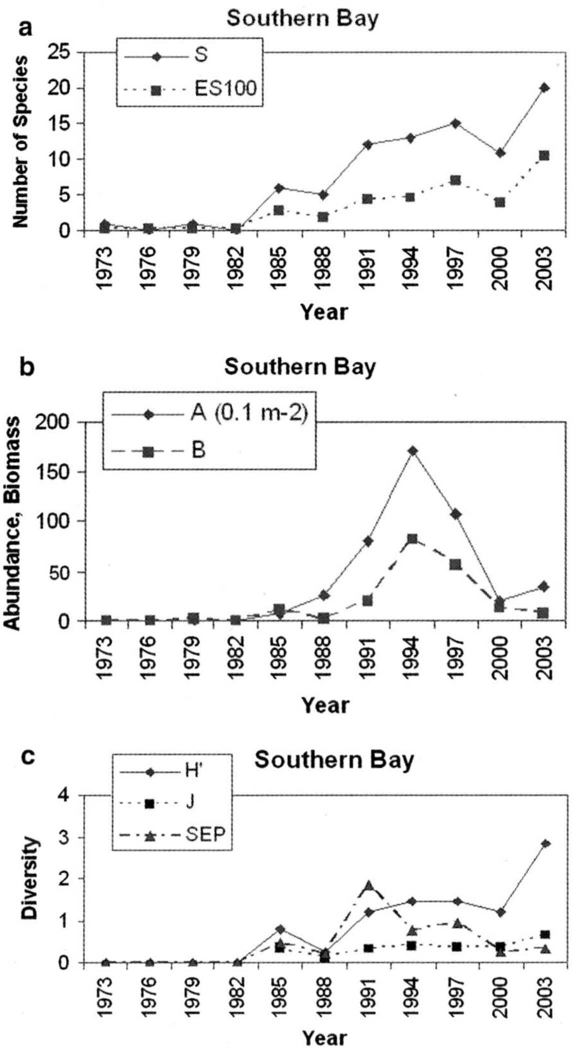

Fig. 3 Southern Bay 1973–2003, showing (**a**) Number of species found (S) and expected number (ES100); (**b**) abundance (A, individuals 0.1 m^{-2}) and biomass (B, g dry wt m^{-2}); (**c**) diversity: Shannon–Weiner index (*H*′), Evenness (*J*) and Shannon–Weiner Evenness Proportion (SEP)

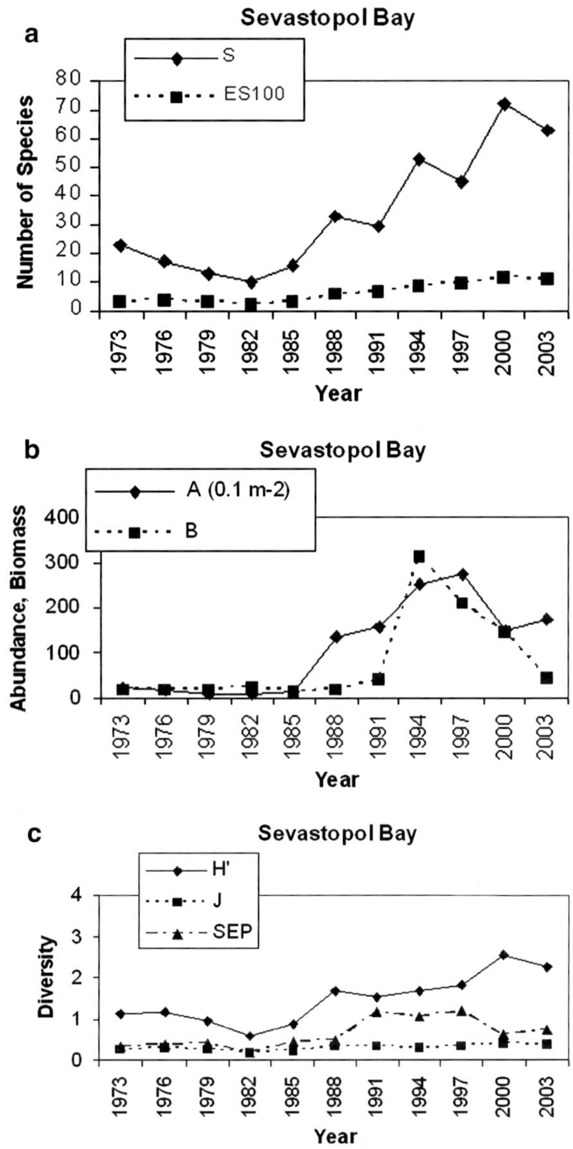

Fig. 4 Sevastopol Bay: legend as Fig. 3

the least, with the remaining Bays in the order Sevastopol > Streletskaya > Kamishovaya.

Overlying the local anthropogenic influences described above are the large-scale climatic factors impacting on the Black Sea as a whole. Of these, two in particular have been identified as of major consequence in system forcing:

- Unusually cold winters in 1976, 1985 and 1993 (Konovalov & Murray, 2001);
- Climate-induced high primary productivity 1988–1992 (Yuney et al., 2002);

There is little evidence from the data presented here that the cold winters had any impact on the inshore benthic macrofauna, since none of the patterns corresponds with the temporal signature of these events. Numbers of species, and to a certain extent abundance of individuals and diversity showed a general trend of increasing from 1982 to 2003, yet this is not mirrored in any temperature trends over that period. The driver of these phenomena must therefore be a combination of factors.

One common feature of all locations was a peak in biomass, although the timing did vary. It was earliest

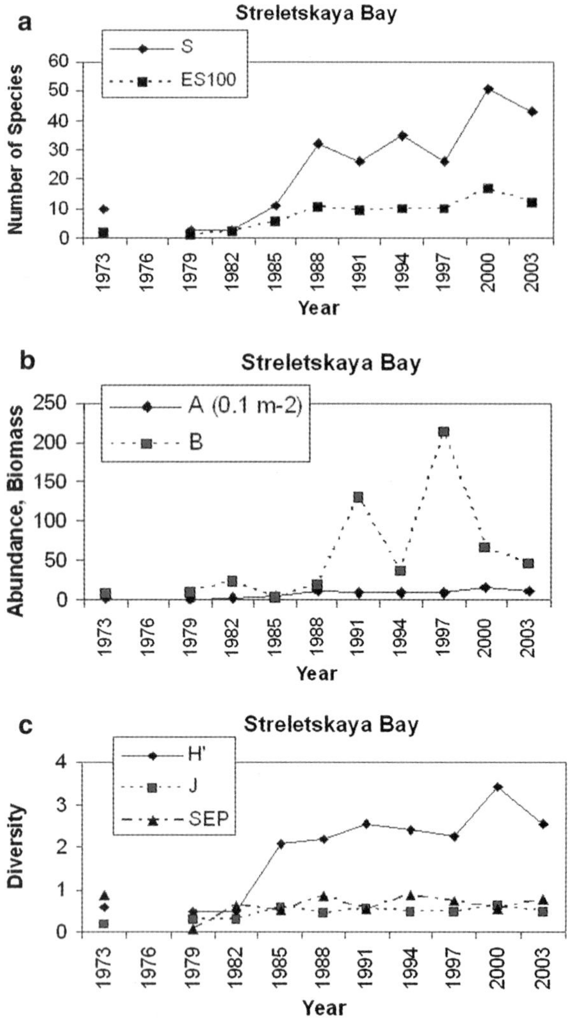

Fig. 5 Streletskaya Bay: legend as Fig. 3

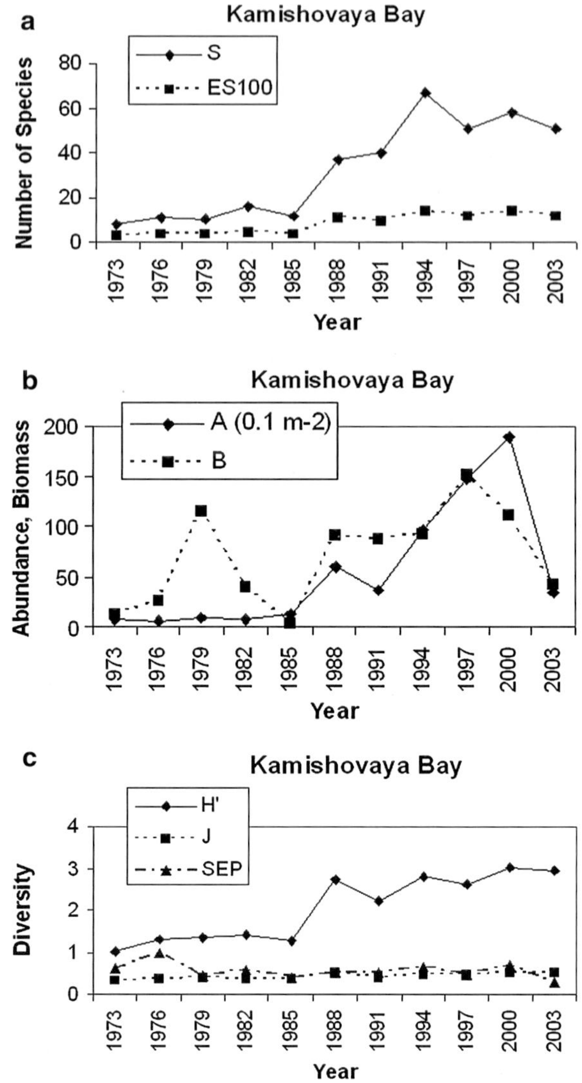

Fig. 6 Kamishovaya Bay: legend as Fig. 3

at the open sea control site (1991, Uchkuevka) and latest at two sites with intermediate levels of contamination (1997, Kamishovaya Bay and Streletskaya Bay). This corresponds well with the observed, and very marked, peak in primary productivity at the end of the 1980s (Yunev et al., 2002), and an obvious mechanism would be that the increased food availability increased the biomass of invertebrate suspensivores and deposivores. Yunev et al. (2002, 2005) clearly ascribed the 1980s peak in primary productivity to climatic forcing of hydrographic conditions. However, they showed in their later paper (Yunev et al. 2005) that the outcome of this forcing was not uniform over the whole of the Black Sea, but rather was modified by basin topography. Some caution needs therefore to be applied in ascribing wide-scale phenomena to climate change. Nevertheless, the locations in this study are located within a relatively small compass, so it is reasonable to assume that they would all benefit more or less equally from any large-scale increase in primary productivity, and what is seen is a local modification in the timing.

Although it is tempting to relate the increases in numbers of species, abundances and diversity to the decreases in contamination over the period studied, two facts argue against such a simplistic explanation.

The first is that practically the same number of species was reported at the end of the period studied

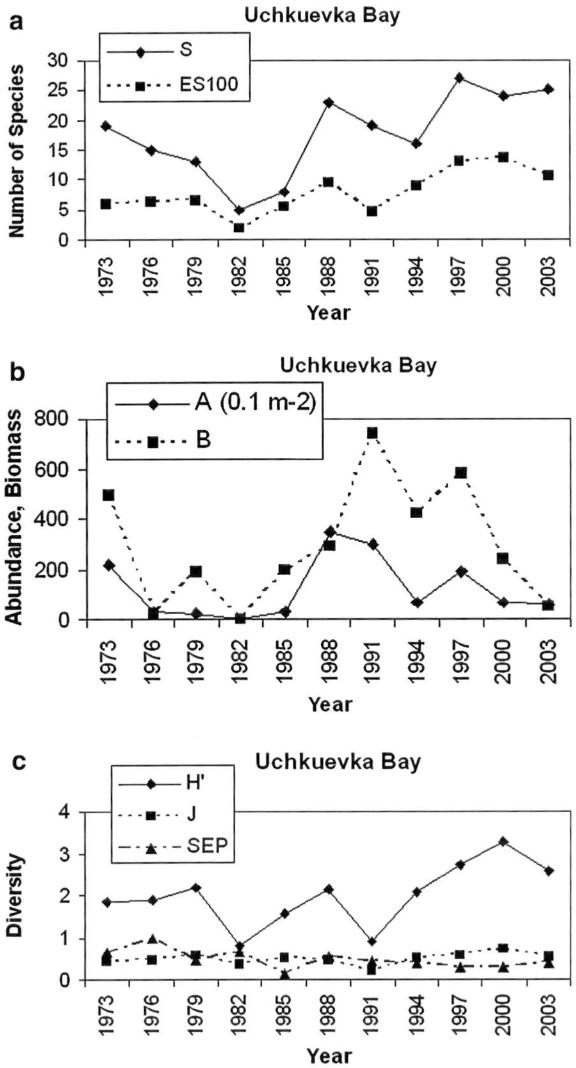

Fig. 7 Uchkuevka Bay: legend as Fig. 3

from each location, despite a wide range (more than one order of magnitude) in contamination. However, it must be noted that the number of species at the start (1973) was very low in Southern Bay, the most-contaminated location, and did continue low for some time before the increase in species numbers became apparent. This suggests that whatever factor was operating, it took some time for the system to respond and that the level of contamination was suppressing the system response.

The second is that the patterns of the increases in all measures (notably in the numbers of species) were broadly similar over all the sites, yet the pattern of changes in contamination over time was not—an apparent mismatch.

The alternative hypothesis is that the increases in the other measures as well as biomass were driven by increasing trophic input. Biomass can increase in response to increased food in three ways:

1. Somatic growth of individuals: i.e. no change in A, *S* or *H*′;
2. Somatic growth increases slightly but the remaining resources are allocated to gametes, following which the chances of recruitment survival are enhanced both by the increased fitness of larvae and also reduced intraspecific competition: i.e. an increase in A as well as B, but not *S* or *H*′.
3. The increase in resources decreases interspecific competition and possibly opens new niches of feeding possibilities, such that not only does the biomass of individual species increase, other species are recruited into the system: i.e. an increase in B, A and *S* and possibly, though not necessarily, *H*′.

From the results presented, here, hypothesis 3 is that which best fits the response of the Sevastopol systems and which suggests the mechanism by which climate change may be propagated through the system, affecting both its composition and its structure.

However, the picture is not as clear as might be. Work on other groups (e.g. meroplankton, Gordina et al., 2001), suggested that the conditions in Sevastopol Bay and the Southern Bay had not measurably improved since the 1970s and that in fact the situation had continued to decline. However, they did note that the recent conditions outside Sevastopol Bay (which was further cut off from the open Black Sea by the construction of barriers across the entrance in the 1970s) were very much better than inside and that there was some evidence that the exterior was acting as a source of larvae for the benthic macrofauna populations within (Gordina et al., 2001).

Examination of the species' composition at the various sites reveals firstly, that the community composition in Southern Bay, which was radically different from the other sites at the beginning of the monitoring period, had by the end of the study converged with those at the other sites. Secondly, the extra species recruited into all the sites comprised a number of similar elements (Table 1), and included to

Table 1 Example of an Ecosystem Evaluation Framework (EEF) for seamounts; a scheme of attributes and status evaluations that may be objectively scored to provide a comparative assessment of seamount food webs, ecology, exploitation status and threats to sustainability and integrity (modified from Pitcher et al., 2007)

Factors EEF	Classification EEF	Knowledge EEF	Function EEF	Notes
Oceanographic factors				
Depth of peak				
Depth of surrounding ocean				
Height of peak				
Slope of seamount				
Proximity to shelf				
Proximity to neighbour seamounts				
Ocean currents link to shelf				
Ocean currents to neighbour seamounts				
Taylor cap forms				
Total oceanographic status				
Ecological factors				
Macrophytes present				
Corals present				
Larval settlement regime				
Nutrient upwelling occurs				
Phytoplankton enhancement				
Zooplankton enhancement				
Deep Scattering layer organisms entrapped				
Settled filter feeders				
Zooplankton migrates in feeding range				
Predators/grazers present				
Detritus build-up present				
Detritivores present				
Small resident invertebrate predators				
Small resident fish predators				
Resident cephalopods				
Aggregating deep sea fish				
Visiting fish predators				
Visiting elasmobranch predators				
Visiting marine turtles				
Visiting mammal predators				
Visiting seabird predators				
Total ecological status				

Status EEF	Classification	Status EEF	(Uncertainty)	Notes
Fisheries				
Trawl fishery				
Longline fishery				
Handline fishery				
Purse seine fishery				
Others				
Total fisheries status				

Table 1 continued

Status EEF	Classification	Status EEF	(Uncertainty)	Notes
Ecosystem concerns (EBM)				
Corals and benthos damage				
Turtle by-catch issues				
Shark by-catch issues				
Dolphin by-catch issues				
Whale by-catch issues				
Seabird by-catch issues				
Others				
Total conservation concern status				

levels in marine food webs. Some have argued to do this deliberately to mitigate global problems of hunger, especially in the face of concerns about the sustainability of terrestrial agriculture, and we can be sure that economic pressures and market values for increasingly scarce protein will lead to tendencies of this type. One example is a possible increased exploitation of krill, which at first sight suggests vastly increased amounts of protein that could be harvested.

First, I set out the rationale for expecting increased krill harvests. Figure 3 shows the world fish catch 1980 to 2004 as reported to the FAO (UN). Despite some issues with these totals being an underestimate of the true fish extractions, such as not including discards of the order of 18–30% (Kelleher, 2004), and 10–40% unreported catches (Pauly & Christensen, 2002), the plot does illustrate trends in the total amount of fish taken for human consumption and agricultural use in animal feeds. The plot also shows the total world demand for fish, projected forwards to 2020 under a set of disaggregated assumptions about per capita fish consumption by region (IFPRI: Delgado et al., 2003), and alternatively under a demand projection disaggregated by product and country (FAO, 2004). Supply projections for capture fisheries by FAO and IFPRI are also shown, and both are stable or fall slightly by 2020. All of the projections agree that, to meet demand, the amount of fish product supplied by aquaculture will have to double in the next 20 years. Since technical developments in aquaculture have been focussed largely on high trophic level carnivorous fish that rely on catches of small pelagic fish for food (on account of the difficulty of supplying a sufficient quantity of double sulphur-bond amino acids that are essential in most fish diets), there is considerable doubt if this can be achieved in time. Moreover, there are serious worries about the sustainability of capture fisheries, which have recently been largely maintained by serial depletion of location and species (e.g. Morato et al., 2006a, b; Berkes et al., 2006), and some of which have exhibited unanticipated and disastrous crashes. Given these uncertainties about both capture fisheries and aquaculture, Fig. 3 implies a considerable pressure to find alternative sources for marine foodstuffs in the near future. Since catching animals from lower trophic levels provides an approximately tenfold gain, per trophic level, in potential harvestable biomass, this has led some to ask if deliberately 'fishing down the food web' should (or will) be performed to meet the unsatisfied world demand for seafood (Sherman, 1991; Pauly & Watson, 2005; Berkes et al., 2006). Unlike most other lower trophic level marine organisms, Antarctic krill is already harvested and, on account of longstanding concerns as a major forage organism (Everson & de la Mare, 1996), is subject to a fishery quota (CCAMLR, 2001; Pitcher & Erfan, 2005). Moreover, a number of markets for krill products have opened up, there have been experimental fisheries in Canada and elsewhere, and relatively minor changes in technology would likely enable greater harvest worldwide (see contributions in Nicol & Endo, 1999; Everson, 2001). However, krill are said to be 'keystone' zooplankters (Laws, 1985) and undoubtedly play a pivotal role in the Antarctic and probably many other ecosystems such as the Bering Sea, Humbolt, Benguela and California currents.

Second, I make preliminary estimates of how much krill might be caught sustainably. One simple estimate might be to take, say, 45% of krill biomass

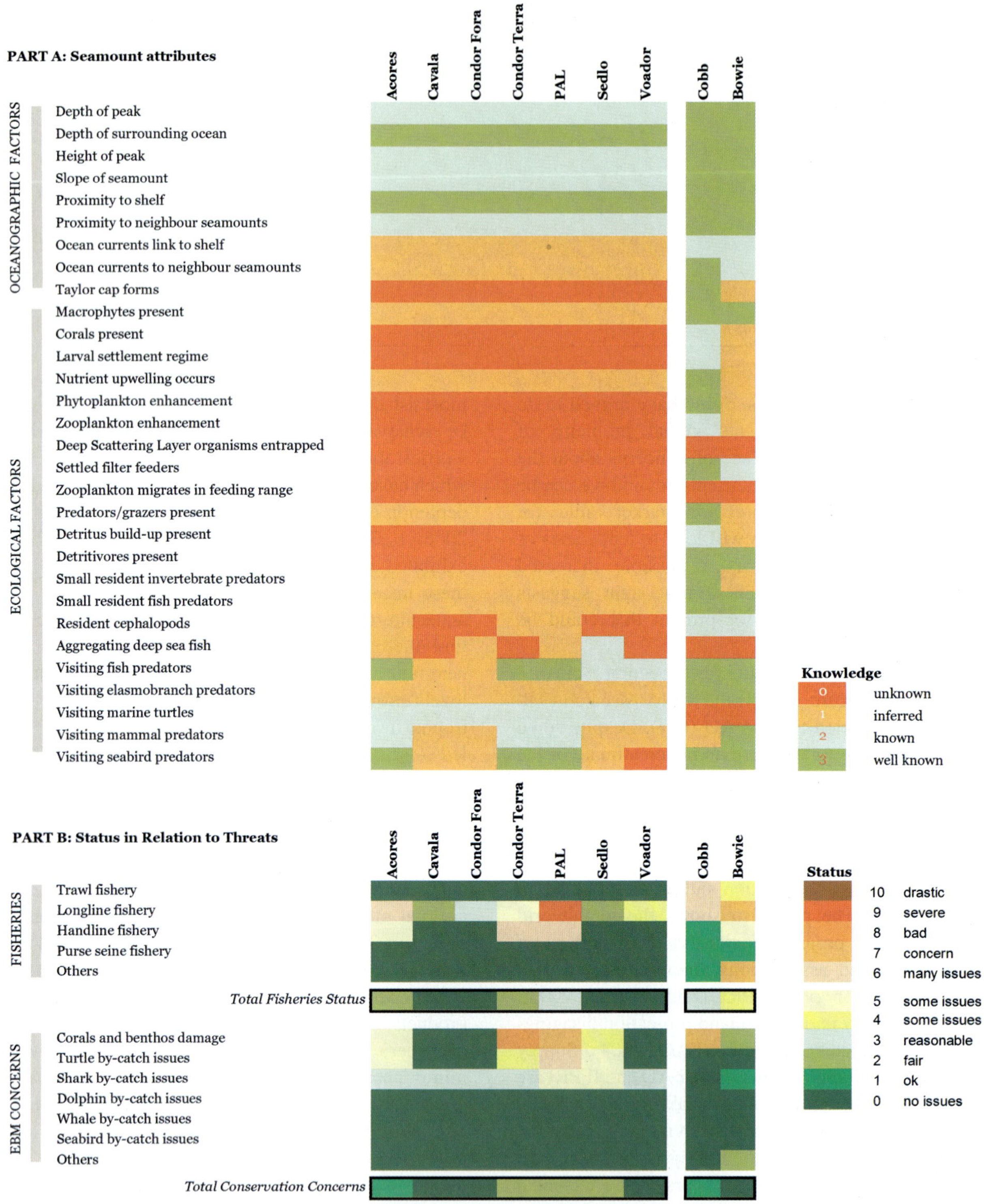

Fig. 2 Example of an EEF applied to nine seamounts; seven named seamounts at left from the North Atlantic (Azores, Portugal) and two at right from the northeast Pacific (Canada and High Seas). See text for discussion. Shading colour indicates, in Part A Knowledge EEF (top), level of knowledge, and in Part B Status EEF (lower), sustainability status of fisheries and level of concern from threats

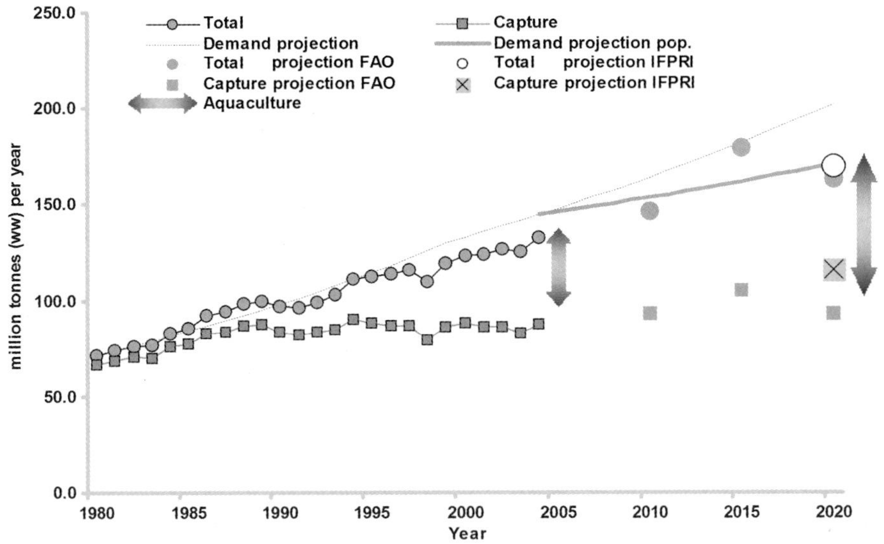

Fig. 3 Projections of demand and yields from world fisheries to 2020 from Food and Agriculture Organization of the United Nations (FAO, 2004) and the International Food Policy Research Institute (IFPRI; Delgado et al., 2003). See text for discussion. Plot shows historical annual total fish yield and capture fisheries 1980 - present (shaded circles and squares, see key); present and project returns from aquaculture (arrows); FAO past and projected future demand based on per capita fish consumption (thin dotted line); FAO alternative demand projection using disaggregated regional human population trends (thick solid line); FAO projected total and capture for (shaded circles and squares); IFPRI 2020 projected total (open circle) and from capture fisheries (crossed square)

(a precautionary amount given that the P/B ratio for adult krill is more than 2 and perhaps as high as 6) predicted from a 10% transfer efficiency of world phytoplankton production. This results in a potential krill catch of around 850 million tonnes per annum, which would represent an astounding and welcome tenfold increase in world fishery yield. Here I use ecosystem simulation modelling (*Ecopath-with-Ecosim, EwE*: Christensen & Walters, 2004) in an attempt to improve upon this simple calculation. The crucial role of krill in many marine ecosystems suggests caution, and indeed, preliminary work with the ecosystem models used here suggested that krill fisheries of this magnitude could result in serious depletions, the collapse of many dependent organisms, and lead to simplified and volatile food webs that could be vulnerable to a wide range of perturbations, including the impacts of climate fluctuations and climate change. Moreover, this is not only an ecological problem since a precautionary krill harvest would hope to have minimal impact on existing fisheries and livelihoods. Hence, modelling was designed to find levels of krill harvest that would be sustainable over the long term (50-year simulations were used), that would not affect ecosystem biodiversity or ecosystem structure, and would not reduce significantly catches from existing fisheries.

I used eight published EwE models that included krill explicitly as a functional group (listed in Table 2). They range in complexity from 24 to 59 functional groups in the food web, with 2 to 15 existing fisheries, each with harvest and bycatch levels that had been fitted to local data by the authors. One of the ecosystem models, of the Antarctic Peninsula, already included a krill fishery. In each case, after a few parameter adjustments to standardize the performance of test Ecosim runs under conditions of zero and extremely high fishing mortalities, a new krill fishery was added to the model, with an initial small harvest set to 0.5 tonnes per square kilometer. The krill fishery was assumed to have no by-catch. All modelling was performed on an annual per km^2 basis and results later raised to actual tonnes.

Simulations were performed using an automated non-linear search procedure that maximized an objective function (Walters et al., 2002). The procedure searched for a maximum level of krill harvest that had little impact on the biomass of other functional groups in the ecosystem, including all those feeding directly on krill, and that did not

Table 2 List of Ecopath-with-Ecosim models employed for krill fishery estimations

Ecosystem location	# Functional groups	# Present fisheries	Time span for model	Source/Citation
West Scotland	37	8	1995–2000	Haggan & Pitcher (2005)
Northern British Columbia	53	17	2000	Ainsworth & Pitcher (2006), Ainsworth et al. (2002)
California current	15	3	1997 (?)	Field (2004), Martell (2006) personal. communication.
North Sea	29	6	1995	Christensen (1995); Christensen (2006) personal communication
Iceland	24	14	1997	Mendy & Buchary (2001)
Antarctic Peninsula	39	2	1995–2001	Erfan & Pitcher (2005)
Falklands	44	6	2000	Cheung & Pitcher (2005)
Norwegian and Barentz Seas	58	19	1950–1997	Skaret & Pitcher (2007)

In each case, minor changes to Ecosim parameters were made to stabilize performance, although author's predator–prey vulnerability parameters were left unchanged wherever possible

significantly reduce the harvest from existing fisheries. All existing fisheries were held at the same fishing rate as current, while algorithms monitored biodiversity (Q90: Ainsworth & Pitcher, 2006) and ecosystem stability. Each 50-year run produced one krill fishery value; I found that between 44 and 113 50-year runs were needed to locate the maximum krill harvest. In order to ensure that genuine global maxima had been found, the procedure was repeated at least four times using random starting values for krill instantaneous fishing mortality between 0.1 and 1.5. In order to ensure equal weighting between the fishery and ecosystem portions of the objective function, four pilot runs provided a sensitivity procedure that adjusted actual weights in the elements of the objective function to produce almost equal changes in the objective function.

The resulting sustainable krill harvests are plotted against primary production levels for each ecosystem in Fig. 4. Although a number of functions would provide likely fits to this plot, I assumed a conservative peaked relationship of yield with increasing primary production using a Ricker-type curve. In fact, 90% of the world's ocean surface has primary production below the peak of the Ricker curve at about 250 g C m^2 yr, and only 1.3% lies above 500 g C m^2 yr, so the Ricker assumption is not very critical. Figure 4 shows the two 95 percentile Ricker curves obtained by using a Monte Carlo procedure on the probability density functions of the main inputs to the curve fitting.

In Table 3 this relationship with primary production is used to predict krill harvest in five major world ocean habitat zones (Pauly & Christensen, 2002). In order to express uncertainty, upper and lower 95 percentile values from the curve fitting have been used. The harvestable proportion in Table 3 expresses operational and technological difficulties in taking the full krill quota over such a vast scale. Figure 5 shows the final result, compared to the simple calculation based on transfer efficiencies (horizontal bar), and projections of 2020 fish catches and aquaculture yields. The analysis suggests that between 260 and 740 million tonnes of krill might be harvested sustainably with little impact on ecosystems or existing fisheries and livelihoods.

We are here assuming that krill fishing technology will be devised to be able to catch a reasonable proportion of krill species (the proportions in Table 3 reflect my own guesses at this figure) in seasons when they are reasonably aggregated, as most in fact are at some stage in their life history (see contributions in Everson, 2000). There have indeed been number of experiments with krill harvesting technology, including bow-mounted scoops. If technological solutions to krill harvest are not so successful as assumed, then sustainable and precautionary catches will likely be lower than suggested here.

Do we believe this forecast of between an approximately 3 and 6-fold increase in seafood yield from the oceans? It is certainly subject to a number of uncertainties and could be improved. In terms of

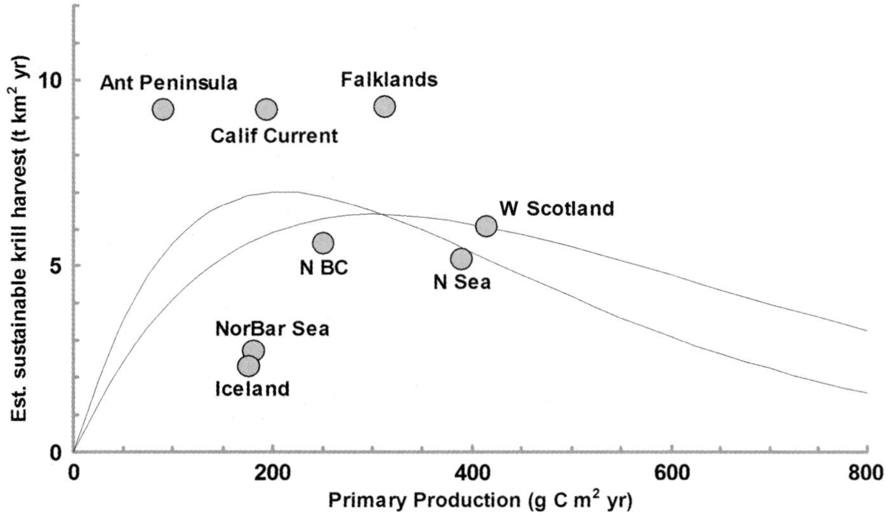

Fig. 4 Estimates of sustainable krill harvest obtained from eight published ecosystem models: Antarctic Peninsula; California Current; Falklands; West Scotland, North Sea; Norwegain and Barebnts Sea (NorBar); Iceland. Details of the ecosystem models are provided in Table y. Sustainable and responsible harvest of krill was obtained from automated searches, using 50-year simulations, for maximum values of an objective function that included krill yield and ecosystem structure. Searches examined fishery mortality parameter values from eight ecosystem models taken from the literature shown in Table y. In each case, searches were constrained to bar any changes larger than 5% in existing (i.e. present day) fisheries and in the biomass of other ecosystem groups. Thin lines show upper and lower ranges for the parameters of a Ricker-type curve fitted to the data by least squares, and then subjected to 500 Monte Carlo simulations using triangular probability density functions for the primary production and optimal krill harvest values

Table 3 Summary of estimation procedure for sustainable krill harvest

Zone	Area 10^6 km^2	PP gC m^2 yr	Estimated Krill SY t/km^2 /yr		Harvestable proportion		Estimated Krill SY tonnes	
			Low	High	Low	High	Low	High
Open ocean	332	103	4.0	6.0	0.15	0.3	198,290,721	597,021,945
Upwellings	0.8	973	0.8	2.4	0.5	0.8	322,422	1,515,732
Trop. shelves	8.6	310	6.1	6.7	0.2	0.6	10,424,404	34,663,578
Shelves	18.4	310	6.1	6.7	0.5	0.8	55,758,442	98,885,247
Coastal/reef	2	890	1.1	2.8	0.5	0.8	1,095,641	4,555,704
Total							265,891,630	736,642,205

PP, average annual primary production. Columns 1, 2 and 3 taken from Pauly and Christensen (2002, Table 2). Krill sustainable krill harvest from calculations described in the text. Low and high values are approximate 95% limits extracted from Monte Carlo error estimation. Harvestable proportions of sustainable yields are guesses. PP, Primary production from phytoplankton; SY, sustainable yield

modelling, there are a number of uncertainties that could be dealt with by adjusting the fitting procedure, or by using more than these eight ecosystem models to determine the relationship with primary production (several more suitable models could likely be located).

However, the important issue for marine biology is that the ecology of krill and links to its predators, food supply and habitats needs to be understood in much more detail for this analysis to be significantly improved. Nicol (2006) shows how Antarctic krill life history has several bottlenecks, especially for the overwintering and juvenile stages feeding on ice algae. There is also critical competition with salps, which are favoured in years with less ice, while years with more ice favour krill. This food web switch is known to have large consequences for penguins, whales and seal species that depend upon Antarctic

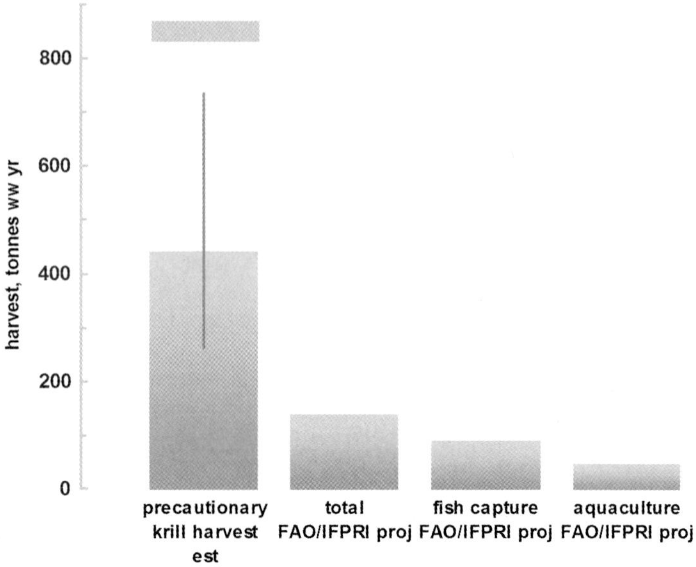

Fig. 5 Sustainable krill harvest compared to 2020 projections of capture fisheries and aquaculture from FAO and IFPRI (see text). Sustainable and responsible harvest of krill was obtained as described in Figure 4 and text. Confidence limits (thin vertical bar) were obtained from Monte Carlo simulations and values shown in Table 3. Thick horizontal bar shows krill harvest estimate derived from assuming a 10% transfer efficiency across trophic levels

krill. Equivalent issues for other krill-eating species outside the Antarctic (for example, humpback whales and shearwaters in the Bering Sea) are almost unknown. Moreover, a clearer ecological perspective on how fisheries alter ecosystems, using evidence from collapsed fisheries is needed (e.g. Smith & Link, 2005). The challenge for marine biologists is to improve insight of krill ecology to a level where much more confident forecasting can be performed with improved models.

This work has used the EwE modelling framework and so it may be asked if alternative modelling techniques would produce different results. Existing generally applicable ecosystem modelling systems include; inverse modelling (e.g. Savenkoff et al., 2004); agent-based modelling (e.g. 'InVitro': Gray et al., 2006); size-spectra models (e.g. Shin & Cury, 2004) and biogeochemical modelling ('Atlantis': Fulton et al., 2004). There are a number of more location-specific models such as the Globally Applicable Area-Disaggregated General Ecosystem Toolbox ('Gadget': Begley & Howell, 2004); Boreal Migration and Consumption (BorMiCon: Stefánnson & Pálsson, 1998); and the European Regional Seas Ecosystem Model (ERMSEM: Baretta-Bekker & Baretta, 1997). It is, however, worth noting that, at present, no general ecosystem simulation modelling systems other than EwE and, possibly Atlantis, have the capacity of performing an equivalent task to that set out here; reviews may be found in Fulton et al. (2007) and Whipple et al. (2000).

A practical restoration agenda for the oceans

My third challenge to marine biology reflects a saying of Spanish-American philosopher George Santayana (1863–1952): "Those who cannot remember the past are condemned to repeat it". The challenge springs from a need to provide new quantitative methods for describing past times when ecosystems were not as depleted as they are today. This includes a deeper understanding of ecological processes, especially the food, nutrient and energy webs of ecosystems, so that we can predict with more confidence with the effects of depleted biomass from fisheries, pollution, habitat destruction and climate variation and shifts. To date, ecology has not been very effective in dealing with changes in the species 'actors' in the ecosystem, such as result from invading species, local extinctions and reintroductions. Nevertheless, the aim is to find rigorous ways to learn from the past to enable forecasts of restoration in the future.

There is wide agreement about the importance of historical ecology, and many have embraced research that has attempted to use a variety of historical surveys, documentary archives, (Jackson et al., 2001), archeology, sediment chemistry, and even catch data from past fisheries (Rosenberg et al., 2005). However, moving beyond historical reconstruction for its own sake, I have attempted to devise a practical restoration agenda for aquatic ecosystems which uses the past as a quantitative guide to what may be restored in the future and hence guide ocean policy: this research has been termed 'Back to the Future' (BTF). BTF research has been described in considerable detail elsewhere (e.g. Pitcher, 2005, 2007; Pitcher et al., 2005a, b; Pitcher & Ainsworth, 2007), and so only a brief summary and some highlights will be described here.

The essence of the BTF idea is to construct simulation models of past ecosystems at a number of points in time, and then use these ecosystem models to evaluate alternative practical targets for future restoration, given risks from climate influences and uncertainty. The credibility of the procedure depends upon the ecosystem models: to date we have used a model framework based on Ecopath-with-Ecosim dynamic simulations (review: Christensen & Walters, 2004), but an alternative modelling system could easily take its place once developed to the same level of utility (see discussion above). In theory, conventional single species models could be used for BTF analysis, provided that doubts about applicability to whole ecosystem dynamics could be assuaged (Walters et al., 2005).

The scientific parts of the BTF analysis for an ecosystem follows six main procedures, which are briefly described below, together with references to sources that provide a more complete account. Figure 6 and its caption summarizes the essential features of the BTF process.

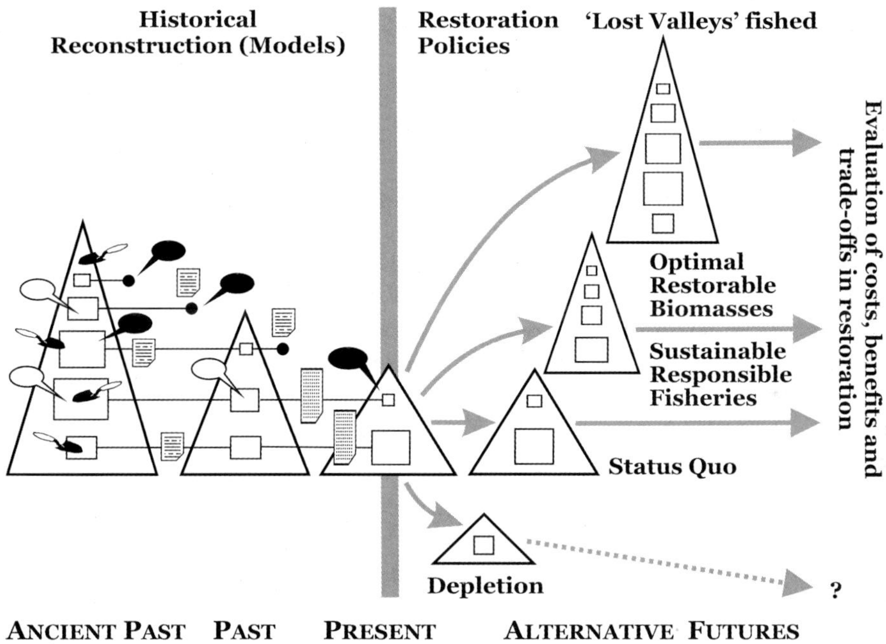

Fig. 6 Diagram illustrating the 'Back-to-the-Future' concept for the adoption of past ecosystems as future policy goals. Triangles at the left represent a time series of historical ecosystem models, constructed at appropriate past times before the present (thick grey vertical line), where the vertex angle is inversely related and the height directly related to biodiversity and internal connectance. Time lines of some representative species in the ecosystems are indicated; size of the boxes represents relative abundance and solid circles represent local extinctions (=extirpations). Sources of information for constructing and tuning the ecosystem models are illustrated by symbols for historical documents (*paper sheet symbol*), data archives (*tall data table symbol*), archaeological data (*trowel*), the traditional environmental knowledge of Indigenous Peoples (*open balloons*) and local environmental knowledge of coastal communities (*solid balloons*). At right are alternative future ecosystems, representing further depletion, the *status quo*, or restoration to 'Lost Valleys' that may be used as alternative policy goals. Restored 'Lost Valleys' may be fished with sustainable, responsible fisheries designed according to specified criteria, and aiming at Optimal Restorable Biomasses determined using objective quantitative policy searches. Final choice of BTF policy goals are made by comparing trade-offs, cost and benefits among possible futures using socio-economic and ecological objectives agreed among industrial and small-scale fishers, government, conservation, coastal communities and other stakeholders in order to maximize compliance. Diagram does not show evaluation of risks from climate fluctuations and model parameter uncertainty (Modified from Pitcher 2007)

Synoptic ecosystem reconstruction

This work includes culling sources of historical data from surveys, archives, historical documents, archaeology, sedimentology, climatology, and local and traditional environmental knowledge (Heymans & Pitcher, 2004). Historical and other sources are used to determine presence and absence, and, where possible relative abundance of the main functional groups in the ecosystem. Mass-balance models (like EwE) are helpful in this respect as they allow only ranges of abundance (in the form of biomass) that can be supported by available food sources (such as plankton). The choice of time periods to model is based on major historical changes in fisheries (for example, the introduction of steam trawlers 1880–1920) and climate (for example, cold and warm ocean periods). The work includes the reconstruction of total fisheries extractions (Christensen et al., 2003; Rose, 2004; Ainsworth & Pitcher, 2005a). This synoptic ecosystem reconstruction work enables the construction of preliminary ecosystem models for chosen time periods with comparison to the present day (e.g. Ainsworth et al., 2002, 2007). An example taken from three historical ecosystem models of changes in the northern Gulf of California, Mexico, is shown in Fig. 7; further description is provided in the caption (Lozano-Montes & Pitcher, 2007).

Modelling non-linear ecological dynamics

The second step is to tune the parameters of the models so that they may simulate dynamic changes over time; for models of more recent periods this can include fitting to time series data. It may also include challenges such as modelling the dynamics of extinct and locally extinct species (Pitcher, 2004a), which in some cases have been reintroduced (e.g. sea otters in British Columbia); emulating trophic cascades and 'wasp-waist' forage fish ecosystems (Pitcher, 2004b); and accounting for climate variability and climate changes (Pitcher & Forrest, 2004). More recent models may be run forwards through time in a dynamic fashion, fitted to time series data from stock assessment and abundance surveys, and driven by suitable climate factors (e.g. Pitcher et al., 2005a, b; Ainsworth et al., 2007; Ainsworth & Pitcher, 2007). Examples from historical reconstructions of the northern Gulf of California, Mexico, for two functional groups (a shrimp and an endemic croaker) are described in Fig. 8A, B and their captions. This step improves the model parameters for past and present ecosystem models in order to provide simulation models that may later be used to forecast responses under changed management conditions.

Designing sustainable and responsible fisheries

The next step is to determine how an ecosystem may be harvested sustainably and responsibly, given that a specified past ecosystem might be restored to a state described in the previous step. The problem here analogous to devising a fleet, fishing with today's technology, that is suddenly allowed to fish a hypothetical remote fjord, a 'Lost Valley', which has been newly discovered (Pitcher et al., 2004). For use in this forecasting process, catch and selectivity parameters of existing fisheries are modified in the model according to a set of criteria that minimise discards, bycatch and damage to habitats, and adopt standards of ethics, equity and transparency (Pitcher et al., 2004). Candidate fisheries are challenged by their likely compliance with the FAO (UN) Code of Conduct for Responsible Fisheries (FAO, 1995), which can be evaluated using a rapid appraisal technique ('Rapfish': Pitcher, 1999; Pitcher & Preikshot, 2001). In practice, rate of change of fleet design towards the ideal may have to be modified by considerations deriving from the availability, practicality and cost-effectiveness of the fishing technology needed to achieve these goals. The impacts of the desired changes in fishing fleets on the sustainability of the livelihoods of existing fishers and their coastal communities needs to be evaluated so that they lie within an achievable envelope. The list of fishery criteria for this step is presented with amplification in Table 4 (modified from Pitcher et al., 2004 and Pitcher, 2005).

Establishing optimal fishing policies and choosing restoration targets

In the EwE model framework, optimal fisheries may be determined by maximizing multi-criterion objective functions using an automated search procedure (Cochrane, 2002; Walters et al., 2002). Usually searches are run over a simulated 50 years, starting with one of the fitted ecosystem models from step 3

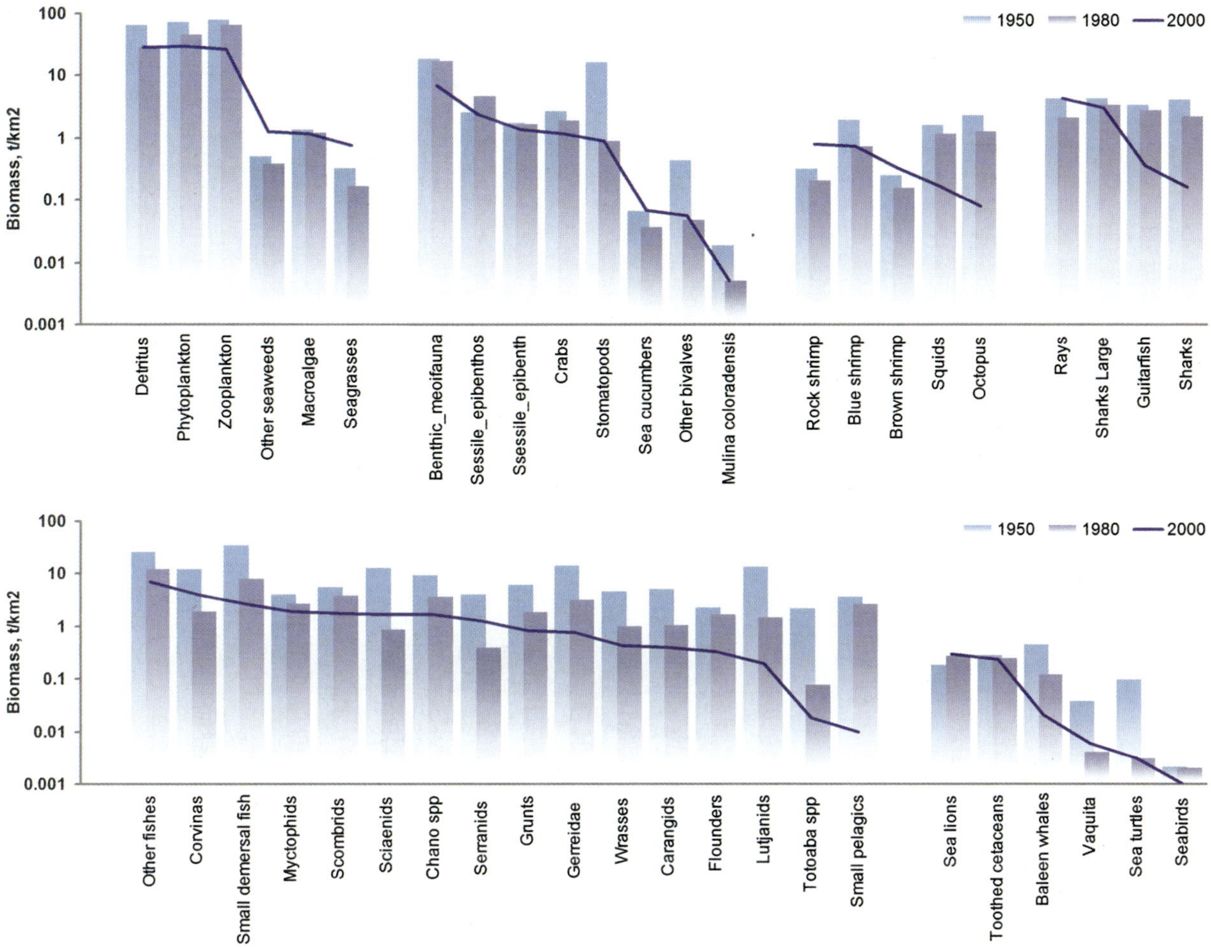

Fig. 7 Example of biomasses (in t/km2) from three historical ecosystem simulation models constructed from archival and historical material, interviews with local fishers, surveys and stock assessments for the northern Gulf of California around the Colorado River delta, Mexico. Present day (2000) biomasses are shown by the thick line; 1980 biomasses by dark blue (dark gray) bars; 1950 biomasses by light blue (light grey) bars. Note logarithmic scale. Biomasses are shown for 45 functional groups in the ecosystem (5 additional juvenile groups have been added to their adults for the plot). Groups on the plot are (upper plot), primary producers, detritus and plankton; benthic invertebrates; shrimps and mollusks; (lower plot), elasmobranchs; bony fish; marine mammals and seabirds. Names of the groups are given on the x-axis: in most cases several species are combined in model functional groups. Further detail of the model construction may be found in Lozano-Montes (2006) and Lozano-Montes & Pitcher (2007). Note major collapses of the endemic porpoise (vaquita), endemic giant croaker (totoaba), groupers (Lutjanids), small pelagics, sharks, whales, seabirds and turtles. Most fish groups exhibit large declines, and only shrimp groups show significant increases in abundance. Most of these changes have been brought about by two factors: changes in river flow and unregulated fisheries. Large reductions in the flow of the Colorado River have been caused by dams constructed in the USA (note also reductions of phytoplankton, zooplankton, and the endemic bivalve *Mulina coloradensis*). The vaquita, totoaba and sharks have been affected especially by net fisheries

(Ainsworth et al., 2004). There is a choice of ecological and socio-economic goals for the objective functions (e.g. Buchary et al., 2003), and hence a range of scenarios for sustainably fishing the 'Lost Valleys' from step 3 above may be investigated. Under the BTF framework, the choice of future restoration targets will resemble, but not fully copy, past ecosystems. The techniques include the adoption of alternative economic discounting procedures, including those based upon intergenerational equity (e.g. Sumaila et al., 2001). An example of results from analyses of the northern British Columbia

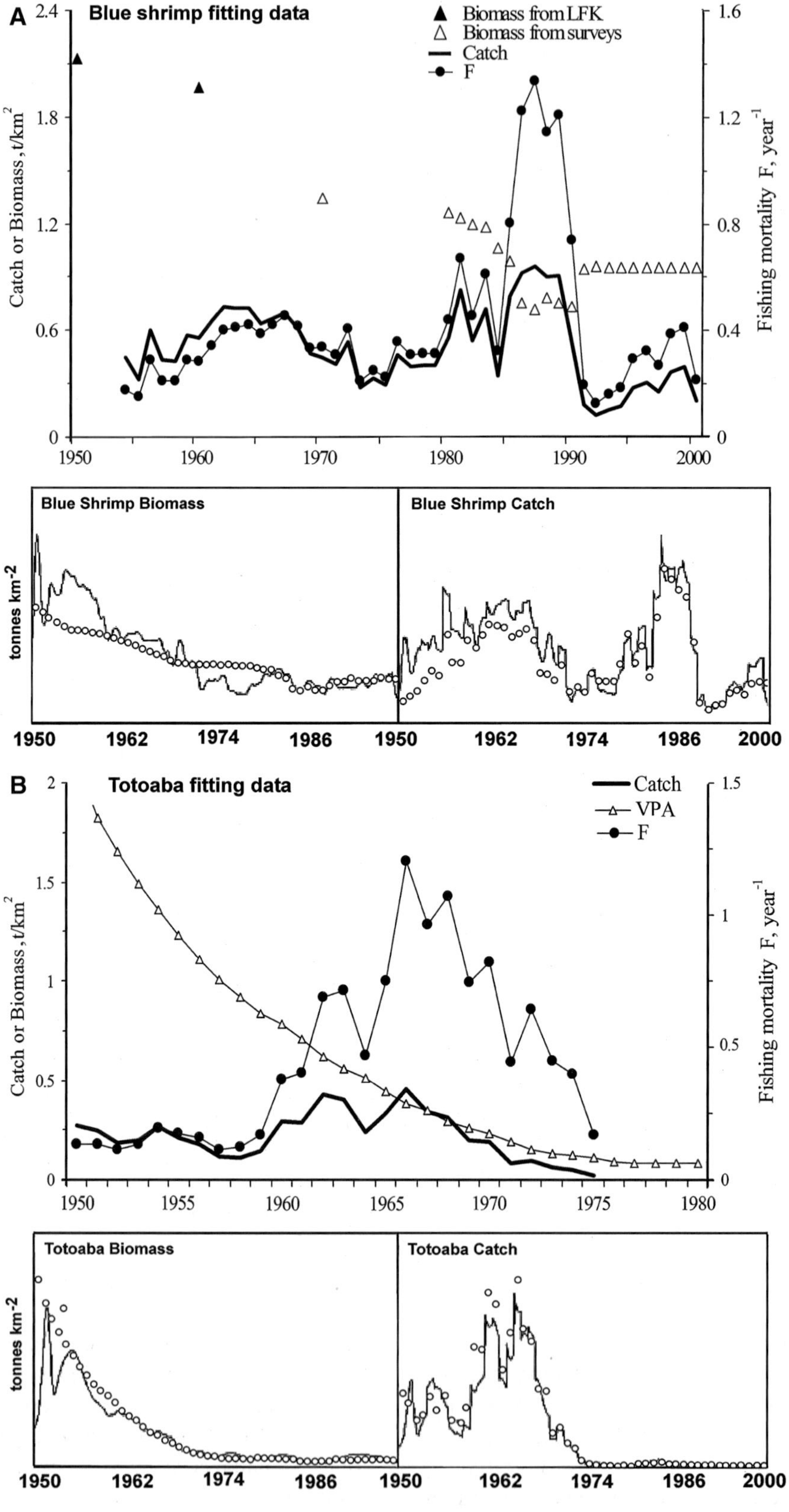
A
Blue shrimp fitting data
Biomass from LFK
Biomass from surveys
Catch
F
Catch or Biomass, t/km2
Fishing mortality F, year-1
Blue Shrimp Biomass
Blue Shrimp Catch
tonnes km-2
B
Totoaba fitting data
Catch
VPA
F
Catch or Biomass, t/km2
Fishing mortality F, year-1
Totoaba Biomass
Totoaba Catch
tonnes km-2

◀**Fig. 8** **A** *Top Panel:* Estimation of annual fishing mortality (circles) of blue shrimp (*Litopenaeus stylirostris*) in the upper Gulf of California from 1954 to 2000 using catch data reported by the government fisheries agency (line), and local biomass surveys (white triangles: see Lozano & Pitcher, 2007 for references) Perception of the average abundances from 1950 to 1960 (black triangles) was estimated from interviews with local fishers (see Lozano-Montes et al., 2007). The annual fishing mortality was used to drive the ecosystem model simulations for blue shrimp. *Lower Panels:* Blue shrimp simulation (gray line) outputs for biomass (left panel) and catch (right panel) from the Gulf of California ecosystem model compared with data (open circles). The Colorado River discharge from 1950 to 2000 was used as a forcing function affecting primary productivity; using this climate factor improved the least squares fit by over 20%. **B**. *Top Panel:* Annual fishing mortality from 1950 to 1978 (black circles) of the Gulf giant croaker or totoaba (*Totoaba macdonaldi*) using stock assessment (VPA: see Lozano and Pitcher 2007 for references) to estimate past biomasses from catches (solid line) until 1978 when the fishery collapsed and was banned. The time-series of fishing mortality of totoaba was one of those employed to tune the upper Gulf of California ecosystem model simulations. *Lower Panels:* Totoaba simulation (gray line) outputs for biomass (left panel) and catch (right panel) from the Gulf of California ecosystem model compared with data (open circles). The Colorado River discharge from 1950 to 2000 was used as a forcing function affecting primary productivity; using this climate factor improved the least squares fit by over 20%

ecosystem, which shows expected economic values from 15 optimal fishing scenarios together with forecast changes in biodiversity, is provided in Fig. 9. The trade-offs that would be associated with each of the four possible restoration periods and 15 policy objectives are made explicit (Ainsworth & Pitcher, 2005a). Ecosystem-based performance criteria may be supplied by comparison with optimal searches based on a standard known as the 'maximum dexterity fishing fleet' (Ainsworth, 2006), a hypothetical set of fishery gear types that catch only single target species with no unintended bycatch or discards.

Evaluation of risks and uncertainties for alternative restoration policies

Repeated simulations using values randomly chosen from parameter uncertainties can enable the output of Probability Density Functions for output catches, economic values and ecological indices (e.g. biodiversity: Ainsworth & Pitcher, 2006). Monte Carlo simulations can also include time series of climate fluctuations, based either on historical series or on forecast climate scenarios. Combined Monte Carlo simulations allow risks of specified levels of depletion (including local extinctions) to be estimated (e.g. Pitcher et al., 2005b). In this way alternative restoration policies may be compared with the status quo: an example from BTF work on northern British Columbia is provided in Fig. 10.

Evaluation of viable economic and social consequences of alternative restoration trajectories

Figure 11 illustrates an important new policy goal that has derived from the BTF work: the Optimum Restorable Biomass (ORB; Ainsworth & Pitcher, 2007). ORB derives from the realisation that there would be little point in going to the trouble to restore a past ecosystem only to then deplete it by fishing, albeit sustainably. In practice, any actual restoration trajectory would have to start from the present depleted system. Hence the policy target of restoration would be a historical ecosystem as it would appear after establishing sustainable fishing according the choice of policy objectives. Comparison of the benefits provided by alternative ORBs allows choice of suitable restoration targets. Moreover, using the ORB as a biomass restoration target entered into an objective function, and a range of permitted economic costs, the simulation model can search for optimal trajectories that approach the target: in a complex non-linear ecosystem model there are often many such. In some cases we have found that comparison of a range of optimal restoration trajectories suggests that some may allow progress towards an ORB restoration target and at the same time generate fishery profits in excess of bank interest, a finding that suggests that BTF may present some win-win options for policy makers (Ainsworth & Pitcher, 2007).

In addition to scientific work in choosing and designing practical restoration policies, underlying the BTF approach is a deeper need for more active stewardship of marine resources and ecosystems. Moreover, if management guided by ORB policies is to work, compliance may be fostered by community and stakeholder involvement at each stage of the BTF analysis process (Power et al., 2004; Pitcher et al., 2005a, b). For example, science, government and coastal community involvement in using local

Table 4 Candidate list of criteria for sustainable and responsible fisheries to be opened in a restored ecosystem (Modified from Pitcher et al. 2003)

	Criteria for sustainable fisheries	Notes
1	Minimal by-catch discards	Technological modifications to gear. Over the past decade, trawl, trap and purse seine fisheries have attained impressive reductions in by-catch through the use of separators, lifters, gates and excluders (review: Kennelly & Broadhurst, 2002), and by altering fishing practices (e.g. dolphins released in tuna purse seine fisheries, Hall, 1988; long-line setting adapted to reduce hook mortality in seabirds, Brothers et al., 1999). These technological advances may be successfully employed to greatly reduce unintended catches of non-target species of fish, marine mammals, reptiles, birds and invertebrates (review: Hall & Mainprize, 2005). Moreover, in some jurisdictions, discards have become illegal (e.g. Norway).
2	No damage to habitat by gear	Technological modifications to gear. Unmodified bottom trawls and dredges may do great harm to sessile benthic invertebrates (e.g. sponges, gorgonids, corals) whose architecture acts as refuge habitat for juveniles of many commercial fish species (Kaiser et al., 2002). To meet this criterion, technological improvements to the fishery will have to be employed to minimise damage, for example by permitting only trawls that fish above the bottom. Where some collateral damage to benthos is inevitable, such as in prawn trawls, large and progressive reductions in damage, say tenfold, might be mandated.
3	Include aboriginal fisheries	Customary rights recognized. Some fisheries by indigenous or aboriginal peoples were sustainable over thousands of years (e.g. eulachon, salmon and halibut in the Pacific Northwest, Richardson, 1992). In terms of equity, they should be included in the candidate fisheries portfolio, provided the take is sustainable, and where such customary rights are recognised. Aboriginal fishers often have an intimate connexion with and knowledge of coastal marine ecosystems and their support for a policy like BTF could enhance compliance with regulations.
4	Include traditional target species	Except where #1 and #2 would bar. Provided criteria 1 and 2 above are satisfied, this category is included because there will be an understandable demand for traditional desirable fish species in local fishing communities. For example, the historic Atlantic halibut fishery has not proven sustainable, but the species would be in demand as a target in a restored ecosystem.
5	Minimise risk to charismatic species	Except as under #3 and #7. Whilst it is evident from the recorded history of seabirds, whales, seals and sirenians that many 'charismatic' species are sensitive to exploitation by humans (e.g. Roman and Palumbi 2003), this criterion may well be in conflict with criteria 3 and 4 above, since coastal peoples traditionally exploited seals, sea lions, whales, dugongs, turtles, ducks, gulls, petrels, auks and other seabirds. Where customary rights are recognised, an aboriginal take of these species would be allowed under criterion 3, with appropriate consent under criterion 7 below. On the other hand, many marine mammal, bird and shark species have recently become 'charismatic' to the conservation movement, and legal bans on killing them reflect public revulsion at their use for human food. But these views are volatile and local, so in the last resort, the choice of whether to exploit these types of animals will be locally or nationally determined. The only rational criterion is avoidance of excessive depletion and minimal risk of extirpation.
6	Exclude fisheries on juveniles	Except where minimal impact is proven. Generally, heavy fishing on juveniles increases the risk of recruitment failure, so such fisheries would not normally be allowed. In some cases, traditional fisheries (criterion 4) include eggs, fry and juveniles of highly fecund species such as herring, anchovy, sardines, milkfish or hake, so such fisheries would be permissible where impacts can be proven to be minimal. Likewise, fisheries might be permitted on restricted numbers of juveniles where adults live and spawn in refuges from fishing, as in traditional Mediterranean fisheries (Caddy, 2000).
7	Participatory vetting of fisheries	By management agency, local community and public. To maximise support, the local fishing community must vet and approve the list of fisheries, notwithstanding criteria 3 and 4. In addition, the management agency must be convinced with science-based evidence that gear is appropriate (criteria 1 and 2), that management and monitoring (criterion 9) are feasible for the chosen fisheries, and that the scientific basis of the forecasting (criterion 8) represents best practice.

Table 4 continued

Criteria for sustainable fisheries	Notes
8 Simulations show fishery sustainable	100-year simulations are satisfactory. Fishery assessments must show that the biomass of the main ecosystem groups, biodiversity, and the fishery catches themselves are likely sustainable and will not fluctuate more than a predetermined and agreed amount over a 100-year period. A tougher criterion would be to insist that the forecasts are robust against climate fluctuations and uncertainty to a specified level of risk. The great importance of ecosystem-based analysis is evident here, since, on their own, single-species stock assessments cannot show risks to charismatic or non-target organisms, or sessile organisms that provide important structural cover. Criterion 8 describes a critical part of the process: examining trade-offs of ecological with social and economic objectives using as a wide a range of indices as possible
9 Adaptive management plan in place	Adaptive changes to the unexpected (e.g. climate change). Because environmental changes (climate, pollution) and our ignorance of fundamental ecology often lead to the unexpected in natural ecosystems, it would be prudent for the restored 'Lost Valley' and its fisheries to be subject to regular monitoring of the indices from criterion 8. This would allow adaptive shifts in fishing, much like the way that catch quotas and fishing locations are regulated today, but driven by an ecosystem approach.

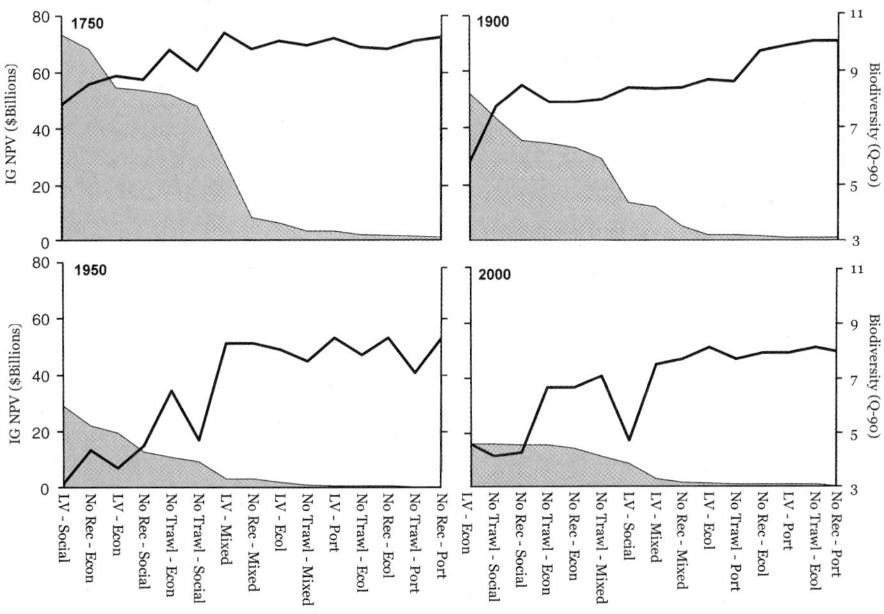

Fig. 9 Diagrams comparing results of optimal fishery searches using fitted ecosystem simulation models of northern British Columbia for four historic time periods that had been fitted to all available time series data (see Ainsworth et al., 2007). Left vertical axes and shaded areas show intergenerational net present value (IG NPV); right vertical axes and solid line show equilibrium biodiversity established after dynamic harvest simulations (Q90: Ainsworth and Pitcher 2005×). Fisheries were set up according to the rules outlined at step 3. Horizontal axes show 15 scenarios that vary harvest objectives for these fisheries; these are social (employment), economic (profit, NPV), ecological (longevity, life history parameters), mixed (social, economic and ecological objectives equally weighted), and 'portfolio log-utility' (a risk-averse policy, Goll & Kallsen, 2003). Several fleet compositions were also included (LV = idealized 'Lost Valley' fleet; NoRec = recreational fishing gear removed; NoTrawl = bottom trawlers removed). Note that the results demonstrate some clear trade-offs revealed by this ecosystem-based fisheries policy analysis: policies that generate large revenues tend to sacrifice biodiversity (left side of plots), whereas policies that maintain biodiversity provide less revenue (right side of plots) (Redrawn from Ainsworth & Pitcher, 2005a)

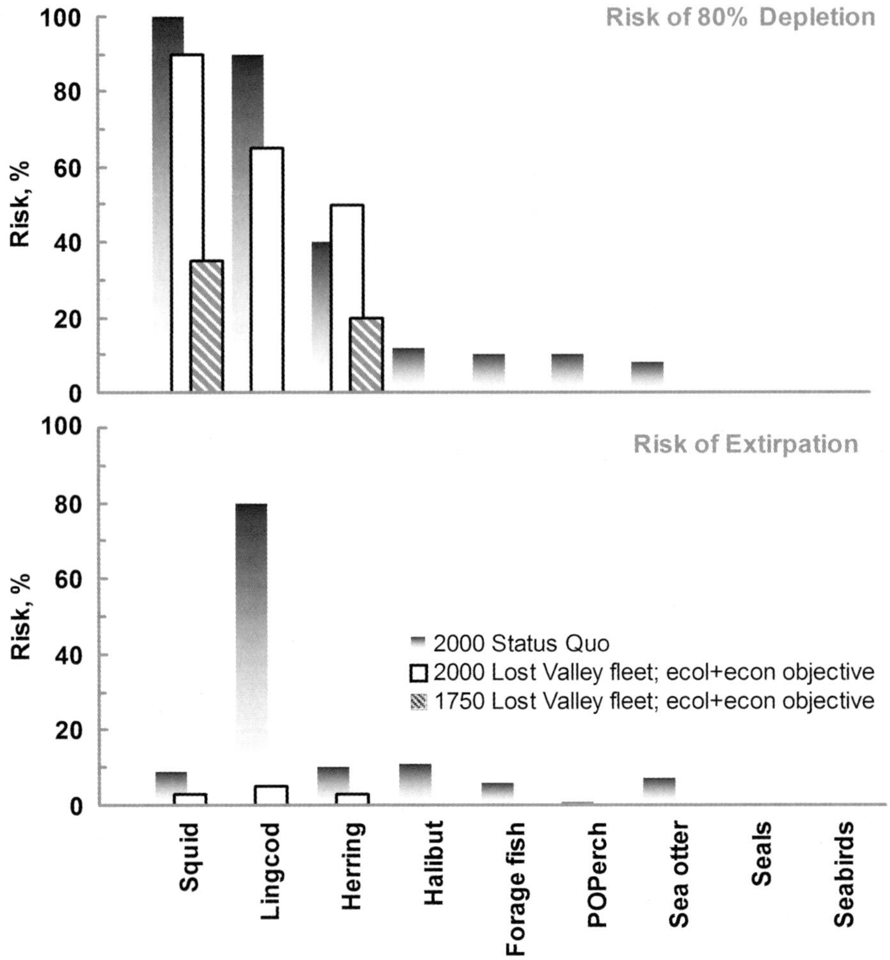

Fig. 10 A comparison of risks under optimized fisheries and a restored ecosystem. An ecosystem model for Northern British Columbia, fitted to time series data, was driven over 50 years with a climate forcing function derived from randomized historical climate data (tree rings) for the area. Uncertainties (coefficients of variation) were added for all major model parameters. Repeated 50 year Monte Carlo simulations used values from these parameters (only sets of parameter values that resulted in mass-balanced solutions were accepted). Risks were assessed from counts of the number of times depletion criteria were exceeded. Plots show the risks of 80% depletion, and of extirpation (local extinction, defined as <0.1% starting biomass), for 9 functional groups in the ecosystem model. Three alternative scenarios are compared: a 'status quo' model based on existing fisheries in Northern British Columbia in 2000; a 2000 model that includes a 'Lost Valley' fishing fleet designed according to the criteria described in step 3 above, and whose fisheries have been optimised for equally-weighted ecological and economic objectives; and an equivalent procedure to the latter, based on a model of a hypothetical restoration of the ancient 1750 ecosystem prior to contact of local peoples with Europeans

environmental knowledge (e.g. Ainsworth & Pitcher, 2005b), input into design of model scenarios (e.g. Haggan & Pitcher, 2002; Power et al., 2004), and the design of policies that may be enacted in legislation (e.g. Power & Pitcher, 2007). Indeed, however desirable the long-term goals, compliance with management regulations that do not receive community assent are likely doomed (e.g. Kaplan & McCay, 2004).

Conclusions

All three areas of challenge to marine biology and the examples I have presented here represent work-in-progress. Their credibility and utility pivot on the quality of our insight into aspects of ecological processes that are critical to the forecasting that has to underlie science-based public policy concerning natural resources. As marine biologists we are living

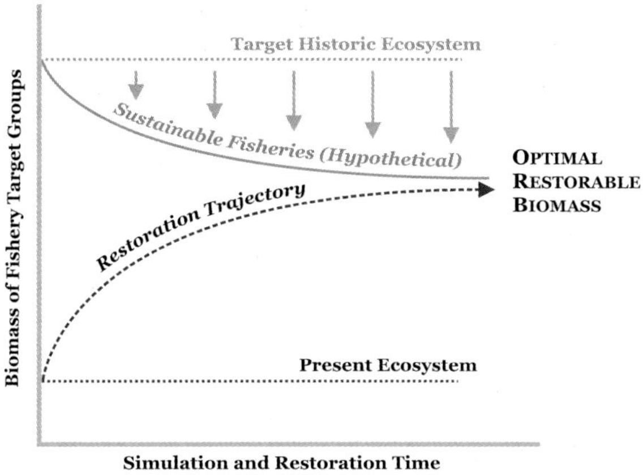

Fig. 11 Diagram illustrating the concept of Optimal Restorable Biomass as an ecosystem-based fishery management target (Redrawn from Ainsworth & Pitcher, 2007)

through an era of unprecedented change through anthropogenic impacts on marine ecology with dire consequences for seafood security. My suggestion is that by focussing research on areas that can help a better understanding of causes and improve the quality of scientific forecasting, we may hope to mitigate, and even reverse, the worst of these changes. We should be brave enough to try to invent new tools and forecast, provided that (Santayana again) "scepticism, like chastity, should not be relinquished too readily".

Acknowledgements This paper comprised one of the keynote addresses at the 41st European Marine Biology Symposium, in Cork, Ireland, September 4–8, 2006. I am most grateful to John and Julia Davenport and the rest of the Organising Committee for the invitation to present this work. For inspiration in tackling this work I am grateful to my colleagues Drs Daniel Pauly and Kevern Cochrane, and for ideas, comments and help with the analyses I thank members of the Fisheries Ecosystem Restoration Research group in the Fisheries Centre at the University of British Columbia, especially Dr Cameron Ainsworth, Dr Telmo Morato, Nigel Haggan, Carie Hoover and Divya Varkey.

References

Ainsworth, C. A., 2006. Strategic marine ecosystem restoration in northern British Columbia. D.Phil. Thesis, University of British Columbia, 422pp.

Ainsworth, C., J. J. Heymans & T. J. Pitcher, 2004. Policy search methods for back to the future. In Pitcher, T.J. (ed.) Back to the Future: Advances in Methodology for Modelling and Evaluating Past Ecosystems as Future Policy Goals, 48–63. Fisheries Centre Research Reports 12(1): 158pp.

Ainsworth, C., J. J. Heymans, T. J. Pitcher & M. Vasconcellos, 2002. Ecosystem models of Northern British Columbia for the time periods 2000: 1950: 1900 and 1750. Fisheries Centre Research Reports 10(4): 41pp.

Ainsworth, C. H. & T. J. Pitcher, 2005a. Evaluating marine ecosystem restoration goals for Northern British Columbia. In Kruse, G. H., V. F. Gallucci, D. E. Hay, R. I. Perry, R. M. Peterman, T. C. Shirley, P. D. Spencer, B. Wilson & D. Woodby (eds), Fisheries Assessment and Management in Data-limited Situations. Alaska Sea Grant, Fairbanks, USA: 419–438.

Ainsworth, C. H. & T. J. Pitcher, 2005b. Using local ecological knowledge in ecosystem models. In Kruse, G. H., V. F. Gallucci, D. E. Hay, R. I. Perry, R. M. Peterman, T. C. Shirley, P. D. Spencer, B. Wilson, & D. Woodby (eds), Fisheries Assessment and Management in Data-limited Situations. Alaska Sea Grant, Fairbanks, USA, 289–321.

Ainsworth, C. H. & T. J. Pitcher, 2006. Modifying Kempton's species diversity index for use with ecosystem simulation models. Ecological Indicators 6(3): 623–630.

Ainsworth, C. H. & T. J. Pitcher, 2007. Back-to-the-future in Northern British Columbia: evaluating historic marine ecosystems and optimal restorable biomass as restoration goals for the future. In Nielson, J. (ed.), Reconciling Fisheries with Conservation: Proceedings of the 4th World Fisheries Congress. American Fisheries Society, Bethesda, USA (in press).

Ainsworth, C., T. J. Pitcher, J. J. Heymans & M. Vasconcellos, 2007. Historical Reconstruction of the Marine Ecosystem of Northern British Columbia from Pre-European Contact to the Present (submitted).

Baretta-Bekker, J. G. & J. W. Baretta (eds.), 1997. Special issue: European regional seas ecosystem model II. Journal of Sea Research 38(3/4).

Baum, J. K., R. A. Myers, D. G. Kehler, B. Worm, S. J. Harley & P. A. Doherty, 2002. Collapse and conservation of shark populations in the Northwest Atlantic. Science 299: 389–392.

Begley, J. & D. Howell, 2004. An overview of Gadget, the Globally applicable Area-Disaggregated General Ecosystem Toolbox. ICES CM 2004/FF: 13pp.

Berkes, F., T. P. Hughes, R. S. Steneck, J. A. Wilson, D. R. Bellwood, B. Crona, C. Folke, L. H. Gunderson, H. M. Leslie, J. Norberg, M. Nyström, P. Olsson, H. Österblom, M. Scheffer & B. Worm, 2006. Globalization, roving bandits, and marine resources. Science 311: 1557–1558.

Buchary, E. A., W. W. L. Cheung, U. R. Sumaila & T. J. Pitcher, 2003. Back to the future: a paradigm shift to restore Hong Kong's marine ecosystem. 3rd World Fisheries Congress, Beijing, November 2000. American Fisheries Society Symposium 38: 727–746.

Caddy, J. F., 2000. Marine catchment basin effects versus impacts of fisheries on semi-enclosed seas. ICES Journal of Marine Science 57(3): 628–640.

Caddy, J. F., 2002. Limit reference points, traffic lights, and holistic approaches to fisheries management with minimal stock assessment input. Fisheries Research 56(2): 133–137.

Caddy, J. F., 2004. Current usage of fisheries indicators and reference points, and their potential application to management of fisheries for marine invertebrates. Canadian Journal of Fisheries and Aquatic Sciences 61: 1307–1324.

CCAMLR Statistical Bulletin, 2001. CCAMLR. Hobart, Australia.

Cheung, W. L. & T. J. Pitcher, 2005. A mass-balance model of the marine ecosystem and fisheries of the Falkland Islands. In Palomares M. L. D., P. Pruvost, T. J. Pitcher & D. Pauly (eds), Modeling Antarctic Marine Ecosystems. Fisheries Centre Research Reports 13(7): 65–85.

Cheung, W. L., R. Watson, T. Morato & T. J. Pitcher, 2007. Change of intrinsic vulnerability in the global fish catch. Marine Ecology Progress Series 333: 1–12.

Christensen, V., 1995. A model of trophic interactions in the North Sea in 1981, the Year of the Stomach. DANA 11(1): 1–19.

Christensen, V., 1998. Fishery-induced changes in a marine ecosystem: Insight from models of the Gulf of Thailand. Journal of Fish Biology 53: 128–142.

Christensen, V., S. Guénette, J. J. Heymans, C. J. Walters, R. Watson, D. Zeller & D. Pauly, 2003. Hundred-year decline of North Atlantic predatory fishes. Fish and Fisheries 4: 1–24.

Christensen, V. & C. J. Walters, 2004. Ecopath with ecosim: methods, capabilities and limitations. Ecological Modelling 172: 109–139.

Cochrane, K., 2002. Using of ecosystem models to investigate ecosystem-based management strategies for capture fisheries: introduction. In Pitcher, T. J. & K. Cochrane (eds), The use of ecosystem models to investigate multispecies management strategies for capture fisheries, 5–10. Fisheries Centre Research Reports 10(2): 156pp.

Delgado, C., N. Wada, M. Rosegrant, S. Meijer & M. Ahmed, 2003. Fish to 2020: Supply and Demand in Changing Global Markets. International Food Policy Research Institute (IFPRI), Washington, DC, USA: 232pp.

Erfan, A. & T. J. Pitcher, 2005. An ecosystem simulation model of the Antarctic Peninsula. In Palomares M. L. D., P. Pruvost, T. J. Pitcher & D. Pauly (eds), Modeling Antarctic Marine Ecosystems. Fisheries Centre Research Reports 13(7): 5–20.

Fulton, E. A., J. S. Parslow, A. D. M. Smith & C. R. Johnson, 2004. Biogeochemical marine ecosystem models II: the effect of physiological detail on model performance. Ecological Modelling 173: 371–406.

Everson I. (ed.), 2000. Krill: Biology, Ecology And Fisheries. Fish and Aquatic Resources Series 6. Blackwell, Oxford, UK: 372pp.

Everson, I. & W. K. de la Mare, 1996. Some thoughts on precautionary measures for the krill fishery. CCAMLR Science 3: 1–12.

FAO, 1995. Code of Conduct for Responsible Fisheries. FAO (UN), Rome, 41pp.

FAO, 2004. Future prospects for fish and fishery products: medium-term projections to the years 2010 and 2015. FAO, Rome. Fisheries Circular FIDI/972–1.

Field, J. C., 2004. Application of ecosystem-based fishery management approaches in the northern California current. Ph.D. Thesis. School of Aquatic and Fishery Sciences, University of Washington, 408 pp.

Fulton, E., T. Morato & T. J. Pitcher, 2007. Modelling seamount ecosystems and their fisheries, pp. 296–332, Chapter 15. In Pitcher, T. J., T. Morato, P. J. B. Hart, M. R. Clark, N. Haggan & R. S. Santos (eds), Seamounts: Ecology, Fisheries and Conservation. Fish and Aquatic Resources Series 12, Blackwell, Oxford, UK. 527pp.

Goll, T. & J. Kallsen, 2003. A complete explicit solution to the log-optimal portfolio problem. The Annals of Applied Probability 13(2): 774–799.

Gray, R., E. A. Fulton, L. R. Little & R. Scott, 2006. Operating model specification within an agent based framework. North West Shelf Joint Environmental Management Study Technical Report, Vol 16. CSIRO, Australia.

Haggan, N. & Pitcher, T. J., 2002. Introduction to the Prince Rupert Community Participation Workshop for 'Back to the Future' in Northern British Columbia. In Pitcher, T. J., M. D. Power & L. Wood (eds), Restoring the past to salvage the future: report on a community participation workshop in Prince Rupert, BC, 6–9. Fisheries Centre Research Reports 10(7): 56pp.

Haggan, N & T. J. Pitcher (eds), 2005. Ecosystem simulation models of Scotland's West Coast and Sea Lochs: Fisheries Centre Research Reports 13(3): 69pp.

Hall, S. J. & B. M. Mainprize, 2004. Towards ecosystem-based fisheries management. Fish and Fisheries 5(1): 1–20.

Hall, S. J. & B. M. Mainprize, 2005. Managing by-catch and discards: How much progress are we making and how can we do better? Fish and Fisheries 6: 134–155.

Halliday, R. G., L. P. Fanning & R. K. Mohn, 2001. Use of the Traffic Light Method in Fisheries Management Planning. Marine Fish Division, Scotia-Fundy Region, Department of Fisheries and Oceans, Bedford Institute of Oceanography, Dartmount, NS, Canada.

Hampton, J., J. R. Sibert, P. Kleiber, M. N. Maunder & S. J. Harley, 2005. Decline of Pacific tuna populations exaggerated? Nature 434: E1–E2.

Harmon, S. Y., 2002. Can there be a science of simulation? Why should we care? Modelling and Simulation 1(1): 2–5.

Hays, G. C., A. C. Broderick, B. J. Godley, P. Luschi & W. J. Nichols, 2003. Satellite telemetry suggests high levels of fishing-induced mortality in marine turtles. Marine Ecology Progress Series 262: 305–309.

Heymans, J. J. & T. J. Pitcher, 2004. Synoptic methods for constructing models of the past. In Pitcher, T. J. (ed.), Back to the Future: Advances in Methodology for Modelling and Evaluating Past Ecosystems as Future Policy Goals, 11–17. Fisheries Centre Research Reports 12(1): 158pp.

Hilborn, R., 2006. Faith-based fisheries. Fisheries 31(11): 554–555.

Holdgate, M. & J. W. L. Beament, 1977. The ecological dilemma. In Duncan R. & M. Weston-Smith (eds), The Encyclopeadia of Ignorance. Pergamon Press, UK, 411–416, 443pp.

Hughes, T. P., D. R. Bellwood, C. Folke, R. S. Steneck & J. E. Wilson, 2006. New paradigms for supporting the resilience of marine ecosystems. Trends in Ecology and Evolution 20(7): 380–385.

Hunter, P., (2005). Is political correctness damaging science? Peer pressure and mainstream thinking may discourage novelty and innovation. EMBO Reports 6(5): 405–407.

Jackson, J. B. C., M. X. Kirby, W. H. Berger, K. A. Bjorndal, L. W. Botsford, B. J. Bourque, R. H. Bradbury, R. Cooke, J. Erlandson, J. A. Estes, T. P. Hughes, S. Kidwell, C. B. Lange, H. S. Lenihan, J. M. Pandolfi, C. H. Peterson, R. S. Steneck, M. J. Tegner & R. R. Warner, 2001. Historical overfishing and the recent collapse of coastal ecosystems. Science 293: 629–638.

Jennings, S., 2005. Indicators to support an ecosystem approach to fisheries. Fish and Fisheries 6(3): 212–232.

Kaiser, M. J., J. S. Collie, S. J. Hall, S. Jennings & I. R. Poiner, 2002. Modification of marine habitats by trawling activities: prognosis and solutions. Fish and Fisheries 3: 114–136.

Kaplan, I. M. & B. J. McCay, 2004. Cooperative research, co-management, and the social dimension of fisheries science and management. Marine Policy 28: 257–258.

Kaplan, R. S. & D. P. Norton, 1996. The Balanced Scorecard: Translating Strategy Into Action. Harvard Business School Press, Boston, USA.

Kelleher, K., 2004. Collateral Damage: Discard discards in the world's marine fisheries. An update. FAO. Technical Report 470: 131pp.

Kennelly, S. J. & M. K. Broadhurst, 2002. Bycatch begone: changes in the philosophy of fishing technology. Fish and Fisheries 3(4): 340–355.

Kennett D. & B. Winterhalder (eds), 2006. Behavioral Ecology and the Transition to Agriculture. University of California Press, USA, 407pp.

Kitchingman, A., S. Lai, T. Morato & D. Pauly, 2007. How many seamounts are there and where are they located? Chapter 2. In Pitcher, T. J., T. Morato, P. J. B. Hart, M. R. Clark, N. Haggan & R. S. Santos (eds), Seamounts: Ecology, Conservation and Management. Fish and Aquatic Resources Series, Blackwell, Oxford, UK (in press).

Laws, R. M., 1985. The ecology of the Southern Ocean. American Scientist 73: 26–40.

Lozano-Montes, H., 2006. Historical ecosystem modelling of the upper Gulf of California (Mexico): following 50 years of change. PhD Thesis, University of British Columbia, 266pp.

Lozano-Montes, H. & T. J. Pitcher, 2007. Ecosystem simulation models of the Gulf of California for 1950: 1980 fnd 2000. Fisheries Centre Research Reports 15(5): (in press).

Lozano-Montes, H., T. J. Pitcher & N. Haggan, 2007. Shifting environmental baselines in the upper Gulf of California (Mexico): Local Fisher's Knowledge reveals a slow motion disaster.

Mendy, A. N. & E. A. Buchary, 2001. Constructing an icelandic marine ecosystem model for 1997 using a mass-balance modelling approach. Fisheries Centre Research Reports 9(4): 182–197.

Millennium Ecosystem Assessment, 2005. Ecosystems and Human Well-being: Synthesis. Island Press, Washington, DC, USA, 155 pp.

Morato, T., W. L. Cheung & T. J. Pitcher, 2006a. Vulnerability of seamount fish to fishing: fuzzy analysis of life history attributes. Journal of Fish Biology 68(1): 209–221.

Morato, T., R. Watson, T. J. Pitcher & D. Pauly, 2006b. Fishing down the deep. Fish and Fisheries 7(1): 23–33.

Morato, T. & T. J. Pitcher, 2005. Ecosystem simulations in support of management of data-limited seamount fisheries. In Kruse, G. H., V. F. Gallucci, D. E. Hay, R. I. Perry, R. M. Peterman, T. C. Shirley, P. D. Spencer, B. Wilson, D. Woodby (eds), Fisheries Assessment and Management in Data-limited Situations. Lowell Wakefield Fisheries Symposium Series 21. Alaska Sea Grant, Fairbanks, USA, 467–486.

Myers, R. A. & B. Worm, 2003. Rapid worldwide depletion of predatory fish communities. Nature 423: 280–283.

Nicol, S., 2006. Krill, currents, and sea ice: *Euphausia superba* and its changing environment. BioScience 56(2): 111–120.

Nicol, S. & Y. Endo, 1999. Krill fisheries: development, management and ecosystem implications. Aquatic Living Resources 12: 105–120.

Parsons, T. R., 1996. The impact of industrial fisheries on the trophic structure of marine ecosystems. In Polis G. A. & K. O. Winnemiller (eds), Marine Food Webs: Integration of Patterns and Dynamics. Chapman & Hall, New York, 352–357.

Pauly, D. & V. Christensen, 2002. Primary production required to sustain global fisheries. Nature 374: 255–257.

Pauly, D. & R. Watson, 2005. Background and interpretation of the 'Marine Trophic Index' as a measure of biodiversity. Philosophical Transactions of the Royal Society of B 360: 415–423.

Pitcher, T. J., 1999. Rapfish, A Rapid Appraisal Technique For Fisheries, And Its Application To The Code Of Conduct For Responsible Fisheries. FAO Fisheries Circular No. 947: 47pp.

Pitcher, T. J., 2001. Fisheries managed to rebuild ecosystems: reconstructing the past to salvage the future. Ecological Applications 11: 601–617.

Pitcher, T. J., 2004a. The problem of extinctions. In Pitcher, T. J. (ed.), Back to the Future: Advances in Methodology for Modelling and Evaluating Past Ecosystems as Future Policy Goals. Fisheries Centre Research Reports 12(1), 21–28, 158pp.

Pitcher, T. J., 2004b. What was the structure of past ecosystems that had many top predators? In Pitcher, T. J. (ed.), Back to the Future: Advances in Methodology for Modelling and Evaluating Past Ecosystems as Future Policy Goals. Fisheries Centre Research Reports 12(1), 18–20, 158pp.

Pitcher, T. J., 2004c. Why we have to open the lost valley: criteria and simulations for sustainable fisheries. In Pitcher, T. J. (ed.), Back to the Future: Advances in Methodology for Modelling and Evaluating Past Ecosystems as Future Policy Goals. Fisheries Centre Research Reports 12(1), 78–86, 158pp.

Pitcher, T. J., 2005. 'Back To The Future': a fresh policy initiative for fisheries and a restoration ecology for ocean ecosystems. Philosophical Transactions of the Royal Society 360: 107–121.

Pitcher, T. J., 2007. Lessons from History: session leader's summary from the 4th World Fisheries Congress. In Nielson J. (ed.), Reconciling Fisheries with Conservation: Proceedings of the 4th World Fisheries Congress. American Fisheries Society, Bethesda, USA (in press).

Pitcher, T. J. & C. H. Ainsworth, 2007. Back-to-the-Future: a candidate ecosystem-based solution to the fisheries problem. In Nielson, J. (ed.), Reconciling Fisheries with Conservation: Proceedings of the 4th World Fisheries

Congress. American Fisheries Society, Bethesda, USA (in press).

Pitcher, T. J., C. H. Ainsworth, H. Lozano, W. L. Cheung & G. Skaret, 2005a. Evaluating the Role of Climate, Fisheries and Parameter Uncertainty using Ecosystem-Based Viability Analysis. ICES CM 2005\M:24: 1–6.

Pitcher, T. J., C. H. Ainsworth, E. A. Buchary, W. L. Cheung, R. Forrest, N. Haggan, H. Lozano, T. Morato & L. Morissette, 2005b. Strategic management of marine ecosystems using whole-ecosystem simulation modelling: the 'Back-to-the-Future' policy approach. In Levner, E., I. Linkov & J.-M. Proth (eds), The Strategic Management of Marine Ecosystems. Springer, NATO Science Series: IV: Earth and Environmental Sciences 50, 199–258, 313pp.

Pitcher, T. J. & C. Bulman, 2007. Raiding the Larder: a quantitative evaluation framework and trophic signature for seamount food webs, Chapter 14. In Pitcher, T. J., T. Morato, P. J. B. Hart, M. R. Clark, N. Haggan & R. S. Santos (eds), Seamounts: Ecology, Conservation and Management. Fish and Aquatic Resources Series, Blackwell, Oxford, UK (in press).

Pitcher T. J. & A. Erfan, 2005. The krill fishery in the Antarctic Peninsula: spatial ecosystem-based simulations addressing conservation concerns for charismatic species. In Palomares, M. L. D., P. Pruvost, T. J. Pitcher & D. Pauly (eds), Modeling Antarctic marine ecosystems. Fisheries Centre Research Reports 13(7): 21–27.

Pitcher, T. J. & R. Forrest, 2004. Challenging ecosystem simulation models with climate change: the 'Perfect Storm', In Pitcher, T. J. (ed.), Back to the Future: Advances in Methodology for Modelling and Evaluating Past Ecosystems as Future Policy Goals. Fisheries Centre Research Reports 12(1): 29–38, 158pp.

Pitcher, T. J., J. J. Heymans, C. Ainsworth, E. A. Buchary, U. R. Sumaila & V. Christensen, 2004. Opening the lost valley: implementing a 'Back To Future' restoration policy for marine ecosystems and their fisheries. In Knudsen, E. E., D. D. MacDonald & J. K. Muirhead (eds), Sustainable Management of North American Fisheries. American Fisheries Society Symposium 43, 165–193.

Pitcher, T. J., T. Morato, P. J. B. Hart, M. R. Clark, N. Haggan & R. S. Santos, 2007. The depths of ignorance: an ecosystem evaluation framework for seamount ecology, fisheries and conservation, Chapter 21. In Pitcher, T. J., T. Morato, P. J. B. Hart, M. R. Clark, N. Haggan & R. S. Santos (eds), Seamounts: Ecology, Conservation and Management. Fish and Aquatic Resources Series, Blackwell, Oxford, UK (in press).

Pitcher, T. J. & D. B. Preikshot, 2001. Rapfish: a rapid appraisal technique to evaluate the sustainability status of fisheries. Fisheries Research 49(3): 255–270.

Polacheck, T., 2006. Tuna longline catch rates in the Indian Ocean: did industrial fishing result in a 90% rapid decline in the abundance of large predatory species? Marine Policy 30: 470–482.

Power, M. D., N. Haggan & T. J. Pitcher, 2004. The community workshop: how we did it and what we learned from the results. In Pitcher, T. J. (ed.), Back to the Future: Advances in Methodology for Modelling and Evaluating Past Ecosystems as Future Policy Goals. Fisheries Centre Research Reports 12(1), 125–128, 158pp.

Power, M. D. & T. J. Pitcher, 2007. Reconciling fisheries and allocation using a justice-based approach: troll fisheries score best. In Nielson J. (ed.), Reconciling Fisheries with Conservation: Proceedings of the 4th World Fisheries Congress. American Fisheries Society, Bethesda, USA (in press).

Richardson, A., 1992. The control of productive resources on the Northwest coast of North America. In Williams, N. M. & E. S. Hunn (eds), Resource Managers: North American and Australian Hunter-gatherers. Australian Institute of Aboriginal Studies, Canberra.

Roman, J. & S. R. Palumbi, 2003. Whales before whaling in the North Atlantic. Science 301: 508–510.

Rose, G. A., 2004. Reconciling overfishing and climate change with stock dynamics of Atlantic cod (*Gadus morhua*) over 500 years. Canadian Journal of Fisheries and Aquatic Sciences 61: 1553–1557.

Rosenberg, A. A., W. J. Bolster, K. E. Alexander, W. B. Leavenworth, A. B. Cooper & M. G. McKenzie, 2005. The history of ocean resources: modeling cod biomass using historical records. Frontiers in Ecology and the Environment 3(2): 78–84.

Sadovy, Y. & W. W. L. Cheung, 2003. Near extinction of a highly fecund fish: the one that nearly got away. Fish and Fisheries 4: 86–99.

Savenkoff, C., C. Castonguay, A. F. Vézina, S.-P. Despatie, D. Chabot, L. Morissette & M. O. Hammill, 2004. Inverse modelling of trophic flows through an entire ecosystem: the northern Gulf of St. Lawrence in the mid-1980s. Canadian Journal of Fisheries and Aquatic Sciences 61(11): 2194–2214.

Schindler, D. E., T. E. Essington, J. F. Kitchell, C. Boggs & R. Hilborn, 2002. Sharks and tunas: fisheries impacts on predators with contrasting life histories. Ecological Applications 12(3): 735–748.

Sherman, K., 1991. The large marine ecosystem concept: research and management strategy for living marine resources. Ecological Applications 1(4): 349–360.

Shin, Y.-J. & P. Cury, 2004. Using an individual-based model of fish assemblages to study the response of size spectra to changes in fishing. Canadian Journal of Fisheries and Aquatic Sciences 61(3): 414–431.

Skaret, G. & Pitcher T. J., 2007. An ecosystem model of the Norwegian Sea and Barents Sea fitted to data. In Pitcher, T. J. (ed.), Fisheries Centre Research Reports 15: (in press).

Smith, T. D. & J. S. Link, 2005. Autopsy your dead … and living: a proposal for fisheries science, fisheries management and fisheries. Fish and Fisheries 6: 73–87.

Southwood, T. R. E., 1977. Habitat, the template for ecological strategies? The Journal of Animal Ecology 46: 337–365.

Stefánsson, G. & Ó. K. Pálsson, 1998. A framework for multispecies modeling of Arcto-boreal systems. Reviews in Fish Biology and Fisheries 8: 101–104.

Sumaila, R. S., T. J. Pitcher, N. Haggan & R. Jones, 2001. Evaluating the Benefits from Restored Ecosystems: A Back to the Future Approach, 1–7, Chapter 18. In Shriver, A. L. & R. S. Johnston (eds), Proceedings of the 10th International Conference of the International Institute of Fisheries Economics and Trade, Corvallis, Oregon, USA, July, 2000.

Walters, C. J., V. Christensen, S. J. Martell & J. F. Kitchell, 2005. Possible ecosystem impacts of applying MSY policies from single-species assessment. ICES Journal of Marine Science 62: 558–568.

Walters, C. J., V. Christensen & D. Pauly, 2002. Searching for optimum fishing strategies for fishery development, recovery and sustainability. In Pitcher, T. J. & K. Cochrane (eds), The Use of Ecosystem Models to Investigate Multispecies Management Strategies for Capture Fisheries. Fisheries Centre Research Reports 10(2), 11–15, 156pp.

Watson, R., A. Kitchingman, A. Gelchu, & D. Pauly, 2004. Mapping global fisheries: Sharpening our focus. Fish and Fisheries 5(2): 168–177.

Wessel, P., 2007. Seamount Characteristics, Chapter 1. In Pitcher, T. J., T. Morato, P. J. B. Hart, M. R. Clark, N. Haggan & R. S. Santos (eds), Seamounts: Ecology, Conservation and Management. Fish and Aquatic Resources Series, Blackwell, Oxford, UK. (in press).

Whipple, S. J., J. S. Link, L. P. Garrison & M. J. Fogarty, 2000. Models of predation and fishing mortality in aquatic ecosystems. Fish and Fisheries 1: 22–40.

White, M., I. Bashmachnikov, J. Arístegui & A. Martins, 2007. Physical processes and seamount productivity Chapter 4. In Pitcher, T. J., T. Morato, P. J. B. Hart, M. R. Clark, N. Haggan & R. S. Santos (eds), Seamounts: Ecology, Conservation and Management. Fish and Aquatic Resources Series, Blackwell, Oxford, UK. (in press).

Worm, B., E. B. Barbier, N. Beaumont, J. E. Duffy, C. Folke, B. S. Halpern, J. B. C. Jackson, H. K. Lotze, F. Micheli, S. R. Palumbi, E. Sala, K. A. Selkoe, J. J. Stachowicz & R. Watson, 2006. Impacts of biodiversity loss on ocean ecosystem services. Science 314(5800): 787–790.

Zeller, D. & D. Pauly, 2005. Good news, bad news: global fisheries discards are declining, but so are total catches. Fish and Fisheries 6(2): 156.

Hydrobiologia (2008) 606:187–193
DOI 10.1007/s10750-008-9335-y

CHALLENGES TO MARINE ECOSYSTEMS

Effects of fishing methods on deep water shark species caught as by-catch off southern Portugal

Rui Coelho · Karim Erzini

© Springer Science+Business Media B.V. 2008

Abstract Deep water sharks are commonly caught as by-catch of longlines targeting bony fishes and trawlers targeting crustaceans in deep water off the southern Portuguese coast. Due to low or no commercial value, these species are most of the times discarded at sea, with only the larger specimens of some species commercialized at very low prices. In this study we present size distributions, maturity distributions, and sex ratios of 2,138 specimens belonging to four different species, namely the lantern sharks *Etmopterus pusillus* and *Etmopterus spinax* and the catsharks *Galeus melastomus* and *Galeus atlanticus*, caught with these two gears. Trawls generally caught smaller-sized specimens, in a wider length range than longlines. Trawls caught mostly immature specimens of all species, namely 83.7% immature of *E. pusillus*, 84.3% of *E. spinax*, 89.5% of *G. melastomus*, and 95.5% of *G. atlanticus*, while longlines caught mostly immature *E. pusillus* (69.2%) and *G. melastomus* (78.6%) and mostly mature *E. spinax* (88.2%) and *G. atlanticus* (87.2%). Trawls tended to catch more males than females of all species except *E. spinax*, while longlines caught more females than males of *E. spinax* and *G. melastomus* and more males than females of the other two species. The main conclusion of this work is that trawls are catching smaller-sized and mostly immature specimens when compared to longlines, meaning that they are probably having a more detrimental effect on these shark populations. The data presented here have significant implications for the conservation of these shark populations since sizes, sexes, and the immature and mature components of the populations are being affected differently by these two fishing gears.

Keywords Size distribution · Sex ratio · Maturity · Deep water sharks · Fishing gears · By-catch

Guest editors: J. Davenport, G. Burnell, T. Cross, M. Emmerson, R. McAllen, R. Ramsay & E. Rogan
Challenges to Marine Ecosystems

R. Coelho (✉) · K. Erzini
FCMA/CCMAR, University of the Algarve, Campus de Gambelas, 8005-139 Faro, Portugal
e-mail: rpcoelho@ualg.pt

Introduction

Due to their typically vulnerable life cycle, characterized by slow growth rates, late maturity, and low fecundity (Cortés, 2000), elasmobranch fishes are highly susceptible to fishing mortality. In these animals, overexploitation can occur even with low levels of fishing mortality (Stevens et al., 2000), and once they start to decline, it can take decades for populations to recover (Anderson, 1990). Examples of well-documented collapses in elasmobranch populations throughout the world include pelagic sharks in the northwest Atlantic (Baum et al., 2003) and several skate species in the northeast Atlantic (Dulvy et al., 2000).

Deep water species are usually even less resilient to fisheries pressure than coastal and epipelagic oceanic species, since productivity is among the lowest of all elasmobranchs (Gordon, 1999). Several deep water elasmobranch populations have been severely depleted, such as the picked dogfish, *Squalus acanthias* Linnaeus, 1758, whose northeastern population has been depleted to about 5% of the virgin biomass (Hammond & Ellis, 2005), or even gone locally extinct such as the blue skate, *Dipturus batis* (Linnaeus, 1758), in the Mediterranean (Abdulla, 2004).

In southern Portuguese waters, several deep water shark species are commonly caught as by-catch by both longliners targeting bony fishes and trawlers targeting crustaceans (Monteiro et al., 2001; Coelho et al., 2005). Although several species are present in this by-catch, two species of lantern sharks (*Etmopterus pusillus* (Lowe, 1839) and *Etmopterus spinax* (Linnaeus, 1758)) and two species of catsharks (*Galeus atlanticus* (Vaillant, 1888) and *Galeus melastomus* Rafinesque, 1810) account for the majority of this shark by-catch. Most of this catch is discarded, given that these species have little or no commercial value, with only the larger catshark specimens (mostly *G. melastomus*) commercialized at very low prices.

Since most of this shark by-catch is discarded and is never accounted for in the official fisheries statistics, it is very difficult to gather information on the impact that these commercial fisheries have on these shark populations. One additional difficulty is that there are still problems with the correct identification of these less known species. As an example, the official fisheries statistics only identify one species of the *Galeus* genus in Portuguese waters (*G. melastomus*), while there is now evidence that another species (*G. atlanticus*) also occurs (Rey et al., 2006) and is probably being misidentified.

Several previous studies have examined or mentioned by-catch of non-commercial elasmobranch species in Portuguese waters. Monteiro et al. (2001) looked into general discards from the trawl fishery, Erzini et al. (2001) studied the hake deep water longline fishery and Coelho et al. (2005) analyzed elasmobranch by-catch from coastal trammel nets and semi-pelagic deep water longlines. These studies presented quantitative values of the by-catch in number or biomass, but did not explore intra-specific characteristics of the by-catch such as length distribution, maturity distribution, or sex ratios of the species caught with the different fishing gears.

Therefore, the objectives of this study are to present and compare size distributions, maturity distributions, and sex ratios of four commonly discarded deep water shark species, namely the lantern sharks *E. spinax* and *E. pusillus*, and the catsharks *G. melastomus* and *G. atlanticus* caught by deep water longlines and deep water trawls off the south and southwestern coasts of Portugal.

Material and methods

Specimens were caught aboard commercial fishing vessels operating off the south and southwestern coasts of Portugal. Overall, data from 46 fishing days, 9 with trawls (corresponding to 29 hauls) and 37 with longlines, from February 2003 to October 2004 were analyzed for this study. Longline catches came from a commercial longliner targeting wreckfish, *Polyprion americanus* (Bloch & Schneider, 1801), and European conger, *Conger conger* (Linnaeus, 1758) (Fig. 1). This fishing vessel usually uses SIAPAL brand number 6 hooks (gape = 17.5 mm, SD = 1.3, $n = 6$), and some 3,200–3,600 hooks are set each fishing day. The catch from the deep water trawls came from the commercial fisheries targeting crustaceans such as Norway lobster, *Nephrops norvegicus* (Linnaeus, 1758), deep water rose shrimp, *Parapenaeus longirostris* (Lucas, 1846), and blue and red shrimp, *Aristeus antennatus* (Risso, 1816) (Fig. 1). The minimum codend mesh size for that fishery is 55 mm, and the hauls analyzed in this study ranged in duration from 2 h 50 min to 10 h 20 min, with boat speed varying from 2.5 to 3 knots. Trawl fishing took place at depths from 199 to 641 m mostly on sandy and muddy bottoms, and longline fishing at depths from 458 to 787 m, mostly near rocky bottoms.

Specimens were taken to the University of the Algarve laboratory, where they were identified to species level, sexed, measured for total length (TL) in a straight line from the tip of the snout to the tip of the upper lobe of the caudal fin in a natural position (nearest lower millimeter) (Compagno, 2001) and the total weight (W) was recorded (nearest lower centigram). The gonads of both males and females were

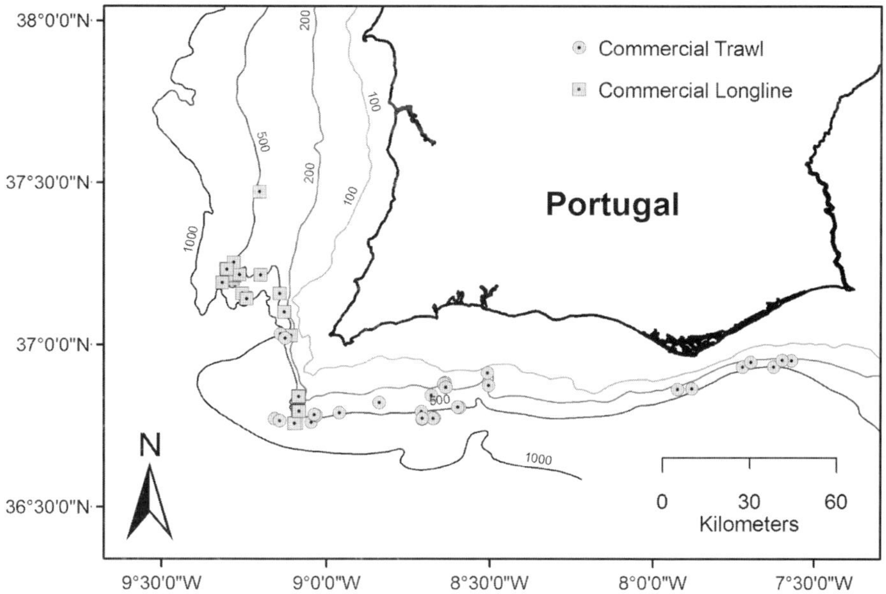

Fig. 1 Map of the south and southwestern coasts of Portugal with the location of the coastline, the bathymetric lines (100, 200, 500, and 1,000 m depth) and the sampling points. Bathymetric lines and coastline adapted from "Atlas do Ambiente Digital—Instituto do Ambiente"

observed macroscopically and classified according to the maturity stages proposed by Costa et al. (2005) for the oviparous species (in this study *Galeus* spp.) and by Clarke et al. (2001) for the aplacental viviparous species (in this study *Etmopterus* spp.). Specimens were then divided into mature or immature, with mature specimens defined as specimens that are ready to produce progeny or have done so in the past.

The Kolmogorov–Smirnov test was used to find differences in the general shape, dispersion, and skewness of the distributions of the catches in length between fishing gears, while the Mann-Whitney statistical test was used to compare catches in length and weight between fishing gears (StatSoft, 2004). The null hypothesis tested by both these statistics was that in each species, specimens caught by the different fishing gears came from the same part of the population. Each species sex ratio per fishing gear was compared by the Chi-square observed versus expected test (StatSoft, 2004), testing the null hypothesis that males and females of each species and in each fishing gear were caught in similar ratios.

Results

A total of 2,138 specimens were caught during this study. In general, trawls caught smaller and lighter specimens than longlines (Table 1), and the observed differences, both in average length and weight, were statistically significant for all species (Mann–Whitney: *P*-values <0.05).

Size distributions by fishing gear showed that trawls tended to catch specimens in a wider length range, while longline catches tended to be more concentrated in the larger size classes (Fig. 2). For all species, the size distribution of the catches with the two fishing gears were significantly different (Kolmogorov–Smirnov: *P*-values <0.05).

Trawls tended to catch more males than females of all species except *E. spinax*, while longlines caught more females than males of *E. spinax* and *G. melastomus* and more males than females of the other two species. Both trawls and longlines caught more male *E. pusillus* than females, although in trawls the differences were more noticeable. On the other hand, both gears caught more females of *E. spinax* than males and this difference was most noticeable in longlines. For *G. atlanticus* the situation was similar to *E. pusillus*, with more males being caught than females by both gears, while relatively more female *G. melastomus* were caught on longlines and more males were caught with trawls (Fig. 3). The observed vs. expected sex ratio frequencies were significantly different for all species and for both gears (χ^2 Observed versus Expected: *P*-values <0.05), except for the case of *G. atlanticus* caught with trawl where no significant differences were detected (χ^2 Observed versus Expected: *P*-value >0.05).

Table 1 Number of specimens (*n*), mean and ranges of TL, and weight (*W*) of the four shark species caught by longline and trawl

		E. pusillus	*E. spinax*	*G. atlanticus*	*G. melastomus*
Longline	*n*	514	279	99	462
	Mean TL (cm)	38.1	33.4	41.8	49.3
	Range TL (cm)	25.3–50.2	23.8–40.7	31.6–45.8	23.2–75.1
	Mean *W* (g)	230.88	165.88	196.20	352.22
	Range *W* (g)	41.61–605.00	42.78–304.33	79.91–298.94	35.14–1219.63
Trawl	*n*	98	407	22	257
	Mean TL (cm)	31.1	23.0	21.0	23.5
	Range TL (cm)	15.8–48.2	9.1–39.6	14.7–41.4	12.5–65.8
	Mean *W* (g)	138.31	56.69	33.48	58.32
	Range *W* (g)	14.44–512.40	2.85–367.05	8.43–198.40	4.66–889.21

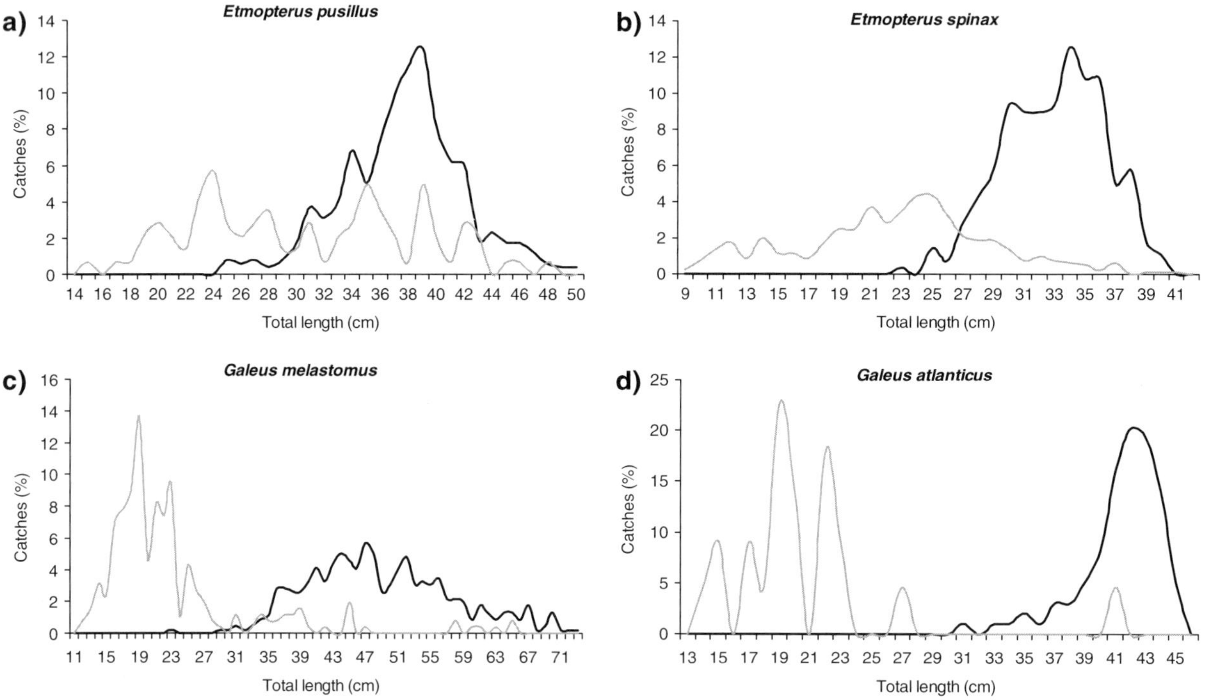

Fig. 2 Size distributions (1 cm TL size classes) in percentage of catches per fishing gear of *Etmopterus pusillus* (**a**), *E. spinax* (**b**), *Galeus melastomus* (**c**), and *G. atlanticus* (**d**) caught both by longline (black line) and trawl (gray line)

In general, more immature than mature specimens were caught. In *E. pusillus*, immature specimens, independent of sex, were always caught in higher percentages than mature specimens, both by trawl (83.7%) and by longline (69.2%). In *E. spinax* there was a discrepancy between fishing gears, with longlines catching mostly mature specimens (88.2%), and trawls mostly immature specimens (84.3%). Immature specimens of both male and female *G. melastomus* were always caught in higher quantities than mature specimens, both by longline (78.6%) and trawl (89.5%), while longlines caught mostly mature *G. atlanticus* (87.2%) and trawls mostly immature specimens (95.5%), independent of sex (Fig. 4).

Discussion

Three of the studied species, *E. pusillus*, *E. spinax*, and *G. melastomus*, are very frequent and are

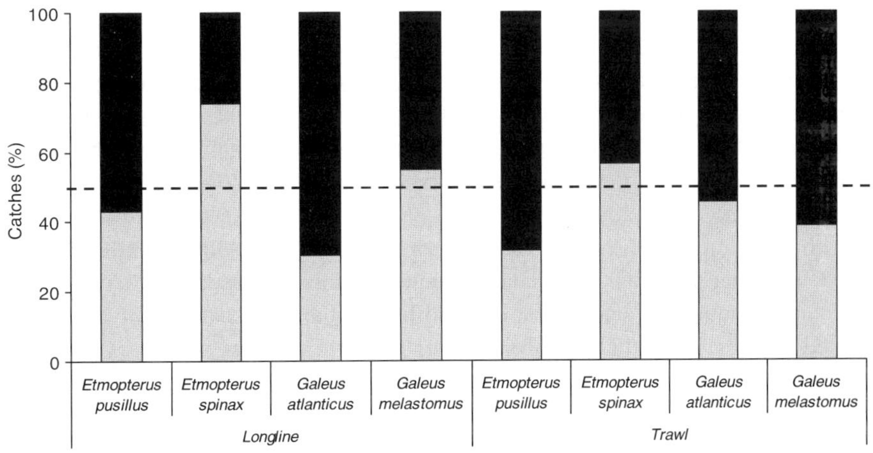

Fig. 3 Sex ratios (females = gray bars and males = black bars) for each species caught with each fishing gear. The dotted line represents 50% of the catches

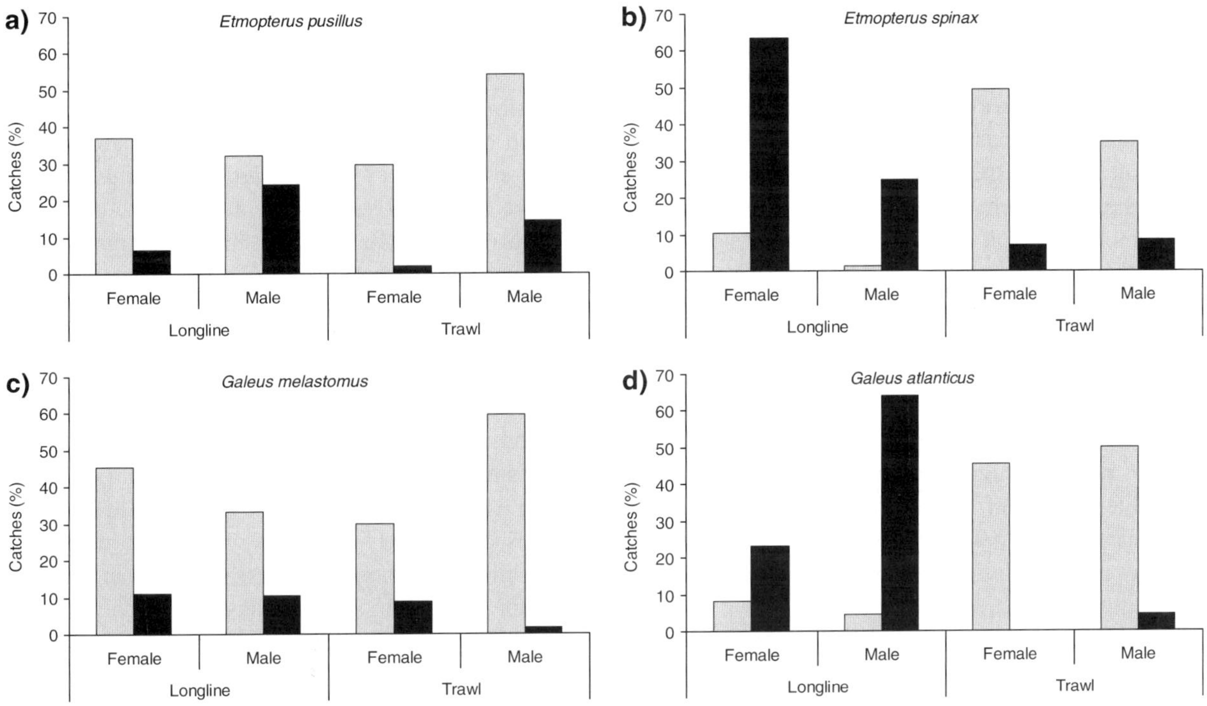

Fig. 4 Catches, in percentage per fishing gear, of immature (gray bars) and mature (black bars) males and females of *Etmopterus pusillus* (**a**), *E. spinax* (**b**), *Galeus melastomus* (**c**), and *G. atlanticus* (**d**)

probably the most common elasmobranch species at the depths fished in this study (Monteiro et al., 2001; Coelho et al., 2005), while the fourth species, *G. atlanticus*, is a relatively rare species, that given its similarity to *G. melastomus* can easily be confounded. In fact this species has only recently been redescribed based on exterior field marks (Rey et al., 2006) and is still not recognized by the official Portuguese fisheries statistics where it is mixed with *G. melastomus*.

Since these species are mostly discarded, only a very small portion is accounted for in the official fisheries statistics. Nonetheless, due to the high fishing mortalities, these populations suffer severe impacts from these fisheries. In fact, even though most of these specimens are discarded, they are returned to sea either dead or with severe injuries that probably affect their survival. Specifically, specimens caught with trawls tend to arrive dead on board, probably due to the trauma of being towed for several hours, while specimens caught with longlines are often still alive, but with injuries caused by the hooks and by the sudden changes in pressure and temperature that might impair their survival. One possible way to effectively reduce by-catch in longlines is to increase the distance for setting the hooks above the seabed (Coelho et al., 2003).

Although longline fishing effort cannot be compared with that of trawls, the data presented in this study can give some idea on the catchability of the studied species. Regarding the *Etmopterus* genus, *E. pusillus* seems to be mostly caught by longlines, while *E. spinax* is mostly caught by trawl. At this point we can put forward some tentative explanations for this observation: (1) these species have different habitats in terms of water column placement, with *E. spinax* being a more benthic species that is more easily caught by trawl, while *E. pusillus* is a more bentho-pelagic species not easily caught by trawl but that is attracted to the baited longlines or (2) these species have different substrate preferences, with *E. spinax* more common on sandy and muddy bottoms, where the trawls operate and *E. pusillus* more associated with rocky bottoms, near where longlines operate. While at this point we can only speculate on the true causes for this observation (Compagno et al., 2005) reports that *E. pusillus* has been recorded in oceanic waters in the southern Atlantic, which might indicate vertical migrations that make them less vulnerable to bottom trawls. For the *Galeus* genus, given that *G. atlanticus* is a rare species, caught in very low numbers by both gears we cannot forward such hypothesis.

Significant differences in sex ratios were detected for all species and for all gears except *G. atlanticus* caught with trawl. For this species, the low sample size may have biased this particular analysis. The differences found in the sex ratios may be correlated with the differences found in the size distribution, since there are differences in maximum sizes of males and females. In fact, in all these species, females tend to grow to larger sizes than males, and given that longlines catch mostly large specimens it would be expected that most of them would be females. This hypothesis seems to hold true for the case of *G. melastomus* where longlines catch more females and trawls catch more males, but not for the other species. In *E. spinax*, although trawls catch smaller sized specimens than longlines, both gears catch mainly females, while in *E. pusillus* the opposite situation occurs with more males than females caught with both gears. As stated before, the fact that these gears are operating in different habitats might be the cause for the different sex composition of the catches, implying that these species are aggregating in different habitats by sex. Although there are no specific studies for these species, there are examples from another squaloid shark, *S. acanthias*, that has for long been known to aggregate by size and sex (Ford, 1921).

Trawls are catching mostly immature specimens of all species while longlines are catching mostly immature *E. pusillus* and *G. melastomus*, and mature *E. spinax* and *G. atlanticus*. Although there is no stock assessment-based evidence (Hilborn & Walters, 1992), the persistent capture of large quantities of small-sized and immature sharks is likely to lead growth overfishing (Bonfil, 1997). In this case, the larger-sized species (*E. pusillus* and *G melastomus*), that are caught largely as juveniles by both gears, are probably more affected by fishing than the smaller sized species (*E. spinax* and *G. atlanticus*.

This work has presented information on the differential population catches of four by-catch and frequently discarded deep water elasmobranch species in southern Portugal. As a conclusion, we showed that when compared to longlines, trawls catch smaller-sized and mostly immature specimens. Given what is known about the quantities of by-catch and discards in the fisheries of southern Portugal, it is likely that trawls are more detrimental than longline gear for these shark populations. Continuous fishing pressure on these populations may lead to their collapse. The use of hooks fished off the bottom (Coelho et al., 2003) would greatly reduce the catch of discarded deepwater sharks. Reduction of shark by-catch in the deepwater crustacean trawl fisheries could be achieved through the use of by-catch reduction devices (BRDs), such as a rigid grid combined with an escape window. Such BRDs have been tested in Portuguese trawl fisheries with mixed results (Campos & Fonseca, 2004).

Acknowledgments The authors wish to thank all the fishermen who contributed with specimens for this study. Special thanks go to Elídio Diogo and Francisco Diogo, skippers of the longliner "Branca de Sagres"; Helder Cavaco, skipper of the bottom trawler "Gamba"; and José Santos, skipper of the bottom trawler "Crustáceo." This study was funded by

POCI 2010 (Programa Operacional Ciência e Inovação 2010) and FSE (Fundo Social Europeu) through a PhD grant from FCT (Fundação para a Ciência e Tecnologia) attributed to Rui Coelho (Ref. SFRH/BD/10357/2002).

References

Abdulla, A., 2004. Status and Conservation of Sharks in the Mediterranean Sea. IUCN Technical Paper, 1–7.

Anderson, E. D., 1990. Fishery models as applied to elasmobranch fisheries. In Pratt, H. L., S. H. Gruber & T. S. Taniuchi (eds), Elasmobranchs as Living Resources: Advances in the Biology, Ecology, Systematics, and the Status of Fisheries. NOAA Technical Report NMFS 90, 473–484.

Baum, J. K., R. A. Myers, D. G. Kehler, B. Worm, S. J. Harley & P. A. Doherty, 2003. Collapse and conservation of shark populations in the Northwest Atlantic. Science 299: 389–392.

Bonfil, R., 1997. Status of shark resources in the southern Gulf of Mexico and Caribbean: implications for management. Fisheries Research 29: 101–117.

Campos, A. & P. Fonseca, 2004. The use of separator panels and square mesh windows for by-catch reduction in the crustacean trawl fishery off the Algarve (South Portugal). Fisheries Research 69: 147–156.

Clarke, M. W., P. L. Connolly & J. J. Bracken, 2001. Aspects of reproduction of the deep water sharks *Centroscymnus coelolepis* and *Centrophorus squamosus* from west of Ireland and Scotland. Journal of the Marine Biological Association of the United Kingdom 81: 1019–1029.

Coelho, R., L. Bentes, J. M. S. Gonçalves, P. G. Lino, J. Ribeiro & K. Erzini, 2003. Reduction of elasmobranch by-catch in the hake semipelagic near-bottom longline fishery in the Algarve (Southern Portugal). Fisheries Science 69: 293–299.

Coelho, R., K. Erzini, L. Bentes, C. Correia, P. G. Lino, P. Monteiro, J. Ribeiro & J. M. S. Gonçalves, 2005. Semi-pelagic longline and trammel net elasmobranch catches in southern Portugal: catch composition, catch rates and discards. Journal of Northwest Atlantic Fishery Science 35: 531–537.

Compagno, L. J. V., 2001. Sharks of the World. An Annotated and Illustrated Catalogue of Shark Species known to Date, Vol. 2: Bullhead, Mackerel and Carpet Sharks (Heterodontiformes, Lamniformes and Orectolobiformes). FAO, Rome.

Compagno, L. J. V., M. Dando & S. Fowler, 2005. Sharks of the World. Collins, London.

Cortés, E., 2000. Life history patterns and correlations in sharks. Reviews in Fisheries Science 8: 299–344.

Costa, M. E., K. Erzini & T. C. Borges, 2005. Reproductive biology of the blackmouth catshark, *Galeus melastomus* (Chondrichthyes: Scyliorhinidae) off the south coast of Portugal. Journal of the Marine Biological Association of the United Kingdom 85: 1173–1183.

Dulvy, N. K., J. D. Metcalfe, J. Glanville, M. G. Pawson & J. D. Reynolds, 2000. Fishery stability, local extinctions, and shifts in community structure in skates. Conservation Biology 14: 283–293.

Erzini, K., J. M. S. Gonçalves, L. Bentes, P. G. Lino & J. Ribeiro, 2001. The hake deepwater semi-pelagic ("pedrabola") longline fishery in the Algarve (southern Portugal). Fisheries Research 51: 327–336.

Ford, E., 1921. A contribution to our knowledge of the life-histories of the dogfishes landed at Plymouth. Journal of the Marine Biological Association of the United Kingdom 12: 468–505.

Gordon, J. D. M., 1999. Management considerations of deep-water shark fisheries. In Shotton, R. S. (ed.), Case Studies of the Management of Elasmobranch Fisheries. FAO, Rome.

Hammond, T. R. & J. R. Ellis, 2005. Bayesian assessment of northeast Atlantic spurdog using a stock production model, with prior for intrinsic population growth rate set by demographic methods. Journal of Northwest Atlantic Fishery Science 35: 299–308.

Hilborn, R. & C. J. Walters, 1992. Quantitative Fisheries Stock Assessment: Choice, Dynamics and Uncertainty. Chapman and Hall, London.

Monteiro, P., A. Araujo, K. Erzini & M. Castro, 2001. Discards of the Algarve (southern Portugal) crustacean trawl fishery. Hydrobiologia 449: 267–277.

Rey, J., B. Séret, D. Lloris, R. Coelho & L. Gil de Sola, 2006. A new redescription of *Galeus atlanticus* (Vaillant, 1888) (Chondrichthyes: Scyliorhinidae) based on field marks. Cybium 30(4) (Supplement): 7–14.

StatSoft, 2004. Electronic Statistics Textbook. StatSoft, Tulsa.

Stevens, J. D., R. Bonfil, N. K. Dulvy & P. A. Walker, 2000. The effects of fishing on sharks, rays, and chimaeras (chondrichthyans), and the implications for marine ecosystems. ICES Journal of Marine Science 57: 476–494.

Hydrobiologia (2008) 606:195–202
DOI 10.1007/s10750-008-9336-x

CHALLENGES TO MARINE ECOSYSTEMS

Catch me in winter! Seasonal variation in air temperature severely enhances physiological stress and mortality of species subjected to sorting operations and discarded during annual fishing activities

Folco Giomi · Saša Raicevich · Otello Giovanardi · Fabio Pranovi · Paolo Di Muro · Mariano Beltramini

© Springer Science+Business Media B.V. 2008

Abstract Several studies have considered the direct and indirect effects of demersal trawling on discarded species in terms of sublethal damages, survival, and stress due to the fishing processes. Nevertheless the effects of air temperature on the physiological stress and the survival of species during sorting operations were only marginally explored. This factor could be particularly important in the context of sustainable fisheries at temperate latitudes where the seasonal variation of air temperature can be particularly pronounced. In this study the seasonal effects of rapido trawling on the non-target species *Liocarcinus depurator* (Portunidae) in the Northern Adriatic Sea (Mediterranean Sea) have been compared by applying survival tests and considering the unbalance in metabolites' concentration as indicators of physiological stress. Results showed consistently higher mortalities during exposure to air in summer (temperature: 28°C), which reached about 96% in 20 min, compared to winter (temperature: 9°C) when only 2% of individuals died. Furthermore trawled and emersed crabs showed significant increase of hemolymph ammonia, lactate, and glucose concentrations as effects of extreme exercise and suffocation, which was more prominent during summer, suggesting that air temperature can play an important role in determining non-target species survival.

Keywords Physiological stress · Mortality · Fishing disturbance · Discard · *Liocarcinus depurator* · Northern Adriatic Sea

Guest editors: J. Davenport, G. Burnell, T. Cross, M. Emmerson, R. McAllen, R. Ramsay & E. Rogan
Challenges to Marine Ecosystems

F. Giomi (✉) · P. D. Muro · M. Beltramini
Department of Biology, University of Padova, Viale G. Colombo 3, 35131 Padova, Italy
e-mail: folco@bio.unipd.it

S. Raicevich · O. Giovanardi
ICRAM—Istituto Centrale per la Ricerca Scientifica e Tecnologica Applicata al Mare, Loc. Brondolo, Chioggia, 30015 Venice, Italy

F. Pranovi
Department of Environmental Sciences, University of Venice, Venice, Italy

Introduction

Fishing activity constitutes one of the major sources of ecological disturbance (sensu White & Pickett, 1985) and alteration among human exploitation of the marine ecosystems (Williams, 1998). The effects of intense and lasting fishing have been widely investigated to assess the possible effects on marine resources and biodiversity (Jennings & Kaiser, 1998) and to delineate effective guidelines for sustainable fisheries (Pikitch et al., 2005). The overexploitation of marine resources has been recognized to cause drastic impairments of the food web structure and the consequent alteration of the productivity (Pauly & Christensen, 1995). Besides, fishing has dramatic effects on the non-target species, and the different techniques employed have been proved to cause a wide range of alterations to biodiversity and community structure (Jennings et al., 2001; Queirós et al., 2006; Tillin et al., 2006). These aspects are reflected, for example, by the pronounced imbalance in terms of damage, physiological stress, and survival among the species subjected to trawling activities. In fact during the fishing process, which includes not only capture by the fishing gear, but also the exposure on deck during commercial catch sorting operations, the individuals may suffer physical damage (Bergmann et al., 2001a; Pranovi et al., 2001) and physiological stress due to thermal shocks and exposure to air (Bergmann et al., 2001b; Gamito & Cabral, 2003) which can lead to high mortality rates (Kaiser & Spencer, 1995; Bergmann & Moore, 2001). Among the discarded species, crustaceans constitute a prominent part of the total biomass and represent an excellent candidate to assess the physical and physiological effects of trawling and sorting procedures (Bergmann et al., 2001b; Pranovi et al., 2001). In particular, emersion represents, for sublittoral species, a traumatic event that induces increased physiological disruption (Burnett, 1988; Durand & Regnault, 1998). The permanence in the emersed state causes a collapse of the gas-exchange organs like the lamellate and the filamentous gills (McMahon, 2001). The exchanges, normally occurring at the level of these organs, are reduced or blocked leading to an increase of concentration of respiratory and metabolic end products in the hemolymph and tissues. The suffocation in air is, in fact, connected with the rapid increase in hemolymph CO_2. This effect, together with pronounced lactate accumulation derived from the anaerobic metabolism, causes the marked respiratory acidosis. As for the gas exchange, during emersion the excretion across gills is reduced leading to an increase of ammonia concentrations in the hemolymph and tissues (Danford et al., 2002). Moreover, the trawling and the emersion events stimulate mechanical exercise and stress for the discarded animals, and these are reflected in a pronounced hyperglycemia caused by the increased glycogen utilization (Taylor & Spicer, 1987; Hagerman et al., 1990; Schmitt & Uglow, 1997). These indicators have been applied to the study of the physiological effects of fishing disturbance on discarded species and, recently, to *Liocarcinus depurator* (Portunidae) in the Clyde Sea (Bergmann et al., 2001b). *L. depurator* is a species distributed in the sublittoral areas from the Mediterranean to the North Sea, and is frequently caught and discarded by demersal fishery all over the continental shelves. *L. depurator* represents a major component of the discard in the Northern Adriatic Sea fishing industry and is usually subjected to high levels of lethal and sublethal injuries during trawling (Pranovi et al., 2001). Some recent papers highlighted the role of air temperature on the survival of crustaceans during the fishing process (Cabral et al., 2002; Gamito & Cabral, 2003).

The aim of this study is to explore the effect of seasonality, reflected by differences of water and air temperatures, as a source of mortality and physiological impairment in individuals of *L. depurator* discarded by commercial trawl fishing activity in the Northern Adriatic Sea. Accordingly, the mortality rate and the physiological disruption have been analyzed and compared during winter and summer trawls and after the subsequent exposure in air, at temperatures of 9°C and 28°C, respectively.

Materials and methods

Experimental trawls and animal sampling

During March 2002 and July 2003 a total number of 8 experimental rapido tows (for fishing gear description see Hall-Spencer et al., 1999) were carried out in the Northern Adriatic Sea (latitude: 45° 03′ N, longitude: 13° 01′ E; depth: 27 m), using a commercial fishing vessel belonging to the port of Chioggia. To reproduce the fishing procedures and to standardize the experimental setting, each tow lasted 40 min at a full speed of 11 km h^{-1}. During the winter experiment, the seabed temperature was 9°C and the air temperature varied between 9 and 12°C; during the summer experiment, the seabed temperature was 14°C, and the air temperature varied between 26 and 28°C. Intermolt adult specimens of *Liocarcinus depurator* were randomly collected from the discard and immediately employed for the survival experiment or for the physiological stress analyses. The designations of molt stage were according to those described by Mangum (1985).

Survival tests

The mortality rates of *L. depurator* were assessed on 100 individuals per tow (3 replicates) immediately after the net was emptied on deck, and after 10 and 20 min of air exposure. In conformity with Bergmann & Moore (2001), individuals were deemed to be dead when no movement of scaphognatite and reaction to external stimulus were observed.

Physiological stress assessment

The physiological stress evaluation was carried out at two different stages of the fishing process: immediately after the end of the tow and during the sorting process. In the first group (hereafter referred to as "trawled"), individuals were immediately collected from the catch (within 3 min) and put into tanks filled with sea water at the same temperature as the seabed water; hemolymph samples were withdrawn within 5 to 15 min of the end of each tow. The second group consisted of individuals (hereafter referred to as "trawled and emersed") which were left on deck exposed to air. Hemolymph samples were withdrawn continuously from randomly selected individuals: 5 to 45 min (winter experiment) and 5 to 25 min (summer), respectively, after the end of the tow. Hemolymph was withdrawn just once from each animal. During the same cruise, to test undisturbed individuals, not subjected to the catch stress induced by the trawl, baited creels were used and six specimens were collected and used as control group. During the summer experiment the same procedures were applied to the trawled group, but, due to the high mortality rate of the emersed crabs, the air exposure was shortened to 5 min.

Hemolymph sampling and analysis

Hemolymph was collected from individuals using a 1 ml syringe whose needle was inserted into the pericardium from the posterior side of the carapace to reduce the duration of handling and to collect post-branchial blood. The samples were immediately frozen and kept in liquid nitrogen until arrival in the lab where they were stored at −20°C until used. Each hemolymph sample was generally used for the determination of all metabolites, but when the sample volume was too small, lactate and glucose measurements were preferred because they need less material to be assessed. The number of individuals sampled for each assay ranged between 23–41 as indicated in Fig. 2.

The concentration of total NH_3/NH_4^+ nitrogen was determined colorimetrically using the method described by Solorzano (1969) with an absorbance reading at 634 nm. Concentration values were determined by reference to a calibration plot obtained by suitable dilutions of a 1 mM NH_4Cl solution. The calibration plot was established in a concentration range so as to include the concentration in the experimental specimens.

L-lactate concentration was determined according to the method of Gutmann & Wahlefeld (1974) based on the substrate oxidation to L-pyruvate catalyzed by lactate dehydrogenase in an assay medium made with 3.75 g/100 ml glycine, 2.0 ml/100 ml hydrazine hydrate (96%) buffered to pH 9.0 with NaOH and containing 6 mM EDTA to remove the interferences due to divalent cations in solution (Engel & Jones, 1978). NAD^+ 40 mM and lactate dehydrogenase (510 U/mg diluted 1:2 in buffer) were also used.

The quantitative assay for D-glucose concentration is based on the hexokinase-catalyzed phosphorylation of the substrate (ATP is required as co-substrate) followed by the one-electron oxidation of glucose-6-phosphate by the dehydrogenase that produces NADPH (Webster, 1996). The assay is carried out in 0.1 M Tris/HCl pH 7.4 containing 5 mM $MgSO_4$. Both L-lactate and D-glucose assays were carried out with the internal standard method.

The concentration of the assayed compounds was calculated after extrapolating to zero the internal standard concentration and correction for dilutions. All spectrophotometric determinations were carried out with a Hewlett-Packard HP-8452 diode array spectrophotometer and all the reagents employed for the analyses were purchased from Sigma-Aldrich.

Statistical analysis

The mean value of the metabolic parameter was pairwise compared by Mann–Whitney U-test. The analyses were applied between the summer and winter trawled samples, and the summer and winter trawled and emersed samples. Due to the low number of specimens collected, the control group was not included in the analyses but presented in the figures simply as references. The limit of significance was considered at $P < 0.001$.

Results

Survival tests

The short-term analyses of mortality rates highlighted the presence of a strong seasonal effect, since during the winter experiment, less than 2% of collected individuals died within 20 min of exposure to air, whereas during the summer experiments up to 96% died in the same time (Fig. 1).

Physiological stress estimation

The alteration of physiological state due to the stress condition occurring during the trawling or/and air exposure can be studied by measuring the concentration levels in the hemolymph of ammonia, L-lactate, D-glucose. In Fig. 2 are shown the data obtained in "trawled" and "trawled and emersed" groups of animals. The creel-caught individuals are used as a control group for comparison with the other conditions, but only as a reference because of the reduced number of specimens. The "reference" data are the average of both the winter and summer values.

The animals collected by rapido trawling in winter showed average ammonia levels of 0.149 ± 0.019 mmol l^{-1}, slightly higher than the value registered in the control group (0.107 ± 0.007 mmol l^{-1}), but in summer the concentration in the hemolymph increased

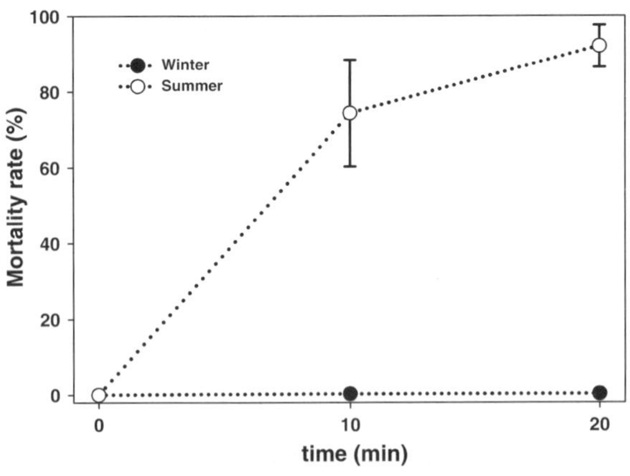

Fig. 1 Cumulative mortality rates (%) in *Liocarcinus depurator* caught by rapido trawl in summer (○) and winter (●), at the end of the tow (0) and after 10 and 20 min of air exposure

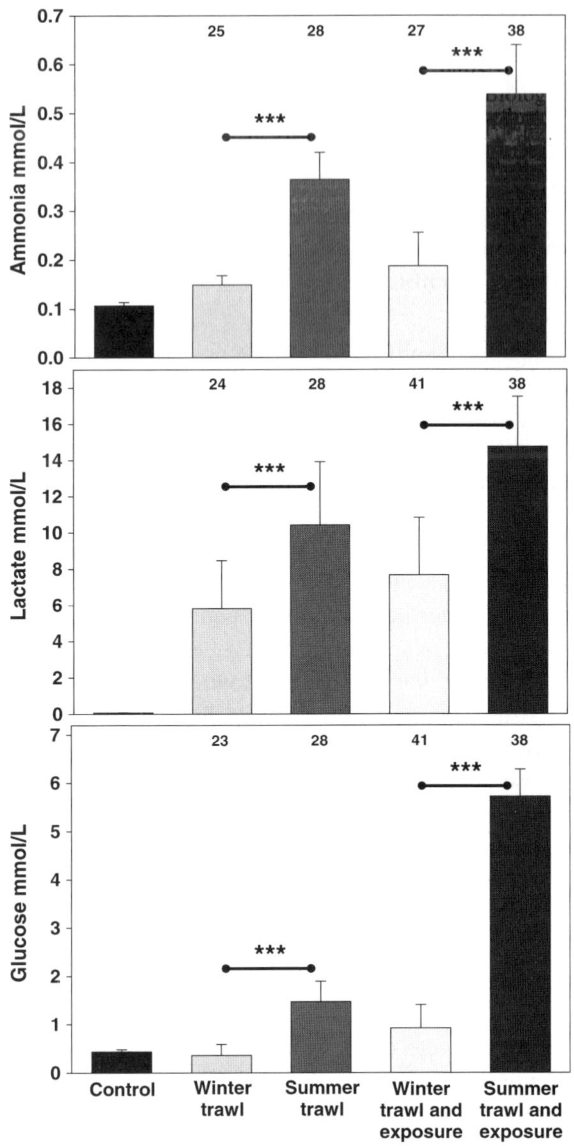

Fig. 2 Hemolymph ammonia, lactate, and glucose concentrations (mmol l^{-1} ± s.d.) measured on experimental crabs. The bars represent the value of concentrations for only trawled and trawled and emersed specimens both in summer and winter catches. The numbers over the bars represent the number of specimens considered while the control group was constantly constituted by six animals. The horizontal bars indicate the pairwise comparisons and the *** stand for the significance values ($P < 0.001$)

up to 0.365 ± 0.055 mmol l^{-1}. In the group of individuals subjected to emersion after trawling, the ammonia value increased with respect to the non-exposed group, both in winter and summer, reaching values of 0.187 ± 0.068 mmol l^{-1} and 0.539 ± 0.101 mmol l^{-1}, respectively.

Lactate was considerably low in the control group (0.068 ± 0.010 mmol l^{-1}) indicating a low stress condition in the animals captured by baited creels. As far as the winter experiment is concerned, the lactate concentration was 5.808 ± 2.654 mmol l^{-1} in trawled individuals, but in air-exposed individuals the levels increased to 7.666 ± 3.160 mmol l^{-1}. A drastic lactate concentration increase was observed in summer: the trawled group showed a value of 10.422 ± 3.497 mmol l^{-1}, about twice the corresponding value in winter while the highest level was achieved in exposed individuals at 14.750 ± 2.755 mmol l^{-1}.

The concentration of glucose in hemolymph of control group was 0.435 ± 0.044 mmol l^{-1}. In winter the variation of glucose levels in trawled animals appears restrained (0.359 ± 0.227 mmol l^{-1}) and there was an increase up to 0.918 ± 0.491 mmol l^{-1} in the air-exposed individuals. In summer the glucose levels exhibited remarkable differences as compared to winter experiments: the trawled group showed a value of 1.475 ± 0.420 mmol l^{-1}, about 300% higher than that in winter experiment, but the exposure to air produced a strong increase reaching the concentration of 5.719 ± 0.559 mmol l^{-1}, 6–7 times higher as compared to the corresponding winter group.

Discussion

The effects of air exposure on crustacean physiology have been accurately described in species that normally experience emersion in the intertidal and coastal fringes (e.g. Bridges & Brand, 1981; DeFur, 1988; Taylor & Spicer, 1988; Stillman & Somero, 1996) as well as for subtidal species that could be forced to emersion as a consequence of commercial fishing activities (Schmitt & Uglow, 1997; Morris & Oliver, 1999a, b; Bergmann et al., 2001b; Danford et al., 2002). From this literature, it emerges that during sorting and shipping procedures benthic crustaceans, which inhabit stable environments and are virtually unable to cope with emersion, undergo a series of circulatory, respiratory, and metabolic impairments. These effects generally produce cumulative consequences on organisms and usually lead rapidly to mortalities.

Results obtained in the present study are consistent with previous findings about the physiological tolerance to emersion in *Liocarcinus depurator* (Bergmann et al., 2001b) and reveal the pronounced sensitivity of this species to mechanical stress during trawling action and to emersion during the sorting process. Moreover, they pointed out the extreme importance of the season in determining the disturbance magnitude on discarded species, underlining the role of the temperature effect in the fishing activities at temperate latitudes. In detail, survival tests showed the extent of seasonal influence, reflected in air and water temperature variations, on the survival chances of individuals involved in the fishing process. On the other hand, the physiological analyses provided the description of the metabolic disruption enhanced by higher temperatures during this stressful and unnatural event. During emersion the collapse of the gills produces a sudden breakdown of gas-exchange and excretion processes, with the consequent accumulation, at the level of tissues and hemolymph, of different metabolic end products (DeFur, 1988; Taylor & Whiteley, 1989). The increase of lactate, CO_2, H^+ concentrations normally occurs during the pronounced anaerobiosis together with the progressive accumulation of ammonia, urate, and amine in the hemolymph (Durand & Regnault, 1998; Danford et al., 2002). Besides, fishing procedures like trawling and sorting constitute stressful events both for commercial and discarded species and generally increase animal activities and metabolism which finally produce addictive effects to the homeostasis disruption (Bergman et al., 2001b). The physiological response of sublittoral organisms to air exposure appears to be markedly species specific and the extent of sublethal levels of metabolites like ammonia and lactate shows a wide range. In portunid crabs, experimentally emersed for 18 h, blood ammonia reaches values of 0.12 mmol l^{-1} in the intertidal *Carcinus maenas* and 1.3 mmol l^{-1} in the sublittoral *Necora puber* (Durand & Regnault, 1998). Similarly, *L. depurator* subjected to ca. 40 min of emersion after fishing trawl presents value of 0.308 mmol l^{-1} of blood ammonia (Bergman et al., 2001b). All these values appear to be sublethal for the investigated species, in consideration with the small mortality rate recorded and with the complete recovery during reimmersion in water. Comparable values of blood ammonia have been measured in the present study both in trawled and trawled and emersed crabs during winter activities, suggesting a sublethal accumulation of nitrogen compounds during emersion. On the contrary, the severe increase of blood ammonia resulting from summer analyses could be a consequence of the accelerated metabolism at the higher temperatures and could partially be responsible for the pronounced mortality rate recorded.

The rapid switch to anaerobic metabolism during emersion produces a high rate of increase in blood lactate level in portunid crabs; however, these species show an efficient mechanism of repayment of the oxygen debt and a rapid restoration of normoxic conditions usually occurs during recovery in water. In particular, when sublittoral species such as *N. puber* are subjected to emersion, individuals show increasing value of blood lactate which reaches 9 mmol l^{-1} and 60 mmol l^{-1} after 4 and 18 h of air exposure, respectively (Durand & Regnault, 1998). *L. depurator* appears to be more sensitive to emersion showing lactate concentrations in hemolymph of 6.2 mmol l^{-1} after 40 min in air following fishing trawling (Bergman et al., 2001b). Similar results are obtained in the present study for the winter samples. In addition, we have observed a prominent effect of the trawl motion rather than the emersion as the cause of stimulation of anaerobic metabolism (compare "trawl" and "trawl and exposure" data of Fig. 2). The mechanical solicitations and the elevated exercise inside the net seem to account for the principal fraction of the total oxygen debt. In summer these effects are significantly enhanced as a consequence of the thermal stress and accelerated metabolism.

In crustaceans, glycogen is the principal reserve of carbohydrates and constitutes the main source of energy during intense and protracted exercise; thus, high levels of circulating glucose represent a reliable indicator of energetic investment (Briffa & Elwood, 2001, 2002). Although glucose concentration could depend on a number of physiological factors and should be carefully adopted as a stress indicator (Kallen et al., 1990; Chang, 1995), the quantification of hemolymph glucose has been used to assess the effect on trawling. Exercise and emersion on *L. depurator*, reveal a significant increase of this metabolite following highly disturbing treatments (Bergman et al., 2001b). Our results agree with the previous studies, but evidenced also for glucose mobilization remarkable differences as a function of the season. The hemolymph concentration in summer experiments, in fact, reached values 6–7 fold higher than that achieved in winter, underlining the effect of temperature on the metabolic disruption.

The evaporative water loss during the emersion constitutes an important cause of homeostasis disturbance and might partially contribute to increased metabolite concentration in crab hemolymph. However, on the basis of the limited reduction in body wet weight reported for *L. depurator* caught in the Clyde Sea and exposed to air for 1 h (Bergmann et al., 2001b), evaporative water loss is likely to be not the main factor in inducing the sharp metabolites' increase observed during the experiments here reported.

The integration of the survival tests with the physiological evaluation of the stress, occurring during trawling and sorting operations, suggests that a decline of *L. depurator* populations should be expected. This conclusion is further supported by the indication that 24 h are necessary for fully recovering the induced physiological impairment (Bergmann et al., 2001b) and that the occurrence of sublethal damages, such as limb loss, hamper the foraging efficiency and agonistic behavior of crabs (Pranovi et al., 2001; Bergmann et al., 2001a). However, data from surveys show a rather stable situation for the Northern Adriatic Sea (Raicevich, 2004). The hypothesis of a trade-off between energetic costs, associated with the fishing disturbance, and energetic benefits, due to discard supply can be put forward: the scavenging behavior of *L. depurator* (Bergmann et al., 2002) would, indeed, allow populations of this species to exploit the supplementary, low-energy expenditure, source of food given by individuals of other species discarded by fishing activities. A similar hypothesis has been proposed also for the scavenging starfish *Asterias rubens* in the North Sea (Ramsay et al., 2000).

The extension of this study to other species could make it possible to rank the non-target species on the basis of their vulnerability to fishing. Such rank could include physiological aspects based on field observation and mortality assessment, integrating former assessments mainly based on life-history and fragility features (Macdonald et al., 1996). In spite of the different mechanisms responsible for the resilience to fishing disturbance in populations of scavengers species, studies showing high discard mortality rates should be considered by the authorities to enforce the legislation aimed to reduce this adverse effect of fishing on non-target species. For instance, the summer fishing ban already implemented (1 month) in the Northern Adriatic Sea could be extended to the whole summer season, avoiding discard production in the period when specimens are more likely to suffer high mortality rates and physiological impairment.

References

Bergmann, M. & P. G. Moore, 2001. Survival of decapod crustaceans discarded in the Nephrops fishery of the Clyde Sea area, Scotland. ICES Journal of Marine Science 58: 163–171.

Bergmann, M., D. J. Beare & P. G. Moore, 2001a. Damage sustained by epibenthic invertebrates discarded in the *Nephrops* fishery of the Clyde Sea, Scotland. Journal of Sea Research 45: 105–118.

Bergmann, M., A. C. Taylor & P. G. Moore, 2001b. Physiological stress in decapod crustaceans (*Munida rugosa* and *Liocarcinus depurator*) discarded in the Clyde *Nephrops* fishery. Journal of Experimental Marine Biology and Ecology 259: 215–229.

Bergmann, M., S. K. Wieczorek, P. G. Moore & R. J. A. Atkinson, 2002. Utilisation of invertebrates discarded from the Nephrops fishery by variously selective benthic scavengers in the west of Scotland. Marine Ecology Progress Series 233: 185–198.

Bridges, C. R. & A. R. Brand, 1981. The effect of hypoxia on haemolymph oxygen consumption and haemolymph lactate levels of some marine Crustacea. Comparative Biochemistry and Physiology A 65: 157–168.

Briffa, M. & R. W. Elwood, 2001. Decision rules, energy metabolism and vigor of hermit crab fights. Proceedings of the Royal Society of London, Series B 268: 1841–1848.

Briffa, M. & R. W. Elwood, 2002. Power of shell-rapping signals influences physiological costs and subsequent decisions during hermit crab fights. Proceedings of the Royal Society of London, Series B 269: 2331–2336.

Burnett, L. E., 1988. Physiological responses to air exposure: acid–base status and the role of branchial water stores. American Zoologist 28: 125–135.

Cabral, H. N., C. M. Teixeria, R. Gamito & M. J. Costa, 2002. Importance of discards of a beam trawl fishery of organic matter into nursery areas within the Tagus estuary. Hydrobiologia 475/476: 449–455.

Chang, E. S., 1995. Physiological and biochemical changes during the molt cycle in decapod crustaceans: an overview. Journal of Experimental Marine Biology and Ecology 193: 1–14.

Danford, A. R., L. Hagerman & R. F. Uglow, 2002. Effects of emersion and elevated haemolymph ammonia on haemocyanin–oxygen affinity of *Cancer pagurus*. Marine Biology 141: 1019–1027.

DeFur, P. L., 1988. Systemic respiratory adaptations to air exposure in intertidal decapod crustaceans. American Zoologist 28: 115–124.

Durand, F. & M. Regnault, 1998. Nitrogen metabolism of two portunid crabs, *Carcinus maenas* and *Necora puber*, during prolonged air exposure and subsequent recovery: a comparative study. Journal of Experimental Biology 201: 2515–2528.

Engel, P. C. & J. B. Jones, 1978. Causes and elimination of erratic blanks in enzymatic metabolite assays involving the use of NAD+ in alkaline hydrazine buffers: improved conditions for the assay of L-glutamate, L-lactate and other metabolites. Analytical Biochemistry 88: 475–484.

Gamito, R. & H. Cabral, 2003. Mortality of brown-shrimp discards from the beam trawl fishery in the Tagus estuary, Portugal. Fisheries Research 63: 423–427.

Gutmann, I. & A. W. Wahlefeld, 1974. L-(+)-lactate: determination with lactate dehydrogenase and NAD+. In Bergmeier, H. U. (ed.), Methods of Enzymatic Analysis. Academic Press, New York: 1464–1468.

Hagerman, L., T. Søndergaard., K. Weile, D. Hosie & R. F. Uglow, 1990. Aspects of blood physiology and ammonia excretion in *Nephrops norvegicus* under hypoxia. Comparative Biochemistry and Physiology 97: 51–55.

Hall-Spencer, J. M., C. Froglia, R. J. A. Atkinson & P. G. Moore, 1999. The impact of Rapido trawling for scallops, *Pecten jacobaeus* (L.), on the benthos of the Gulf of Venice. ICES Journal of Marine Science 56: 111–124.

Kaiser, M. J. & B. E. Spencer, 1995. Survival of by-catch from a beam trawl. Marine Ecology Progress Series 126: 31–38.

Kallen, J. L., S. L. Abrahamse & F. Van Herp, 1990. Circadian rhythmicity of the crustacean hyperglycemic hormone CHH in the hemolymph of the crayfish. Biological Bulletin 179: 351–357.

Jennings, S. & M. J. Kaiser, 1998. The effects of fishing on marine ecosystems. Advances in Marine Biology 34: 201–351.

Jennings, S., T. A. Dinmore, D. E. Duplisea, K. J. Warr & J. E. Lancaster, 2001. Trawling disturbance can modify benthic production processes. Journal of Animal Ecology 70: 459–475.

MacDonald, D. S., M. Little, M. C. Eno & K. Hiscock, 1996. Disturbance of benthic species by fishing activities: a sensitivity index. Aquatic Conservation 6(4): 257–268.

Mangum, C. P., 1985. Molting in the blue crab *Callinectes sapidus*: a collaborative study of intermediary metabolism, respiration and cardiovascular function, and ion transport. Preface Journal of Crustacean Biology 5: 185–187.

McMahon, B. R., 2001. Respiratory and circulatory compensation to hypoxia in crustaceans. Respiration Physiology 128: 349–364.

Morris, S. & S. Oliver, 1999a. Circulatory, respiratory and metabolic response to emersion and low temperature of *Jasus edwardsii*: simulation studies of commercial shipping methods Comparative Biochemistry and Physiology A—Molecular and Integrative Physiology 122: 299–308.

Morris, S. & S. Oliver, 1999b. Respiratory gas transport, haemocyanin function and acid–base balance in *Jasus edwardsii* during emersion and chilling: simulation studies of commercial shipping methods. Comparative Biochemistry and Physiology A—Molecular and Integrative Physiology 122: 309–321.

Pauly, D. & V. Christensen, 1995. Primary production required to sustain global fisheries. Nature 374: 255–257.

Pikitch, E. K., C. Santora, E. A. Babcock, A. Bakun, R. Bonfil, D. O. Conover, P. Dayton, P. Doukakis, D. Fluharty, B. Heneman, E. D. Houde, J. Link, P. A. Livingston, M. Mangel, M. K. McAllister, J. Pope & K. J. Sainsbury, 2005. Ecosystem-based fishery management. Science 305: 346–347.

Pranovi, F., S. Raicevich, G. Franceschini, P. Torricelli & O. Giovanardi, 2001. Discard composition and damage to non-target species in the "rapido" trawl fishery. Marine Biology 139: 863–875.

Queirós, A. M., J. G. Hiddink, H. Hinz & M. J. Kaiser, 2006. The effects of chronic bottom trawling disturbance on biomass, production and size spectra of invertebrate infauna communities from different habitats. Journal of Experimental Marine Biology and Ecology 335: 91–103.

Raicevich, S., 2004. Assessment of ecological disturbance in marine environments. Ph.D. thesis, University "Ca' Foscari", Venice.

Ramsay, K., M. J. Kaiser, A. D. Rijnsdorp, J. A. Craeymeersch & J. Ellis, 2000. Impact of trawling on populations of the invertebrate scavenger Asterias rubens. In Kaiser, M. J. & S. J. de Groot (eds), The effects of fishing on non-target species and habitats: biological, conservation and socio-economic issues. Fishing News Books: 151–162.

Schmitt, A. S. C. & R. F. Uglow, 1997. Haemolymph constituent levels and ammonia efflux rates of *Nephrops norvegicus* during emersion. Marine Biology 127: 403–410.

Solorzano, L., 1969. Determination of ammonia in natural waters by the phenolhypochlorite method. Limnology and Oceanography 4: 799–801.

Stillman, J. H. & G. N. Somero, 1996. Adaptation to temperature stress and aerial exposure in congeneric species of intertidal porcelain crabs (genus *Petrolisthes*): correlation of physiology, biochemistry and morphology with vertical distribution. Journal of Experimental Biology 199: 1845–1855.

Taylor, A. C. & J. I. Spicer, 1987. Metabolic responses of the prawns *Palaemon elegans* and *P. serratus* (Crustacea: Decapoda) to acute hypoxia and anoxia. Marine Biology 95: 521–530.

Taylor, A. C. & J. I. Spicer, 1988. Functional significance of a partial-emersion response in the prawn *Palaemon elegans* (Crustacea: Palaemonidae) during environmental hypoxia. Marine Ecology Progress Series 44: 141–147.

Taylor, E. W. & M. G. Whiteley, 1989. Oxygen transport and acid–base balance in the haemolymph of the lobster, *Homarus gammarus*, during aerial exposure and resubmersion, Journal of Experimental Biology 144: 417–436.

Tillin, H. M., J. G. Hiddink, M. J. Kaiser & S. Jennings, 2006. Chronic bottom trawling alters the functional composition of benthic invertebrate communities on a sea basin scale. Marine Ecology Progress Series 318: 31–45.

Webster, S. G., 1996. Measurement of crustacean hyperglycaemic hormone levels in the edible crab *Cancer pagurus* during emersion stress. Journal of Experimental Biology 199: 1585–1597.

White, P. S. & S. T. A. Pickett, 1985. Natural disturbance and patch dynamics: an introduction. In White P. S. & S. T. A. Pickett (eds), The Ecology of Natural Disturbance and Patch Dynamics. Academic Press, New York: 1–3.

Williams, N., 1998. Overfishing disrupts entire ecosystems. Science 279: 809.

Hydrobiologia (2008) 606:203–211
DOI 10.1007/s10750-008-9334-z

Grazing and assimilation rate estimates of hydromedusae from a temperate tidal creek system

Daniel Marshalonis · James L. Pinckney

© Springer Science+Business Media B.V. 2008

Abstract Previous research implicates gelatinous zooplankton grazing for reducing microzooplankton populations and influencing pelagic carbon fluxes. The principal objective of this study was to determine in situ ingestion rates and assimilation efficiencies for seasonal hydromedusae populations in North Inlet, SC, USA. Hydromedusae collected during July 2005, 2006 and January 2006 were exposed to natural plankton communities that were previously radiolabeled with ^{14}C. Ingestion rates and assimilation efficiencies were calculated from measurements of hydromedusae tissue radioactivity after ingestion and egestion, respectively. Net tows determined that the numerically dominant gelatinous predators in this system were *Bougainvillia muscus* (July 2005, 2006) and *Nemopsis bachei* (January 2006). Summer ingestion rates for *B. muscus* significantly increased with increased temperature, but there was no effect of temperature on winter ingestion rates for *N. bachei*, suggesting temperature responses are species specific. Moreover, there was a significant species effect on ingestion rates for July 2006. Hydromedusae assimilation efficiencies are relatively constant throughout the year and are not significantly affected by temperature, species, or season. Seasonal differences in grazing are likely a function of the pelagic community and temperature-dependent metabolism. These results coupled with seasonal hydromedusae abundance data imply summer populations of hydromedusae graze more heavily on prey than winter populations, while also assimilating more carbon into predator tissue. Effects of hydromedusae predation may include modification of trophic structure, shifts in algal community composition, and decline in ecosystem stability, leading to impacts on sustainable commercial fisheries.

Keywords Hydromedusae · Ingestion rate · Assimilation efficiency

Abbreviations

LTER	Long-Term Ecological Research
NERR	National Estuarine Research Reserve
DO	Dissolved oxygen
DIC	Dissolved inorganic carbon
DPM	Disintegrations per minute

Guest editors: J. Davenport, G. Burnell, T. Cross, M. Emmerson, R. McAllen, R. Ramsay & E. Rogan
Challenges to Marine Ecosystems

D. Marshalonis (✉) · J. L. Pinckney
Department of Biological Sciences, University of South Carolina, Columbia, SC 29208, USA
e-mail: dino@biol.sc.edu

J. L. Pinckney
e-mail: jpinckney@biol.sc.edu

Introduction

Gelatinous zooplankton (hydromedusae, ctenophores, and scyphomedusae) are prominent members of pelagic communities and major predators of other zooplankton species. Gelatinous zooplankton blooms have become more prevalent in coastal waters in recent years and have caused widespread concern among marine ecologists, fishermen, beachgoers, and the general public (Mills, 2001; Sullivan et al., 2001). Large blooms of gelatinous zooplankton have been associated with negative estuarine ecosystem events, such as eutrophication (Arai, 2001), and may be harbingers of major ecosystem change. Their capability to reproduce rapidly coupled with their extremely high prey consumption rates enables them to exert control over energy flow and nutrients in estuarine waters (Purcell et al., 2001; Gordina et al., 2005).

Predation effects by gelatinous zooplankton are often substantial in environments where they are abundant, consuming more than 30% of available zooplankton daily (Purcell & Arai, 2001). Predation by hydromedusae has been shown to alter zooplankton populations (Larson, 1987). The hydromedusa *Sarsia gemmifera* can ingest up to 3 μg C h^{-1} with daily carbon values exceeding individual body weight (Stibor & Tokle, 2003). In a food web model of Chesapeake Bay, Baird & Ulanowicz (1989) determined that gelatinous zooplankton exerted primary control on the planktonic food chain. They found that the controlling organisms for recycling of material within the plankton community shifted from crustacean zooplankton in the spring to the jellyfish *Chrysaora quinquecirrha* in the summer. Similar seasonal shifts are likely to occur in other estuarine systems.

Numerous studies have investigated the role of gelatinous zooplankton grazing in stressed environments [low dissolved oxygen (DO), collapsing fisheries, eutrophication], but few have examined the importance of gelatinous predators in environments receiving minimal anthropogenic stressors. North Inlet, SC, is located in the southeastern US and has received protection status as one of the 27 National Estuarine Research Reserves (NERR) by the National Oceanographic and Atmospheric Administration. North Inlet is considered a pristine tidal creek estuary with low pollutant and low nutrient input (Allen et al., 1997; Buzzelli et al., 2004; White et al., 2004). Examination of gelatinous zooplankton trophic pathways in North Inlet can yield baseline estuarine system information regarding the role of gelatinous predators under normal or "healthy" environmental conditions.

The known trophic importance of gelatinous zooplankton species in other estuarine systems suggests that gelatinous zooplankton may be a major component of the food web in the North Inlet estuary. In this study we examined seasonal trends in hydromedusae grazing and assimilation efficiencies. Our goal was to determine if temperature or hydromedusae species had any effect on grazing and assimilation efficiencies. Moreover, we considered the potential impacts of hydromedusae grazing on zooplankton populations for the North Inlet ecosystem.

We hypothesized that elevated summer temperatures would raise ingestion rates and lower assimilation efficiencies for hydromedusae compared to winter rates, with summer species having higher ingestion rates and lower assimilation efficiencies compared to winter species. There is limited knowledge concerning metabolic assimilation efficiencies for hydromedusae. Hydromedusae implement different predation strategies based upon morphology, with oblate medusae feeding while swimming and prolate medusae feeding only at rest (contracted tentacles while swimming) (Costello & Colin, 2002). Thus, different species with different activity levels may have different capabilities to assimilate ingested prey. Moreover, ambient temperatures may have an effect on assimilation rates. In North Inlet, seasonal changes in abiotic parameters are limited to temperature. Tidal flushing keeps salinity rather constant due to limited freshwater input into the system. Increases in temperature have been shown to increase respiration rate for gelatinous zooplankton (Arai, 1997), which could decrease assimilation rate. Consequentially, an organism would require increased ingestion rates to compensate for lower assimilation efficiencies.

Methods

Study site

The study location was Town Creek in North Inlet, South Carolina. The inlet is a bar built tidal creek estuary that is shallow (~3 m in the channel) with a semidiurnal tide that has an average tidal range of 1.4 m, which fosters a short hydraulic residence time of the estuary (1–2 days). The water column is well mixed with no salinity or density stratification, and circulation is driven by tidal pumping (Vernberg et al., 1992). Seasonal abundances of hydromedusae have been recorded for the North Inlet estuary during the last 15 years as a part of their The Long-Term Ecological Research (LTER) and NERR Programs. These monitoring programs have shown hydromedusae to be the dominant gelatinous predator in this system in terms of numerical abundance (Allen et al., 2006). The LTER and NERR Programs' sampling site in the middle of Town Creek, a major branch of the North Inlet estuary, receives a large percentage of the tidal influx and much of the water flushing because the smaller tributaries of the estuary pass through this location. As such, sampling results should be considered representative of what the Inlet may receive on a typical tidal cycle in terms of hydromedusae and zooplankton community structure and densities.

Hydromedusae were collected using short-duration (<1 min) plankton tows at a depth of 0.5–1.5 m with a 1 m dia 1,000 μm mesh net to minimize damage to fragile tissue. Organisms were examined under a dissecting microscope for signs of damage and identified to species prior to all bioassays. To minimize stress, bioassays were initiated within 2 h of collection. This delay also allowed the test organisms to fully egest any food available to them during collection and transportation. Because the gelatinous community in both July 2005 and January 2006 was dominated by a single species, experiments did not test species differences. We tested species effects in July 2006 even though the community was composed of primarily one species (*B. muscus* >90% of gelatinous organisms).

Quantitative net tows for zooplankton and physical measurements were made biweekly during the experimental periods. Simultaneous replicate 30 dia cm 153 μm mesh closing nets were towed for 1 min intervals at the bottom, mid-depth, and surface. Samples were fixed with seawater-buffered formalin and specimens larger than copepod nauplii were enumerated. Measurements taken at the surface and bottom were salinity, temperature, DO, and dissolved inorganic carbon (DIC).

Radiolabeling

To radiolabel plankton for the grazing incubations, collected seawater with natural assemblages of phytoplankton/zooplankton was incubated in situ in 1 l polycarbonate Nalgene bottles for 24 h with $NaH^{14}CO_3$ to yield a final seawater concentration of 0.05 μ Ci ml^{-1}. Grazing incubation water consisted of radiolabeled zooplankton retained with a 63 μm filter added to 1 l of 0.2 μm filtered seawater. Egestion incubation water consisted of 0.2 μm filtered seawater. Yurista (2004) used ^{14}C as a tracer for cladoceran bioenergetics studies and found incubations of radiolabeled phytoplankton and *Daphnia lumholtzi* lasting 10 min were adequate for detecting tracer signals in zooplankton samples. All incubations took place in 1 l polycarbonate bottles placed in circulating water baths set at the monthly averaged in situ temperature.

The grazing rate protocol used in this study is modified from Stibor & Tokle (2003) and similar methods have been successfully employed in other studies (Katechakis et al., 2004; He & Wang, 2006). Duplicate incubations were performed to find the proportion of zooplankton carbon radiolabeled during our incubations and convert ingested ^{14}C into total carbon ingested. Half the incubations included radioisotope tracer while the other half contained sample seawater alone. At the end of the incubation, radioactivity was measured in the >63 μm fraction for the tracer bottles while the non-radiolabeled samples were filtered onto Whatman GF/F filters and total carbon was quantified by CHN analysis. Qualitative microscopy upon retained food material showed the most abundant items were copepods, copepod nauplii, and large chain-forming diatom species.

Collected hydromedusae (two hydromedusae per experimental container) were allowed to feed in the grazing incubation water for 2 h. Hydromedusae were observed in the water column of experimental containers to confirm that they were healthy and actively feeding or swimming. After grazing, hydromedusae were rinsed with GF/F filtered seawater to remove non-ingested radioactive material. One hydromedusa was placed in a scintillation vial, with 1 ml of tissue solubilizer (Soluene 350) and heated to 60°C for 24 h, with a subsequent 5 ml addition of scintillation reagent (Hionic Fluor). The other hydromedusa was placed in a new container of non-radioactive food (1.0 l of 0.2 μm filtered seawater) and allowed to clear its digestive tract of radioactive material for 1 h. Previous observations showed hydromedusae egestion occurred within 0.5 h after ingestion. Hydromedusae placed in egestion incubation water were prepared for analysis as described above. Samples were analyzed using a Beckman–Coulter LS 6500 Liquid Scintillation Counter. Disintegrations per minute (DPM) were converted to ingested carbon using the following equation:

$$I = \frac{R}{T} \times \left(\frac{Z_C}{Z_{DPM}} \right) \quad (1)$$

where I is the ingested carbon (μg C ind^{-1} h^{-1}), R is the quench corrected DPM for each sample, T is the duration of the grazing experiment (h), Z_C is the total zooplankton carbon present to be grazed during experiments (g C l^{-1}), and Z_{DPM} is the amount of measured radioactivity after 24 h incubation with radioisotope (DPM l^{-1}); Z_C/Z_{DPM} is a scalar correction factor converting ingested radioactive C to total C. Ingestion was measured from the amount of radioactivity eaten before the hydromedusae started to egest undigested radioactive material. Assimilated material was measured by the amount of radioactivity left in the organism after clearing its gut.

Our experimental design tested the effects of two factors on grazing rates and assimilation efficiencies: temperature (with levels of 12 and 21°C for January 2006 and 28 and 34°C for July 2006) and species (*Clytia hemisphaerica* and *Bougainvillia muscus* for July 2006). Each factor level combination was replicated 12–18 times per season except *C. hemisphaerica* ($n = 3$).

Results

Community composition and physical environment

Monthly averaged ambient temperatures for North Inlet for July 2005, January 2006, and July 2006 were 29, 12, and 28°C, respectively while average ambient DIC values were 12.5, 21.7, and 15.8 mg C ml^{-1}, respectively (Table 1). Physical parameters recorded for North Inlet show relative stability throughout the study period. Current velocity and salinity show negligible variability between sampling episodes. There is little variability between the two July sampling times in terms of temperature, DO, and DIC. January 2006 had elevated DO and DIC values and depressed temperatures relative to the summer sampling periods. Net tows determined that the numerically dominant (>95%) gelatinous predators in North Inlet were *B. muscus* (July 2005) and *Nemopsis bachei* (January 2006). During July 2006, *B. muscus* individuals dominated the gelatinous community (>90%) with *C. hemisphaerica* present in low numbers. Microscopic examination of plankton communities used as prey in these experiments demonstrated a clear dominance of copepod adults, copepodids, and copepod nauplii, with occasional other taxa of limited abundance (decapod crab zoea, decapod shrimp larvae, larvaceans, and nematodes). Thus, it is likely that the radiolabeled food presented to hydromedusae in this study was almost exclusively crustacean prey.

Table 1 Natural densities for predator and prey and physical parameters for Town Creek, North Inlet, SC, during this study

Date	Density (ind m^{-3})		Salinity	Temperature (°C)
	Copepod	Hydromedusae		
July 2005	14191.5	9.57	31.9	29.2
January 2006	1890.1	0.33	30.7	11.6
July 2006	3460.8	1.95	31.3	28.0

Data are means from biweekly NERR sampling during the month

Ingestion rates

The highest ingestion rate measured was for the summer species *C. hemisphaerica* at 28°C (20.05 μg C ind^{-1} h^{-1}), while the lowest measured ingestion rate was for the other summer species *B. muscus* at 28°C (4.60 μg C ind^{-1} h^{-1}) (Fig. 1). A single factor model I analysis of variance with species (*B. muscus*, *C. hemisphaerica*) as factor levels was performed to determine the effects of species on grazing rates during July 2006. The ANOVA indicated that *C. hemisphaerica* ingestion rates were significantly higher than those for *B. muscus* ($P < 0.001$).

To test the effects of temperature on ingestion rates, two separate single factor model I ANOVAs with temperature as the main factor (12, 21°C during January 2006 and 28, 34°C during July 2006) were performed. The January ANOVA indicated that there was no temperature effect on the ingestion rates for *N. bachei* ($P > 0.05$), while the ingestion rates for *B. muscus* in July were significantly greater at 34°C than those at 28°C ($P < 0.05$).

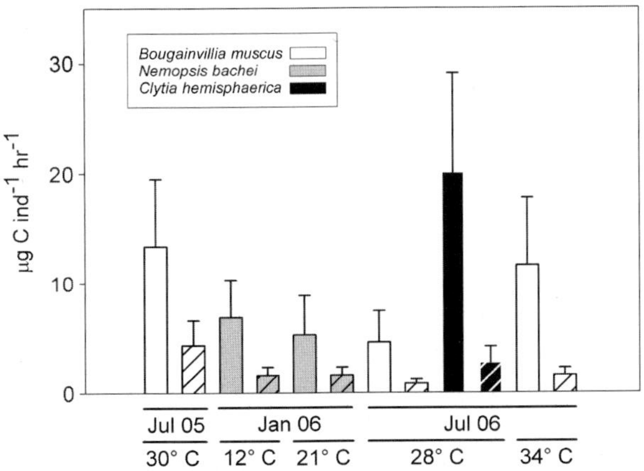

Fig. 1 Grazing rates for hydromedusae during seasonal bioassays. White bars indicate *B. muscus*, gray bars indicate *N. bachei*, black bar indicates *C. hemisphaerica*. Solid bars indicate measured radioactivity after 2 h ingestion period; striped bars indicate measured radioactivity in organisms after complete egestion. Note additional temperature factor level for January 2006 and July 2006, as well as additional species level for July 2006

Assimilation efficiencies

Mean assimilation efficiency values ranged from 12.6% to 37.9% with the winter hydromedusae *N. bachei* at 21°C having the highest measured value (Fig. 2). Assimilation values for July 2006 tended to be lower than the other measurements. A three-way ANOVA indicated that none of the main factors (species, temperature, or season) had a significant effect on assimilation efficiencies ($P > 0.05$).

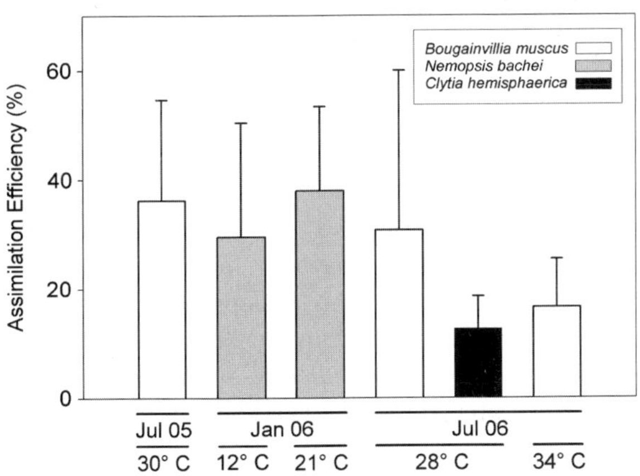

Fig. 2 Assimilation efficiencies for hydromedusae during seasonal bioassays. White bars indicate *B. muscus*, gray bars indicate *N. bachei*, black bar indicates *C. hemisphaerica*. Note additional temperature factor level for January 2006 and July 2006, as well as additional species level for July 2006

Discussion

Protocol and collection

Most previous studies examining cnidarian feeding rates have used gut content analysis or changes in prey density to quantify ingestion. Feeding estimates based on metabolic demands are typically much lower than other methods and clearance rate estimates cannot be used on multiple types of prey due to differential prey capture dynamics (Purcell, 1997). Gut content analysis is laborious, prone to identification mistakes, and dependant on the assumption that all particles ingested have similar digestion times, which has been shown repeatedly to be false. Factors influencing digestion times include prey size (Suchman & Sullivan, 2000), number of prey in gut (Martinussen & Bamstedt, 1999), temperature (Martinussen & Bamstedt, 2001), and the predator species consuming the prey (Martinussen & Bamstedt, 1999; Suchman & Sullivan, 2000). Such results typically lead to over-estimation of particular taxa (usually crustaceans) relative to others and require the use of differential digestion time scaling. Finally, gut content analysis is dependent on accurate measurements of zooplankton prey density, which is known to be patchy and difficult to adequately quantify under most sampling regimes.

There are numerous advantages to the ^{14}C radiolabeling method for determination of ingestion rates. Radioactive tracers are extremely sensitive and allow researchers to quantify small rates with accuracy. Moreover, incubations can be short in duration, significantly reducing laboratory artifacts caused by prey settling, container adhesion, etc. The current study used natural seawater planktonic communities as prey for our feeding experiments. The authors recognize hydromedusae may prefer particular prey due to dietary niches (Costello & Colin, 2002) and believe natural communities of prey represent the most realistic feeding estimate compared to those measured in situ.

Our experimental protocol does suffer from drawbacks. Our 1 l incubation bottles created container volume: predator volume ratios of ~7100:1 which are lower than those suggested by Toonen & Chia (1993) and indicative of potential container effects. However, our results are well within the range of other observations of gelatinous zooplankton ingestion rates. Incomplete radiolabeling of prey is a reality, but we estimated the proportion of labeling in supplemental bioassays. Incubation effects on community structure can be substantial, especially on fragile flagellated prey or prey readily consumed by other prey species before being consumed by hydromedusae. The small, transparent nature of hydromedusae makes direct observation of feeding while in incubation chambers impossible without undue stress from manipulation. As a result, we chose to rinse hydromedusae gently with filtered seawater to remove unincorporated isotope. Finally, we are unaware of the energetic state of collected organisms at the start of the experiments; however, hydromedusae tend to retract tentacles, i.e. not feed, when agitated (in a net tow), and the two-hour buffer zone between collection and experiment start should be sufficient for full manubrium egestion.

Ingestion rate

Other studies have shown hydromedusae grazing values similar to the results presented in this study. Maximal hydromedusae grazing of copepods by *Moerisia lyonsi* peaked at 1 copepod medusa^{-1} h^{-1} (Purcell et al., 1999), which when converted to C biomass matches the experimental results of this study. Similarly, our results for *N. bachei* are in agreement with those found by Purcell & Nemazie (1992) using the same species. Matsakis & Nival (1989) found ingestion for *Phialidium* (=*Clytia*) spp. to be proportional to prey density, with maximal rates of 20.86 μg C h^{-1}, which is nearly identical to our findings for *Clytia*. Matsakis and Conover (1991) determined ingestion for *Aglantha digitale* to vary from 0.62 to 3.16 μg C h^{-1} and Matsakis (1990) measured C uptake by *Clytia* spp. to be 2.7 to 5.0 μg C h^{-1}. Stibor & Tokle (2003) used radiotracer methods to quantify the grazing rate of *Sarsia gemmifera*, which was as high as 3 μg C h^{-1}. Our results are higher than their results but are still on the same order of magnitude.

Differences in grazing rates may be attributable to morphological and feeding strategy differences between hydromedusae species. Anthomedusae primarily consume crustacean prey and are equipped with both desmonemes, used to entangle and adhere to prey, and stenoteles or euryteles, capable of puncturing exoskeleton, while Leptomedusae primarily consume soft-bodied prey and utilize microbasic mastigophores to penetrate prey (Purcell & Mills, 1988). Both *B. muscus* and *N. bachei* are prolate anthomedusan species, while *C. hemisphaerica* is an oblate leptomedusan species. Oblate hydromedusae swimming concurrently produces favorable prey entrainment currents and this mode of propulsion decreases energy expenditure relative to prolate medusae, whose jet propulsion swimming requires more energy despite their low drag and small bell contraction times (Colin et al., 2003). Swimming activity and tentacle behavior help determine trophic impacts and dietary niches in hydromedusae. Dietary differences are created because oblate rowing entrains prey while prolate jetting allows ambush predators to drift the majority of the time and deploy their tentacles. It is plausible that the in situ zooplankton prey community composition dictated that *C. hemisphaerica* would be better suited to grazing due to its mode of prey capture.

The effects of temperature on hydromedusae ecology are ambiguous. Fulton & Wear (1985) found no significant effect of water temperature on the clearance rates of hydromedusae species found in New Zealand waters. However, Matsakis (1993) determined that growth rates for *Clytia* spp. significantly increased with increasing temperature and with increasing prey density. However, these data may be population specific because their organisms also showed high mortality above 25°C. In our study, medusae thrived under temperatures as high as 34°C. The lack of seasonal pattern in our data suggests that in situ hydromedusae communities are well acclimated to their ambient temperature and maximally consume available prey.

At high concentrations Matsakis & Conover (1991) found ingestion was limited by digestion. Purcell et al. (1999) also found an increase in predation rate with increasing prey density but predation began to level off at densities higher than 40 prey l^{-1}. Ingestion rates may be limited at high prey densities due to mechanical saturation, in which tentacles are already in use to continue capturing prey (Clifford & Cargo, 1978; Reeve & Walter, 1978). The short duration of incubation, natural prey community used as food, and the container volume: predator volume ratio used in this study preclude concern regarding prey saturation or substantial reduction of prey.

Assimilation efficiency

Estimates of hydromedusae assimilation efficiency from previous studies are also similar to the efficiencies determined in the present study. Matsakis (1993) estimated *Clytia* spp. growth efficiencies to be 18%, which is higher than the value for *Clytia* determined in this study (12.6%). Other studies have calculated assimilation efficiencies for hydromedusae of 9–17% (Daan, 1986). Assimilation efficiency estimates (10–30%) from Stibor & Tokle (2003) agree with estimates from this study.

Assimilation efficiencies for other gelatinous organisms are variable but high. Others (Reeve & Walter, 1978; Reeve et al., 1989) have found high variability in *Pleurobrachia* and *Mnemiopsis* assimilation, ranging from 18.2% to 82.8%. Siphonophore carbon assimilation efficiencies ranged from 87% to 94%, presumably an adaptation to low prey abundances (Purcell, 1983). The superfluous feeding theory of Becklemishev (1962) has been used to explain rather low assimilation efficiencies found in gelatinous predators. Under high prey densities and favorable feeding conditions, hydromedusae may experience sloppy feeding due to their single opening digestive tract design. When food is less available, digestion times may be longer and allow for more complete digestion. Thus, food availability may play a role in assimilation efficiencies for hydromedusae. Studies have shown digestive efficiency for gelatinous organisms to be inversely proportional to ingestion rate (Kremer, 1979). Our data do not support this claim; summer assimilation efficiencies were not significantly lower than winter values as predicted, although assimilation values for July 2006 are some of the lower values determined in this study. Regardless, there is no significant effect of season, temperature, or species on assimilation efficiency. Moreover, superfluous feeding may be energetically costly compared to a constant feeding rate and assimilation efficiency (Conover, 1966). Accordingly, greater ingestion rates do lead to increased net C assimilation and can be used as a metric for C sequestered by gelatinous predators.

Other factors have been attributed to the variability in assimilation efficiencies. Conover (1966) determined assimilation efficiency was inversely related to the quality (% ash) in food. Schindler (1971) found food type, i.e. prey species, influenced assimilation efficiencies, and there was an inverse relationship between assimilation efficiency and ingestion rate when predators are given a variety of food types. These data suggest that hydromedusae may behaviorally alter grazing when prey densities are high, either through decreasing prey capture effort and frequency of prey capture or increasing prey selectivity, rather than altering assimilation efficiency.

Ecosystem impacts

Estimates of gelatinous predation impacts on zooplankton prey populations can be calculated from current data. We estimate hydromedusae communities in North Inlet estuary to substantially graze upon zooplankton standing stock (Table 2), and we suggest hydromedusae are a major predator upon microzooplankton secondary production in this system. Our estimates of daily consumption of zooplankton by hydromedusae communities are conservative because they ignore bloom events or other times of maximal hydromedusae density. Regardless, there is a very clear seasonal difference in grazing impacts, which has important implications to seasonal spawning commercial fisheries (Lynam et al., 2005, 2006), which include crabs, penaeid shrimps, bivalves, and finfish (Dame et al., 2000).

Table 2 Estimated ecosystem impacts of hydromedusae predation upon copepod prey populations and copepod production for North Inlet during this study

Calculated impacts	
Hydromedusae-ingested carbon (kg C d^{-1})	
July 2005	852.9
January 2006	126.4
July 2006	295.0
Copepod standing stock reductions (%)	
July 2005	36.3
January 2006	40.3
July 2006	51.4
Copepod production reduction (%)	
July 2005	28.1
January 2006	70.8
July 2006	42.5

Copepod abundances were converted to biomass using calculations found in Durbin & Durbin (1992); copepod production was estimated using the temperature-corrected factors specified in White (1991)

Conclusion

As gelatinous predators become more widespread and conspicuous in the world ocean (or at least in the public eye) (Mills, 2001), their effects on ocean food webs could cause mass changes in conventional fish dominated food webs (Mills, 1995). Yet little is known of the ecology and trophodynamics of many "gelata" because of their fragile nature and the associated difficulty in performing robust laboratory experiments. Hydromedusae are particularly under-represented in terms of ecological knowledge, presumably due to their small size (<10 mm). However, they represent a major portion of gelatinous organism diversity (Colin et al., 2005). In this study, we quantified the grazing impacts of hydromedusae communities in a pristine temperate tidal creek system. Our data indicate hydromedusae grazing rates are species specific and not a function of seasonality. Moreover, we suggest that predator behavior (feeding strategy) dictates assimilated carbon, not temperature-dependent metabolism. These data will provide baseline values for comparison between other estuarine systems while also elucidating the major factors influencing gelatinous control over carbon fluxes in pelagic food webs. A clear understanding of gelatinous ecology is requisite if we hope to appreciate their importance in the pelagic coastal region or mitigate their negative effects.

Acknowledgements We appreciate the technical assistance and input of Tammi Richardson, Dennis Allen, Jean Marie Buschur, and Brock Renkas. This work was supported by The National Oceanic and Atmospheric Administration/National Estuarine Research Reserve Graduate Research Fellowship (CFDA: 11.420, Opportunity Number: NOS-OCRM-2005-2000046) and The Slocum Lunz Foundation. Contribution number 1471 of the Belle W. Baruch Institute for Marine and Coastal Sciences, University of South Carolina.

References

Allen, D. M., D. L. Barker, V. Ogburn-Matthews & S. E. Stancyk, 2006. Long-term mesozooplankton (153 micron) data for the North Inlet estuary, Georgetown, South Carolina: 1981–2003. Belle W. Baruch Institute for Marine and Coastal Sciences, University of South Carolina, Data Archives Website: http://links.baruch.sc.edu/Data/MESOZPK/index.html.

Allen, D. M., D. Edwards, R. J. Feller, S. E. Hutchinson & V. Ogburn-Matthews, 1997. Detection and analysis of unusual events in long-term zooplankton and nekton data sets from North Inlet Estuary, South Carolina, USA. Oceanologica Acta 20: 165–175.

Arai, M. N., 1997. A Functional Biology of Scyphozoa. Chapman & Hall, London.

Arai, M. N., 2001. Pelagic coelenterates and eutrophication: a review. Hydrobiologia 451: 69–87.

Baird, D. & R. E. Ulanowicz, 1989. The seasonal dynamics of the Chesapeake Bay ecosystem. Ecological Monographs 59: 329–364.

Beklemishev, C. W., 1962. Superfluous feeding of marine herbivorous zooplankton. Rapp. PV. Reun. Cons. Tnt. Explor. Mer 153: 108–113.

Buzzelli C., O. Akman T. Buck E. Koepfler J. Morris & A. Lewitus, 2004. Relationships among water-quality parameters from the North Inlet-Winyah Bay National Estuarine Research Reserve, South Carolina. Journal of Coastal Research 45: 59–74.

Clifford, H. C. & D. G. Cargo, 1978. Feeding rates of sea nettle, *Chrysaora quinquecirrha*, under laboratory conditions. Estuaries 1: 58–61.

Colin, S. P., J. H. Costello, W. M. Graham & J. Higgins, 2005. Omnivory by the small cosmopolitan hydromedusa *Aglaura hemistoma*. Limnology and Oceanography 50: 1264–1268.

Colin, S. P., J. H. Costello & E. Klos, 2003. In situ swimming and feeding behavior of eight co-occurring hydromedusae. Marine Ecology-Progress Series 253: 305–309.

Conover, R. J., 1966. Factors affecting assimilation of organic matter by zooplankton and question of superfluous feeding. Limnology and Oceanography 11: 346–354.

Costello, J. H. & S. P. Colin, 2002. Prey resource use by coexistent hydromedusae from Friday Harbor, Washington. Limnology and Oceanography 47: 934–942.

Daan, R., 1986. Food-intake and growth of *Sarsia tubulosa* (Sars, 1835), with quantitative estimates of predation on copepod populations. Netherlands Journal of Sea Research 20: 67–74.

Dame, R., M. Alber, D. Allen, M. Mallin, C. Montague, A. Lewitus, A. Chalmers, R. Gardner, C. Gilman, B. Kjerfve, J. Pinckney & N. Smith, 2000. Estuaries of the south Atlantic coast of North America: their geographical signatures. Estuaries 23: 793–819.

Durbin, E. G. & A. G. Durbin, 1992. Effects of temperature and food abundance on grazing and short-term weight change in the marine copepod *Acartia hudsonica*. Limnology and Oceanography 37: 361–378.

Fulton, R. S. & R. G. Wear, 1985. Predatory feeding of the hydromedusae *Obelia geniculata* and *Phialella quadrata*. Marine Biology 87: 47–54.

Gordina, A. D., J. A. Zagorodnyaya, A. E. Kideys, L. Bat & H. H. Satilmis, 2005. Summer ichthyoplankton, food supply of fish larvae and impact of invasive ctenophores on the nutrition of fish larvae in the Black Sea during 2000 and 2001. Journal of the Marine Biological Association of the United Kingdom 85: 537–548.

He, X. J. & W. X. Wang, 2006. Releases of ingested phytoplankton carbon by *Daphnia magna*. Freshwater Biology 51: 649–665.

Katechakis, A., H. Stibor, U. Sommer & T. Hansen, 2004. Feeding selectivities and food niche separation of *Acartia clausi, Penilia avirostris* (Crustacea) and *Doliolum denticulatum* (Thaliacea) in Blanes Bay (Catalan Sea, NW Mediterranean). Journal of Plankton Research 26: 589–603.

Kremer, P., 1979. Predation by the ctenophore *Mnemiopsis leidyi* in Narragansett Bay, Rhode-Island. Estuaries 2: 97–105.

Larson, R. J., 1987. Trophic ecology of planktonic gelatinous predators in Saanich Inlet, British-Columbia—diets and prey selection. Journal of Plankton Research 9: 811–820.

Lynam, C. P., M. J. Gibbons, B. E. Axelsen, C. A. J. Sparks, J. Coetzee, B. G. Heywood & A. S. Brierley, 2006. Jellyfish overtake fish in a heavily fished ecosystem. Current Biology 16: 492–493.

Lynam, C. P., M. R. Heath, S. J. Hay & A. S. Brierley, 2005. Evidence for impacts by jellyfish on North Sea herring recruitment. Marine Ecology-Progress Series 298: 157–167.

Martinussen, M. B. & U. Bamstedt, 1999. Nutritional ecology of gelatinous planktonic predators: digestion rate in relation to type and amount of prey. Journal of Experimental Marine Biology and Ecology 232: 61–84.

Martinussen, M. B. & U. Bamstedt, 2001. Digestion rate in relation to temperature of two gelatinous planktonic predators. Sarsia 86: 21–35.

Matsakis, S., 1990. Behavior and metabolism of medusae from the Ligurian Sea and the Bedford Basin. Ph.D. Dissertation, University of Paris, Paris.

Matsakis, S., 1993. Growth of *Clytia* spp. hydromedusae (Cnidaria, Thecata): effects of temperature and food availability. Journal of Experimental Marine Biology and Ecology 171: 107–118.

Matsakis, S. & R. J. Conover, 1991. Abundance and feeding of medusae and their potential impact as predators on other zooplankton in Bedford Basin (Nova-Scotia, Canada) during spring. Canadian Journal of Fisheries and Aquatic Sciences 48: 1419–1430.

Matsakis, S. & P. Nival, 1989. Elemental composition and food-intake of *Phialidium* hydromedusae in the laboratory. Journal of Experimental Marine Biology and Ecology 130: 277–290.

Mills, C. E., 1995. Medusae, siphonophores, and ctenophores as planktivorous predators in changing global ecosystems. Ices Journal of Marine Science 52: 575–581.

Mills, C. E., 2001. Jellyfish blooms: are populations increasing globally in response to changing ocean conditions? Hydrobiologia 451: 55–68.

Purcell, J. E., 1983. Digestion rates and assimilation efficiencies of siphonophores fed zooplankton prey. Marine Biology 73: 257–261.

Purcell, J. E., 1997. Pelagic cnidarians and ctenophores as predators: selective predation, feeding rates, and effects on prey populations. Annales De L Institut Oceanographique 73: 125–137.

Purcell, J. E. & M. N. Arai, 2001. Interactions of pelagic cnidarians and ctenophores with fish: a review. Hydrobiologia 451: 27–44.

Purcell, J. E., U. Bamstedt & A. Bamstedt, 1999. Prey, feeding rates, and asexual reproduction rates of the introduced oligohaline hydrozoan *Moerisia lyonsi*. Marine Biology 134: 317–325.

Purcell, J. E. & C. E. Mills, 1988. The correlation of nematocyst types to diets in pelagic Hydrozoa. In Hessinger, D. A. & H. M. Lenhoff (eds), The Biology of Nematocysts. Academic Press, San Diego: 463–485.

Purcell, J. E. & D. A. Nemazie, 1992. Quantitative feeding ecology of the hydromedusan *Nemopsis bachei* in Chesapeake Bay. Marine Biology 113: 305–311.

Purcell, J. E., T. A. Shiganova, M. B. Decker & E. D. Houde, 2001. The ctenophore *Mnemiopsis* in native and exotic habitats: US estuaries versus the Black Sea basin. Hydrobiologia 451: 145–176.

Reeve, M. R., M. A. Syms & P. Kremer, 1989. Growth dynamics of a ctenophore (*Mnemiopsis*) in relation to variable food supply. 1. Carbon biomass, feeding, egg production, growth and assimilation efficiency. Journal of Plankton Research 11: 535–552.

Reeve, M. R. & M. A. Walter, 1978. Laboratory studies of ingestion and food utilization in lobate and tentaculate ctenophores. Limnology and Oceanography 23: 740–751.

Schindler, J. E., 1971. Food quality and zooplankton nutrition. Journal of Animal Ecology 40: 589–595.

Stibor, H. & N. Tokle, 2003. Feeding and asexual reproduction of the jellyfish *Sarsia gemmifera* in response to resource enrichment. Oecologia 135: 202–208.

Suchman, C. L. & B. K. Sullivan, 2000. Effect of prey size on vulnerability of copepods to predation by the scyphomedusae *Aurelia aurita* and *Cyanea* sp. Journal of Plankton Research 22: 2289–2306.

Sullivan, B. K., D. Van Keuren & M. Clancy, 2001. Timing and size of blooms of the ctenophore *Mnemiopsis leidyi* in relation to temperature in Narragansett Bay, RI. Hydrobiologia 451: 113–120.

Toonen, R. J. & F. S. Chia, 1993. Limitations of laboratory assessments of Coelenterate predation: container effects on the prey selection of the Limnomedusa, *Proboscidactyla flavicirrata* (Brandt). Journal of Experimental Marine Biology and Ecology 167: 215–235.

Vernberg, F. J., W. B. Vernberg, E. Blood, A. Fortner, M. Fulton, H. Mckellar, W. Michener, G. Scott, T. Siewicki & K. Elfigi, 1992. Impact of urbanization on high salinity estuaries in the southeastern United States. Netherlands Journal of Sea Research 30: 239–248.

White, J. R., 1991. Seasonal study of zooplankton dynamics in the mesohaline Chesapeake Bay. Ph.D. Dissertation, University of Maryland, College Park, Maryland.

White, D. L., D. E. Porter & A. J. Lewitus, 2004. Spatial and temporal analyses of water quality and phytoplankton biomass in an urbanized versus a relatively pristine salt marsh estuary. Journal of Experimental Marine Biology and Ecology 298: 255–273.

Yurista, P. M., 2004. Bioenergetics of a semi-tropical cladoceran, *Daphnia lumholtzi*. Journal of Freshwater Ecology 19: 681–694.